ENVIRONMENTAL ACTIVISTS

ENVIRONMENTAL ACTIVISTS

Edited by John Mongillo
and Bibi Booth

Greenwood Press
Westport, Connecticut • London

To the memory of David R. Brower and Hazel Wolf, who gave meaning to the word *activists*. They will be greatly missed.

Library of Congress Cataloging-in-Publication Data

Environmental activists / edited by John Mongillo and Bibi Booth.
    p.   cm.
    Includes bibliographical references and index.
    ISBN 0–313–30884–5 (alk. paper)
    1. Environmentalists—United States—Biography.   I. Mongillo, John F.   II. Booth, Bibi.
GE55.E57   2001
363.7'0092'273—dc21   00–034129
    [B]

British Library Cataloguing in Publication Data is available.

Library of Congress Catalog Card Number: 00–034129
ISBN: 0–313–30884–5

First published in 2001

Greenwood Press, 88 Post Road West, Westport, CT 06881
An imprint of Greenwood Publishing Group, Inc.
www.greenwood.com

Printed in the United States of America

The paper used in this book complies with the
Permanent Paper Standard issued by the National
Information Standards Organization (Z39.48–1984).

10 9 8 7 6 5 4 3 2 1

Copyright Acknowledgments

The editors and publisher gratefully acknowledge permission for use of the following material:

Cover photos (*clockwise from upper left*): Rachel Carlson, by Bob Hines, used by permission of Rachel Carson History Project; Helen Kim, courtesy of Helen Kim; Denis Hayes, courtesy of Tara Middlewood; Fabiola Tostado, courtesy of James A. Brown; Robert Bullard, courtesy of Robert Bullard and the Enviromental Justice Resource Center, Clark Atlanta University; and Lois Gibbs, courtesy of Lois Marie Gibbs.

Material for Ansel Adams, Gloria Anable, Archie Carr, Rosalie Edge, George Grinnell, Olaus Murie, and Wallace Stegner entries taken from *Biographical Dictionary of American and Canadian Naturalists and Environmentalists*. Edited by Keir B. Sterling, Richard P. Harmond, George A. Cevasco, and Lorne F. Hammond. Copyright © 1997 by Keir B. Sterling, Richard P. Harmond, George A. Cevasco, and Lorne F. Hammond. Reproduced with permission from Greenwood Publishing Group, Inc., Westport, CT.

Contents

Introduction

Environmental Activists profiles a diverse array of American men and women whose work for environmental causes spans the period from the early 1800s to the present. These revolutionaries, reformers, and crusaders have devoted themselves to issues ranging from global deforestation and air pollution to environmental racism and the dumping of toxic wastes in our own communities and backyards. Undoubtedly, you will recognize some names from previous studies, but you may find yourself even more intrigued by the stories of activists who are new to you. The editors and contributors have provided a large selection of activists, though the assortment of featured profiles is not all-inclusive.

A typical dictionary might define an activist as "one who takes action to bring about change." But how activists set about causing change and what inspires them to do so vary widely from one individual to another, and from one historical period to the next. The approaches of the environmentalists profiled here have ranged from low-key and businesslike to aggressive, combative, and occasionally even outside the law. However much these activists have differed in the methods they have used, they nevertheless have much in common as a group. Caring, concerned advocates and devoted earth stewards all, they respect the natural world and place supreme value on the welfare of humans and other species. Field biologist George Schaller once stated that a conservationist was someone who was a "politician, diplomat, educator, sociologist, fundraiser, and, almost incidentally, a biologist." His description also aptly describes some of the many faces of environmental activism over the course of U.S. history.

The primary intent of this book is to present a broad variety of past and present activists involved in the diverse domain of environmental issues. The profiles encompass well-known activists from the early 1800s to the early 1900s, including Henry David Thoreau, Gifford Pinchot, Rosalie Edge, John James Audubon, and John Muir. In addition to such environmental pioneers, this book features some of the twentieth-century activists who built upon earlier philosophies and groundwork. This latter group includes Hugh H. Bennett, Ynes Mexia, Aldo Leopold, J. N. "Ding" Darling, and Rachel Carson. Many contemporary activists are profiled as well, including David Brower, Stephanie Mills, Ralph Nader, Winona LaDuke, Ruth Patrick, and Kirkpatrick Sale.

Environmental Activists also embraces men and women who have been part of the evolution and growth of the environmental justice movement. Their lifework has been devoted to achieving equality for people of color and for the disenfranchised, and to eliminating the disproportionate environmental burdens that have historically been placed on these communities. These activists include Robert Bullard, Helen Sunhee Kim, Chellis Glendinning, and Juana Beatríz Gutiérrez.

Since activism is fortunately not limited to any particular age group, this book also features the accomplishments of youth organizers such as Ocean Robbins, Nevada Dove, Fabiola Tostado, and Maria Perez, all of whom first involved themselves in environmental campaigns as teenagers. At the other end of the spectrum are those whose activism has lasted over long, astonishing lifetimes; David Brower (age 88), Ruth Patrick (92), Hazel Wolf (101), and Marjory Stoneman Douglas (108) are all wonderful examples. Finally, to provide historical context, we also present a timeline that locates the activities of individual activists in relation to the founding of environmental organizations and the enactment of important environmental laws and regulations.

What kinds of influences and experiences help create an environmental activist? If the activists profiled here are any indication, there must be a hundred different answers to this question, representing a hundred different ways "to get from here to there." Many of our activists were provoked into action when they were threatened with the loss of something they loved, perhaps a cherished patch of wilderness or the clean water of a favorite stream. Others responded when an environmental threat literally showed up in their workplaces, communities, or homes. Many of our activists credited the important people in their lives—family, teachers, mentors, and heroes—with helping them focus their energies on the natural world and environmental issues.

For example, aquatic scientist Ruth Patrick recalled that she was seven when she received her first microscope from her father. Together, they spent countless hours examining microorganisms gathered from local rivers, streams, and ponds. Robert Bullard's inspirations included his parents and the civil rights activists Frederick Douglass, Sojourner Truth, Malcolm X, and Martin Luther King, Jr. As a budding zoologist, George Schaller was convinced to attend the University of Alaska when his cousin described a campus surrounded by wilderness, whose black bears and moose were welcomed as frequent campus visitors.

When naturalist Roger Tory Peterson was a New York schoolchild, his

seventh-grade teacher enrolled him and his classmates in the Junior Audubon Club, which introduced them to the wonders of bird life. The
teacher often led the class on walks to a nearby forest, where she used
nature to teach writing, art, and science. Great Lakes activist Lee Botts
grew up in Oklahoma, where she loved to spend time on her grandparents'
wheat farm, a homestead acquired at the turn of the twentieth century. Her
wanderings about the land made her keenly aware of her relationship to
the natural world and its subtle changes from season to season, year to
year. To this day, Botts still keeps a little book containing the pressed
wildflowers she collected on that farm as a child.

Today, as the environmental consequences of industrialization, overconsumption, and population growth become increasingly clear, environmental activists are fighting on more fronts than ever. The long list of
contemporary environmental issues—from global warming to pesticides,
from hazardous waste to habitat loss—might seem to amount to an insurmountable obstacle. But thanks to the courage, devotion, and persistence
of today's activists, there is much reason to believe that we will find solutions to the earth's environmental problems. America's environmental activists are living proof that each of us can truly make a difference.

The editors and contributors made a prodigious effort to research and
interview a diversity of environmental activists for this volume in order to
enable each reader to find inspiration and individual motivation to become
more active in his or her community. We believe that the resulting assortment represents a variety that is unique among works of this type; you
might be pleasantly surprised to recognize a bit of yourself among the profiles on these pages. *Environmental Activists* provides an excellent starting
point to understand how individuals not only identify environmental problems and wrongdoings, but also develop the courage to do something about
them.

As we were putting the finishing touches on this book, we received a
thoughtful letter from Nevada Dove, Fabiola Tostado, and Maria Perez,
youth organizers for Concerned Citizens of South Central Los Angeles. In
their correspondence, the young women expressed gratitude that we were
acknowledging their efforts on behalf of their community. We were particularly gratified to read the following excerpt:

In recent times, the only role models children are exposed to seem to be adults.
This false reality gives youth an unrealistic view of role models. Growing up, had
we known that there were other youth leaders, we may have obtained the courage
and motivation to start earlier. Being among the first youth pioneers fighting for
environmental justice is an adventure all in itself and we hope that students who
use this textbook will realize that they too can make a difference.

ENVIRONMENTAL ACTIVISTS

EDWARD ABBEY
(1927–1989)

The love of wilderness is more than a hunger for what is always beyond reach; it is also an expression of loyalty to the earth, the earth which bore us and sustains us, the only home we shall ever know, the only paradise we ever need—if only we had the eyes to see.

—Edward Abbey

A wilderness writer, militant wildlands preservationist, and radical environmentalist, Edward Abbey was one of the American West's most passionate advocates and a spiritual father of the modern American environmental movement. Known to friends as "Cactus Ed," through his novels, essays, letters, and speeches, Abbey consistently voiced the belief that the West was in danger of being developed to death, and that the only solution lay in the preservation of wilderness. He was the author of more than twenty books that eloquently attacked the defiling of the West by builders of dams, dwellers in mobile homes, and developers of ski resorts. Many environmental activists say that it is Abbey's humor that keeps them motivated and prevents them from becoming too angry and burning themselves out. And it is Abbey's humor that keeps new readers flocking to his books more than ten years after his death.

The first of five children, Edward Abbey was born in the town of Indiana, Pennsylvania. Abbey's mother, Mildred Postlewait, was a schoolteacher and church organist who taught her son classical music and literature. His father, Paul Revere Abbey, was a noted anarchist, socialist, and atheist whose political radicalism rubbed off on his oldest son at an early age. (Later in life, Abbey rejected his father's socialism in favor of his own version of anarchism, yet in doing so, he was actually emulating his father's own independent streak.) As a young man, Paul Abbey had spent some time in the West as a ranch hand. His memories and mementos of the West were Edward's earliest boyhood incentives to head west himself.

While Abbey was growing up, his family moved frequently, living in poverty in the midst of the Great Depression; there were at least eight scattered residences in Indiana County in which they lived from just after Abbey's birth through his adolescence. They finally moved to a farmhouse nestled near a cornfield and acres of forest that Abbey later memorialized as "the Old Lonesome Briar Patch." This was the scene of everything that mattered most to Abbey about his boyhood. Near here were the "Big Woods" that the Abbeys loved.

Because Abbey's family's many moves in his earliest years evidently preceded or escaped his memory, and because he loved to link himself to appealing place names, Abbey always referred to Home, Pennsylvania, as his birthplace. This was the case not only in print, but also in his private diary, much of which was ultimately published as *Confessions of a Barbarian*. Later, while living in the West, Abbey claimed "Wolf Hole" or "Oracle" as his place of residence; though both are real Arizona locations, Abbey never lived in either of them. He simply liked the sound of their names and also preferred to misdirect fans after he became an environmentalist cult figure.

Abbey's future career and interest in writing began during high school. His first known publications appeared when he was in his mid-teens: an anti-Hitler editorial, "America and the Future," in the *Marion Center Independent* in December 1941 and "Another Patriot," a short story in the spring 1942 Marion Center High School compendium of student writings. But the life adventure that was most frequently recounted in Abbey's writings was his hitchhiking and rail-riding trip west during the summer of 1944, between his junior and senior years of high school. In numerous places in his writings and in interviews, Abbey cited this as the key formative experience of his life. "I became a Westerner at the age of seventeen, in the summer of 1944, while hitchhiking around the USA. For me it was love at first sight—a total passion which has never left me" (Abbey's Web online). Abbey first saw the beautiful red-rock country and canyons of the Southwest while riding the rails in boxcars. He was arrested in Flagstaff and finally abandoned his travels in Albuquerque and took a bus home. Abbey wrote about his trip in a series of seven striking articles in Indiana High School's *High Arrow* during his senior year there. "Abbey Walks 8,000 Miles by Adroit Use of Thumb," declared the lead headline. The *High Arrow* articles hint at a writer beginning to find his true subject and voice.

After high school, Abbey served as a military policeman from 1945 to 1947 in Naples, Italy. In 1947, he continued his education at Indiana University of Pennsylvania (known then as Indiana State Teachers College), where he enrolled in writing classes and published his first adult short story, the experimental piece "A Fugue in Time," in the 1948 issue of *The Indiana Student Writes*. The Federal Bureau of Investigation started its file on Abbey after he posted a letter against the draft while he was in college.

In 1948, at the age of twenty-one, Abbey left the town of Home and set off to see the West. He again hitchhiked and rode the rails across the Midwest to the Rocky Mountains and the west coast, returning by way of the Southwest. He fell in love with the desert, a love that shaped his life and art for the next forty years. Abbey earned a bachelor's degree in 1951 and a master's degree in 1956 at the University of New Mexico in Albuquerque. He also won a Fulbright Fellowship to Edinburgh University, Scotland, in 1951. The subject of his master's thesis was "Anarchism and the Morality of Violence." He entered a second graduate program at Yale University, but hated the constraints of the Ivy League and dropped out after two weeks.

Abbey subsequently worked for fifteen years as a part-time park ranger and fire lookout for the National Park Service in the Southwest, developing an intimacy with the region's landscape that would shape his writing career. Even as early as the 1950s, he depicted the Southwest not as a virgin utopia inhabited by rugged individualists, but as a region at risk from government and corporate greed, its peoples threatened with separation from the primary source of their spiritual strength: wild places. Abbey's two seasons as a ranger at Arches National Monument in Utah provided the inspiration and raw material for his first book, *Desert Solitaire* (1968), an extensive meditation on the wilderness of southeastern Utah and the human intrusions on it, which has been one of the most enduring works of contemporary American nature writing. Though Abbey has been called an "environmental writer," he preferred not to categorize his style at all, but rather to "let his prose do his talking" for him (Abbey's Web online).

The Monkey Wrench Gang, Abbey's infamous 1975 novel, is a tribute to both individual freedom and activism to save wildlands from development. The book follows the exploits of a band of ecological guerrillas who sabotage dams and bridges in the Southwest. Abbey's story captured the magnitude of the ecological, emotional, and spiritual damage that resulted from the erection of the Glen Canyon Dam on the Colorado River. Abbey contended, "The love of wilderness is more than a hunger for what is always beyond reach; it is also an expression of loyalty to the earth, the earth which bore us and sustains us, the only home we shall ever know, the only paradise we ever need—if only we had the eyes to see" (Abbey 1990, 167). The book inspired the establishment of many radical environmental groups, including Earth First!, the hard-core group of "ecowarriors" that has drawn attention to the plight of wilderness through direct action that is sometimes illegal.

Through both his fiction and nonfiction writings, Abbey constantly warned that all people should be troubled by the prospect of a nation that lacked wilderness. At the same time, he expressed optimism that the "fires of revolt may be kindled, which means hope for us all" (Yozwiak 1999). As he scribbled away during his early years in remote fire lookouts and in his later years at his Arizona home, his goal was always the same: to rescue the beauty that remained. Abbey was a genuine rebel who simply did not believe in the modern industrial way of life. "Resist much, obey little," a favorite quotation from Walt Whitman, was his ecowarrior motto.

Abbey believed that the New York literary establishment frequently misunderstood or ignored his efforts; at the same time, he could be uncooperative about receiving recognition. When he was invited to a banquet inducting him into the American Academy of Arts and Letters in 1987, Abbey's written response stated, "I appreciate the intended honor but will not be able to attend the awards ceremony on May 20th: I'm figuring on going down a river in Idaho that week. Besides, to tell the truth, I think that prizes are for little boys. You can give my $5,000 to somebody else. I don't need it or really want it" (Abbey's Web online).

Edward Abbey died in 1989 at his home near Tucson, Arizona, from surgical complications after four days of esophageal hemorrhaging. His stubborn disposition and his contempt for medical technology were likely factors in his failure to seek early attention for his condition. Decades of rough living may also have contributed to his premature death.

Before his death, Abbey had written a message to his fifth wife, Clarke, detailing what he wanted done for him in the event of his demise. According to his message, he wanted his body transported in the bed of a pickup truck, he wanted to be buried as soon as possible, and he wanted no undertakers, no coffin, and no embalming—just an old sleeping bag. Abbey also requested that his family and friends disregard all state laws concerning burial. "I want my body to help fertilize the growth of a cactus or cliff rose or sagebrush or tree," he had written. "If my decomposing carcass helps nourish the roots of a juniper tree or the wings of a vulture—that is immortality enough for me. And as much as anyone deserves" (Abbey's Web online). He is buried—illegally, as requested—somewhere in the Cabeza Prieta Desert in southern Arizona. "No death's ever affected me that much," remembers Dave Foreman, one of the founders of Earth First! Abbey, he says, was "a visionary, the greatest inspiration my generation of conservation activists in the West ever had" (Bergman 1998). Said another friend, author William Eastlake, "Ed decided early in life to save the planet, and he damn near did. It's up to us to carry on, so Ed's will not be the last voice heard crying in the wilderness" (Yozwiak 1999).

The Abbey Collection at the University of Arizona contains large quantities of archival material, so future generations of Abbey scholars will have ample resources for many more articles and books. All of Abbey's books have remained in print except for his first novel, *Jonathan Troy*, which Abbey disdained and refused to have reprinted.

BIBLIOGRAPHY

Abbey, E. (1976). *The Monkey Wrench Gang*. New York: Avon Books.

Abbey, E. (1990). *Desert Solitaire: A Season in the Wilderness*. New York: Touchstone.

Abbey, E., & Peterson, D. (Eds.). (1994). *Confessions of a Barbarian: Selections from the Journals of Edward Abbey, 1951–1989*. Boston: Little, Brown and Co.

Abbey's Web online. *http://www.utsidan.se/abbey/bio/index.html*

Bergman, B. J. 1998. Wild at heart: Environmentalist Dave Foreman. *Sierra, 83*, pp. 24–33.

Biography online. Abbey, Edward. *http://www.biography.com/cgi-bin/biomain.cgi*

Encyclopaedia Britannica online. Abbey, Edward. *http://www.britannica.com*

Magnuson, E. (1986, June 16). This land is our land: From craggy Atlantic shores to the western vistas, America's natural beauty is a precious heritage, preserved by a model system of national parks. *Time*, pp. 40–41.

Milestones: Edward Abbey. (1989, March 27). *Time*, p. 85.

Shealey, T., Gorman, J. Harlin, J., Hogan, T., Hostetter, K., McGirney, A., Morris, M., & Dorn, J. (1998). In a league of their own: *Backpacker* Silver Service Awards. *Backpacker, 26*, p. 66.

Temple, E. (1982). Interview with Edward Abbey. Canyon Productions, Arizona. Available: *http://www.utsidan.se/abbey/articles/etemple/index.html*

Whitman, D. (1999, June 21). The grand parking lot? *U.S. News & World Report*, p. 18.

Yozwiak, S. (1999, March 7). Legacy on the land: Environmentalists still follow author Abbey's call to action. *Arizona Republic*, p. E1.

—*Bibi Booth*

ANSEL EASTON ADAMS
(1902–1984)

Ansel Adams, an American nature photographer, educator, mountaineer, and environmentalist, was born in San Francisco, California, on February 20, 1902. The only child of Olive Bray and a prosperous businessman, Charles Hitchcock Adams, Adams grew up in California. His father strongly influenced the young Adams, teaching him to appreciate the natural beauty of California and instilling in him a strong Puritan work ethic, sense of duty, and spirit of self-reliance. A hyperactive child in grammar school, Adams required the help of private tutors to complete grade school, which he did in 1917. In 1928, Adams married Virginia Best, the daughter of the proprietor of a photographic studio in Yosemite, California. The Adamses had two children.

Early in his career, Adams worked as a custodian at the Sierra Club headquarters at Yosemite National Park in California, where he led mountaineering trips in the 1920s. The Sierra Club then began to publish Adams's photographs and articles in the *Sierra Club Bulletin*. Adams opened his own commercial studio in 1930 and attracted a diverse clientele. In 1934, Adams and his colleagues formed "Group fl64," an association dedicated to the advancement of photography as an art form. Important early exhibitions of Adams's work took place during the 1930s in major New York studios.

During World War II, Adams taught practical photography and printed top-secret military negatives. Also during this period, Adams became the official photographer of the Mural Project, documenting Native American reservations, national parks, and other facilities controlled by the U.S. Department of the Interior. Photography, a favorite hobby of his father, became his passion and career. Throughout his life, Adams maintained an enduring admiration for his father and cherished his early outdoor experiences. In his words, "I constantly return to the elements of nature that surrounded me in my childhood, to both the vision and the mood."

In the early 1940s, Adams developed a photographic "zone system" that encompassed a range from total black (zone zero) to pure white (zone ten). Adams's major professional accomplishment was his masterful use of black-and-white photography to reflect his love of nature and to plead for the preservation of wilderness. Adams called this technique "visualization," wherein the photographic artist intuitively searches visual and emotional qualities of a completed picture. Adams believed that through his work he had "been able to have had some small effect on the increasing awareness of the world situation through both my photographs and my social assertions." In 1934, he was elected to the Board of Directors of the Sierra Club, which he served throughout his career. He was elected an honorary vice president of the Sierra Club in 1979.

Adams published throughout his career, beginning about 1931 with his photography reviews for *The Fortnightly*. In 1935, Adams published *Making a Photograph: An Introduction to Photography*. His photographs of the Sierra Mountains and the Southwest United States were published in the first historical survey of photography by New York's Museum of Modern Art. In 1938, Adams published *Sierra Nevada: The John Muir Trail*. His World War II photo-essay collection *Born Free and Equal* (1944) reflected his view of imprisoned Japanese Americans.

Adams undertook several special assignments in his career. In 1946, 1948, and 1950, he received Guggenheim Fellowships, which permitted him to work in Alaska and Hawaii. He joined President Lyndon Johnson's environmental task force in 1965. In 1966, Adams was elected a fellow of the American Academy of Arts and Sciences. In 1967, he founded Friends of Photography in Carmel, California. In 1974, he made his first trip to

Europe, to teach at the Arles, France, photography festival. In the following year, he established the Center for Creative Photography at the University of Arizona, Tucson. President Jimmy Carter awarded him the Presidential Medal of Freedom in 1980. Adams received honorary doctorates from the University of California (1961) and Harvard University (1981). After Adams died in San Francisco in 1984, a peak in Yosemite National Park was named Mt. Ansel Adams in his honor.

BIBLIOGRAPHY

Adams, A. (1990). *The American Wilderness*. Ed. A. G. Stillman. Boston: Bulfinch Press.
Adams, A., & Alinder, M. S. (1985). *Ansel Adams: An Autobiography*. Boston: Bulfinch Press.
Sterling, K. B., Harmond, R. P., Cevasco, G. A., & Hammond, L. F. (Eds.). (1997). *Biographical Dictionary of American and Canadian Naturalists and Environmentalists*. Westport, CT: Greenwood Press.

—*John Mongillo*

GLORIA ELAINE "GLO" HOLLISTER ANABLE
(1901–1988)

Gloria Anable, a conservationist, zoologist, and ichthyologist, was born in New York City on June 11, 1901. She was the daughter of Frank Canfield Hollister and Elaine Shirley Hollister. Her father, a physician and diagnostician, encouraged her to explore nature and guided her in this process. Anable attended Miss Rayson's School in New York until 1919 and then spent one year at Hillside School in Norwalk, Connecticut. She received a B.S. in zoology from Connecticut College in 1924 and an M.S. in zoology from Columbia University in 1925. Gloria married Anthony Anable, an engineer, in October 1941.

During her early career, Anable worked as a cancer research assistant under Alexis Carrel. She then joined William Beebe, head of the Department of Tropical Research at the New York Zoological Society, and began her research in ichthyology. On her first scientific trip with Beebe, the Bermuda Oceanographic Expedition in 1928, the objective was to discover and categorize the numerous forms of aquatic life found in the waters off the coast of Bermuda. During the expedition, Anable studied deep-sea life from a steel bathysphere. Anable accompanied Beebe's research team on the West Indies Oceanographic Expedition (1932–1933) and the Pearl Islands Oceanographic Expedition (1934), both of which concentrated on the study and collection of marine fishes. On June 11, 1932, Anable set the women's record of 410 feet for the deepest oceanic descent.

Anable's research focus changed to zoology when she led three expeditions for the Department of Tropical Research. Two Trinidad zoological expeditions (1926, 1936) afforded her the opportunity to explore the Arma Gorge and study and photograph the unusual guácharo, or "oilbird." As a participant on the 1936 British Guiana Expedition, Anable had to traverse dense, tropical jungle to reach the Kaieteur Plateau and Falls. On this trip, she discovered the half-inch golden frog and the Canje pheasant, or hoatzin, a bird having both avian and reptilian features. During World War II, Anable left the Zoological Society to help the American National Red Cross establish the nation's first blood-donor center in Brooklyn, New York.

Referring to the desire to experience the raw natural beauty of the environment, Anable wrote, "Man, whether he realizes it consciously or not, has an urge to run away from what he has made for himself in order to renew his contact with the world as it was before he put his mechanical mark on it" (Sterling et al.). Anable dedicated her life and talents to the exploration of nature, for which she maintained a strong personal reverence. Among her important accomplishments in the fields of ichthyology and zoology were her discoveries of the guácharo, the golden frog, and the hoatzin. As a result of her chairmanship of the Mianus River Gorge Conservation Committee and her successful challenge of the proposed dam construction and subsequent development, she ultimately created a 555-acre nature preserve heralded as one of the finest examples of northeastern flora and fauna.

Anable wrote prolifically about her expeditions and discoveries, chiefly in *Bulletin of the New York Zoological Society, Zoologica*, and *The American Girl*. Important career positions Anable held included chairman and chairman emeritus of the Mianus River Gorge Conservation Committee of the Nature Conservancy from 1953 to 1988 and director of the Stamford, Connecticut, Museum and Nature Center. She served as national vice chairman of Conservation Committee of the Garden Club of America and participated actively in the Girl Scouts of America. Anable died in Connecticut in 1988.

BIBLIOGRAPHY

Sterling, K. B., Harmond, R. P., Cevasco, G. A., & Hammond, L. F. (Eds.). (1997). *Biographical Dictionary of American and Canadian Naturalists and Environmentalists*. Westport, CT: Greenwood Press.

—*John Mongillo*

JOHN JAMES AUDUBON
(1785–1851)

Nature must be seen first alive and well studied before attempts are made at representing it.
—John James Audubon (Princeton Audubon online)

A Haitian-born, French-American ornithologist and ardent conservationist, John James Audubon was noted for his sketches and paintings of birds and was one of the first U.S. naturalists. His intricately detailed compositions depicted scenes he had actually witnessed in nature, though he has been criticized during the twentieth century for having killed the animals he used as models. His admirers note that before the invention of photography and high-resolution binoculars, killing animals was the only possible way to render accurate images of them. Audubon is credited with being the first to band the legs of American wild birds and was one of the earliest naturalists to pioneer conservation.

Audubon was born in Les Cayes, Santo Domingo (now Haiti), the illegitimate son of a successful French navy captain, sugar planter, and slave dealer. Audubon's biological mother, a young Creole woman who was a plantation employee, died within a short time after his birth. In 1789, Audubon's father took him and his half sister back to France, where the captain and his childless wife legally adopted both children in 1794. Audubon's early education among the upper classes resulted in a deep interest in natural history, art, and music. By the age of fifteen, he was sketching French birds, and at seventeen, he studied drawing with the famous neoclassical painter Jacques-Louis David in Paris.

In 1803, to avoid the French military draft, Audubon moved to a family-owned plantation outside Philadelphia. He began to experiment with bird banding and the investigation of avian migration; largely through his own observations and self-study, he developed into a top-notch field ornithologist. In 1808, he married Lucy Bakewell, with whom he eventually had two sons. During the early years of his marriage, Audubon was so passionately absorbed in his nature studies that he failed in the businesses he had attempted to establish along the Kentucky border. In 1819, he was even imprisoned for debt and was declared bankrupt.

In 1820, Audubon set a goal to publish a comprehensive anthology of life drawings of animals. In search of new birds to study and collect, in the years immediately following, he explored the Ohio and Mississippi rivers and the Great Lakes. Before Audubon, most painters of birds had used highly stylized drawing techniques, often employing stuffed specimens as subjects. Audubon's approach was very different. To produce his life-sized paintings, Audubon first intensively studied and sketched his subjects in the field, carefully noting how they moved, their behaviors, and the habitats they preferred. He then shot or trapped the animals to take with him. At his home studio, he bent his subject into lifelike, characteristic poses after running support wires through its body. Audubon's understanding of birds, along with his artistic talent, lent his drawings and paintings a compelling, dramatic realism that made the stiff depictions of other bird artists seem lifeless and dated. He also interspersed his bird studies with writings on American life during the nation's tumultuous beginnings.

Audubon moved to New Orleans in 1821, where his wife soon was providing much of the family income through tutoring. Audubon contributed some funds by painting portraits and teaching art. When he was unable to find patrons or a publisher in America for the paintings he had produced, he traveled to London in 1826. His paintings—more than 400—were published in four volumes in huge "double elephant folio" format in London as *Birds of America* between 1827 and 1838. The work enjoyed immediate success. In Paris, for example, the eminent scientist Georges Cuvier declared that the paintings were "the greatest monument ever erected by art to nature." In the course of promoting his book, Audubon educated many people in both North America and Europe about the New World and its fauna, becoming a reluctant icon for naturalists of the time. Audubon never accepted offers for the individual prints that comprised *Birds of America*; he insisted that the book be sold intact, which is one reason that his prints are so rare today. William MacGillivray, a Scottish naturalist, collaborated with Audubon and supplied most of the scientific data for the accompanying text, *Ornithological Biography*, which described the habits of Audubon's birds. That work appeared in five volumes between 1831 and 1839.

On his return to the United States in 1839, Audubon published a bound edition of his *Birds of America* plates, with some additions. In 1841, he settled on a rural estate along the Hudson River (now known as Audubon Park) in New York City. Audubon set off again in 1843, when he traveled from St. Louis up the Missouri River to Fort Union, Montana, and then overland to the Yellowstone River. By the time of this excursion, Audubon had already become well known in the United States for his *Birds of America*. The purpose of the trip, which was his last great adventure before his death, was to gather specimens for a new series of paintings of American mammals. His mammal project proved more difficult than he had anticipated, however, as he found that many of his intended subjects were nocturnal and their habits were hard to learn.

Audubon returned to St. Louis spectacularly attired in Native American hunting dress and accompanied by live animals to serve as models for his new series of paintings. With the help of his sons and the naturalist John

Bachman, Audubon began preparation of *The Viviparous Quadrupeds of North America* in 1845. He died in New York on January 27, 1851; the work was completed by his sons and Bachman and published in three volumes in 1846–1854. In 1886, the National Audubon Society, America's first bird-preservation organization, was founded and named in his honor.

BIBLIOGRAPHY

Audubon Society online. John James Audubon, 1785–1851. *http://www.audubon.org/nas/jja.html*

Audubon's wilderness palette: The birds of Canada. (1998, September 16). *Canadian Corporate News*, Available: *http://www2.cdn-news.com/scripts/cnn-release.pl?1998/09/16/0916016n.html*

Biography online. Audubon, John James. *http://www.biography.com/cgi-bin/biomain.cgi*

Colby, J. H. (1994, September 22). Traveling exhibit reopens debate on Audubon's values. Gannett News Service.

The Columbia Encyclopedia online. Audubon, John James. Available: *http://www.elibrary.com*

Di Silvestro, R. (1994, October/November) Stories behind the paintings. *National Wildlife, 32* (6), pp. 52–58.

Drawn to nature. (1994, November 3). *St. Louis Post-Dispatch*, p. 1G.

Encyclopaedia Britannica online. Audubon, John James. *http://www.britannica.com*

Encyclopedia.com. Audubon, John James. *http://www.encyclopedia.com*

Funk and Wagnalls Encyclopedia online. Audubon, John James. *http://funkandwagnalls.com*

Gray, P. (1997, June 9). Inspired naturalist: A Smithsonian show celebrates the science and art of America's birdman John James Audubon. *Time*, p. 85.

Hume, C. (1998, September 20). Birds of a feather. *Toronto Star*, p. D3.

The Hutchinson Dictionary of Scientific Biography online. Audubon, John James Laforest. Available: *http://www.elibrary.com*

Princeton Audubon online. *http://www.princetonedition.com/audubon.htm*

The Reader's Companion to American History online. Audubon, John James. Available: *http://www.elibrary.com*

Shaw, L. (2000, January 1). Leaving home, family to pursue America's birds and mammals. *Washington Times*, p. 8B.

Shotwell, R. (1998). Audubon: Life and art. *Sierra, 83,* 78.

Tuleja, T. (1994). Audubon, John James. The New York Public Library Book of Popular Americana online. Available: *http://www.elibrary.com*

University of Nebraska Museum of Nebraska Art online. *http://monet.unk.edu/mona/artexplr/audubon/audubon.html*

Wallach, A. (1993, November 28). Is it a bird? Is it a cliché? Is it politics? Is it art? A John James Audubon painting can be all the above—and more. *Newsday*, p. 16.

Young Students Learning Library online. Audubon, John James. Available: *http://elibrary.com*

—*Bibi Booth*

Photo courtesy of National Archives and Records Administration.

HUGH HAMMOND BENNETT
(1881–1960)

> Even in times of great agricultural prosperity, land destroyed by erosion
> will fail to support a prosperous agriculture.
>
> —Hugh Hammond Bennett

In an article written in 1944, Herbert Corey, in *Nation's Business*, described Hugh Hammond Bennett as "the foremost soil conservationist in
the world. So say other soil scientist." Many others today refer to him as
the "father of soil conservation." Bennett was born in 1881 in what was
called the Lick Skillet neighborhood near Wadesboro, North Carolina. As
one of nine children, Bennett lived, worked, and grew up on his father's
cotton farm. The name Lick Skillet was given by the local farmers. At one
time, it was a good area for farming—a place of fertile soil and prosperous
times. But overfarming, the removal of trees, and the erosion of topsoil
caused many farmers in the area to move away or go into bankruptcy. The

young Bennett observed the problems of soil erosion. But while other farms failed, the Bennett farm was successful and well managed.

Bennett's father seemed to know a thing or two about preventing soil erosion on his farm. Herbert Corey reported a story that when Bennett was seven years old, he tagged along one day at his father's heels. The older Bennett was working through his fields with a spade. "What are you doing that for?" asked the boy. "Making all those little dams." His father responded, "To keep the good earth from washing away." Bennett always remembered those eight words his father spoke.

All of the children of William Osborne Bennett had the opportunity to go to college. Hugh Bennett went to the University of North Carolina at Chapel Hill, but took a few years off to earn enough money to continue his education. In 1903, after earning a degree in chemistry from the University of North Carolina, Bennett became a soil surveyor with the U.S. Department of Agriculture's Bureau of Chemistry and Soils. Bennett specialized in conducting soil surveys in the southern part of the United States. He also had the opportunity to go abroad to study farming conditions. He visited Alaska and various Central and South American countries as well as the Panama Canal. During World War I, he served as a lieutenant of engineers.

Working outdoors for many years, Bennett observed the effects of erosion on soil and the need to conserve soil and water resources. In the 1920s and early 1930s, he gave numerous speeches on the radio and in person at agricultural events, on college campuses, and to farmers and scientists. He wrote hundreds of articles and several books about the threat of soil erosion. One of the best-known government articles he coauthored was *Soil Erosion: A National Menace*. He soon became a leading soil-conservation advocate and crusader recognized by legislators and the general public. In 1929, he used his expertise to advise the U.S. Congress to fund a series of soil-erosion experiment stations under his supervision.

In 1933, Bennett was given the opportunity to put his soil-conservation ideas into practice to help reduce soil erosion. He became the director of the Soil Erosion Service (SES) created by the Department of Interior. By this time, a severe drought was occurring in the southwestern section of the United States. In time, the area would be known as the Dust Bowl. It covered parts of Oklahoma, Texas, Colorado, New Mexico, and Kansas. Since the 1900s, farmers had been overfarming. The soil was thinning out. As a result, the light topsoil, stripped of its protective cover that anchors the soil to the landscape, was easily picked up by the winds and formed into dense sandstorms that could be seen hundreds of miles away. (Lee Botts, who is profiled in this book, observed the dust storms firsthand as a third-grader.) The wind erosion was so intense that most farm families and about 50 percent of the population in the Great Plains moved to other places. Estimates based on a reconnaissance erosion survey of the United

States made in 1935 show that approximately fifty million acres of formerly productive agricultural land had been virtually ruined for further cultivation. Another fifty million acres were in bad condition.

In 1935, the SES was moved to the U.S. Department of Agriculture and renamed the Soil Conservation Service (SCS). The Soil Conservation Service was set up by Congress primarily to demonstrate to the farmers of the country how soil erosion could be controlled or prevented. Bennett continued as the chief of this new agency.

Bennett and his colleagues devised an entirely new approach to land management: putting every acre to its best use and treating every acre according to its needs. Thus was born the conservation plan based on soil resources. Bennett's concept involved the ecology of the entire farm—woods, crops, pastures, and wildlife. Bennett also believed in a hands-on approach to solving problems. He advocated that his SCS staff work directly with farmers rather than from some distant office.

By December 1935, the Soil Conservation Service was employing from relief funds more than 32,000 workers (not including Civilian Conservation Corps [CCC] enrollees) on erosion-control demonstration and work projects. The workers were engaged in such outdoor tasks as seeding, tree and shrub planting, and other improvements on depleted pastures and steep cultivated fields. They built dams and dikes to divert water for irrigation and constructed diversion waterways to stop the soil erosion of expanding gullies. They installed systems for contour cultivation and strip cropping. Other work included resodding worn-out fields for permanent pasture. Most historians agree that Bennett's ideas changed the face of America's landscape. On farmland scarred by gullies and blown by winds, lush fields of grain once again waved. Conservation practices like strip cropping, terraces, and waterways had stopped the erosion of the soil and had returned the land to its former productivity.

Along with the soil-conservation plan, the revolutionary idea of soil-conservation districts was born. In 1937, President Franklin D. Roosevelt sent to the governors of all states legislation that would allow the formation of soil-conservation districts. The first district established was the Brown Creek Soil Conservation District, which included Bennett's home county, Anson.

Bennett's success as a soil-conservation advocate was due to his great skills as a communicator. As mentioned earlier, he gave a number of speeches. In the speech *Erosion, an Old Process*, Bennett discusses how early civilizations practiced soil conservation.

Not all men have been so blind to the effects of erosion. The process is an old one. It began in the dim past, probably in the first sloping field cultivated by man. In various parts of the world the evil has been strenuously opposed for many centuries. There are terraces in the Mediterranean basin on which olive trees a thousand years old are still growing. Some of these ancient walled benches were constructed before the time of Christ, built to hold the land in place for continuing agricultural use. In the hinterlands of the Philippine Islands the aborigines were building terraces on steep mountain slopes at the time of the first visit by civilized man. The same thing was true of the ancients of the continent to the south of us, as in parts of Peru. Washington gave up the growing of tobacco, planted clover and practiced crop rotations on his Mount Vernon estate in order to check erosion. You can still see ancient gullies on the lands that once were his. Their proximity to venerable cedars and oaks indicate that they were there in his time.

In another speech, Bennett described the soil-conservation ideas of Thomas Jefferson:

In 1813 Thomas Jefferson, writing about his farm in Albemarle County, Virginia, said: "Our country is hilly and we have been in the habit of plowing in straight rows, whether up or down hill, in oblique lines, or however they lead, and our soil was all rapidly running into the rivers. We now plow horizontally following the curvature of the hills and hollows on dead level, however crooked the lines may be. Every furrow thus acts as a reservoir to receive and retain the waters, all of which go to the benefit of the growing plant instead of running off into the streams."

Bennett summarized Jefferson's observation by stating that many of Jefferson's fields are still in good shape, but those of some of the farms adjacent to his (Jefferson's) estate are terribly gullied. The contrast is enough to suggest that some of his neighbors were not convinced of his assertion that the soil in that part of the country was rapidly running off into the rivers.

In 1951, Bennett retired as chief of the Soil Conservation Service. His influence had spread worldwide as many other nations adopted the programs he had established. He was one of the founders of the Soil Conservation Society of America (now the Soil and Water Conservation Society).

Bennett believed that Americans had never really learned to love the land but only regarded it as an enduring resource. Bennett believed that on a long list of nature's gifts, none is so essential to life as soil. The topsoil lying at an average depth of seven to eight inches is the principal feeding zone of the plants that provide food for human or livestock consumption, fiber for clothing, and timber for shelter. Soil constitutes the physical basis of our agricultural enterprise.

On July 7, 1960, Hugh Hammond Bennett died of cancer in Burlington, North Carolina, at the age of seventy-nine. He is buried in Arlington National Cemetery. His legacy lives on in the National Resources Conservation Service (formerly the SCS), the nation's 3,000 soil-conservation districts, and the land he worked so hard to protect and enrich.

BIBLIOGRAPHY

Brink, Wellington. (1951). *Big Hugh: The Father of Soil Conservation*. New York: Macmillan.

Corey, Herbert. (1944, April). Scientist with his feet in the topsoil. *National's Business.*

Hugh Hammond Bennett, Father of Soil and Conservation. U.S. Department of Agriculture. *http://www.nc.nrcs.usda.gov/History/bennett.htm*

United Resources Conservation Service. U.S. Department of Agriculture. (1938, March 9). Erosion and rural relief. Speech before the Special Senate Committee to Investigate Unemployment and Relief.

—John Mongillo

PETER BERG AND JUDY GOLDHAFT
(1937–   and 1939–   )

> Ask the next person you meet where they live. They'll probably reply by giving a house number, street name, town, state, or country. Most people won't mention anything about local natural systems.
>
> —Peter Berg

Planet Drum Foundation (*www.planetdrum.org*), founded in 1973 in San Francisco, California, is an effective grassroots organization that emphasizes sustainability, community self-determination, and regional self-reliance. Peter Berg, the director and founder, and Judy Goldhaft promote a bioregional orientation and encourage the general public to rethink their attitudes about the places where they live. The two conduct seminars and workshops on bioregional topics for various groups. They define a bioregion as an interconnected geographic area with similar patterns of plant and animal life and climatic and geological characteristics. Bioregions also include unique cultures that have developed out of living there in a sustainable way.

Peter Berg was born in 1937 in New York. At an early age, he moved to Florida and was exposed to a completely new environment. He experienced hurricanes, hunted in the Everglades, and in his early teens worked aboard a fishing boat in the Gulf of Mexico and southern Atlantic. He witnessed a wide variety of natural phenomena. Judy Goldhaft grew up in agricultural South (New) Jersey. Her father was a veterinarian as were her grandfather, aunt, and uncle. They worked together and pioneered and developed vaccines for poultry and other animals. Goldhaft grew up in an extremely rural area learning about the native plants, birds, and animals and walking to a two-room elementary school. She began formal ballet classes when she was nine.

Berg went to college in the segregated South, at the University of Florida, and he was part of a group that opposed segregation. He put up signs that said, "Integrate in '58." He went into work of various kinds in the civil rights movement, including sit-ins. Berg felt that the oppression of racial

Photo by R. D. Deines, New
Settler Interview.

Photo © Erik Weber 1999.

segregation was so strong that opposing it influenced the next ten years of
his life. After high school, Goldhaft went to Cornell University and graduated with a bachelor's degree. Because of her continuing interest in dance,
she went on to receive a master's degree in it from Mills College. Goldhaft
began working with the San Francisco Mime Troupe, a radical theater
group, where she met Berg. They both wrote, directed, and performed
"guerrilla theater." They decided to start Planet Drum Foundation after
travelling across North America visiting back-to-the-land groups and as a
result of Berg's attending the 1972 United Nations' Conference on the Environment in Stockholm.

At Planet Drum, Berg and Goldhaft conduct workshops to help attendees, especially young people, understand the natural systems that are local
to their bioregions. According to Berg and Goldhaft, this concept is not an
easy one for people to understand because they are too immersed in industrial society; people are also disinhabited from the places in which they
live by the electronic media and the educational system. These sources tend
to provide information in an abstract, dislocated way. As a result, people
do not receive much information about (1) the native species in their region,
(2) the origin of the water they drink, (3) the watershed where they live,
and (4) other natural features that are indigenous to their bioregion.

In one of their workshops, Berg and Goldhaft use a book, *Discovering
Your Life-Place*, written by Berg, that is primarily a workbook with a map-
making exercise. Participants are asked to draw the place where they live
using natural characteristics. The exercise starts with the body of water
that is nearest to them. They draw and label the hills and the other features

that cause the flow of water to the lake or river near them; in other words, they diagram the watershed they live in, and go on to identify its soils as well as the native plants and animals. Then they are asked to indicate what is the best thing people are doing to harmonize with these elements and what is the worst thing. The point of the workshop is to create study groups and activist groups around bioregional issues. Adults have participated in these exercises as well as teenagers, primary school children, and college students.

Berg was a teacher for two years in a school that taught mathematics by using examples from nature. Students also did water testing to learn the chemistry of pollutants. In their biology class, they were actively involved in the rearing of salmon to restore them to the local stream.

Planet Drum is also interested in urban sustainability projects. In 1986, Planet Drum brought together representatives of environmental working groups, small businesses, government agencies, and activist organizations to recommend needed changes. The conference resulted in a book, *A Green City Program for the San Francisco Bay Area and Beyond*, that is a comprehensive platform for urban areas in general. Goldhaft describes some elements of a green city anywhere. As an example, green cities would incorporate native vegetation and native species, fewer cars and a good public transportation system that would not use fossil fuels but alternative energy, other city uses of solar or wind power, and the establishment of community gardens where people could grow their own food. Green cities would also have events and celebrations in which music, dance, poetry, and the cultural history of the area going back to indigenous people would be featured.

Another program is called Education + Action. The Planet Drum staff take schoolchildren and expose them for at least a day to some project that has to do with urban sustainability, for example, a garden, a mural of native plants and animals, or stenciling words on the drains in their neighborhood. They write, "Don't dump, goes to the bay" with a fish and some waves in blue paint.

Berg has further commented that urban sustainability would include water conservation and treatment of wastewater. He particularly likes the way that sewage is treated in Arcata, California, where the tertiary or final form of sewage flows into sloughs and ponds in the actual living bay but doesn't pollute the water. Because it is the final form of sewage and because of the abundance of marsh plants, the sewage, rather than being a pollutant, provides food for native birds and fish. This is a very successful sewage system that doesn't use toxic chemicals and produces usable waste as the end product. Any city at this point should be seriously considering whether it wants to continue to use fresh, pure water to flush its toilets. The shortage of water is probably one of the biggest of the human ecology issues that are going to loom up in the next ten years, especially in the American West. At this point, cities should be seriously converting to dual water systems, with water that has been used once but is still usable, for example, shower water, used to flush the toilet rather than what we are doing, which is the equivalent of taking pure, bottled water and upending it each time one uses the toilet.

Planet Drum also encourages the use of local currencies. Berg reported how Ithaca, New York, uses an Ithaca dollar. In Berkeley, California, a similar kind of money is called Bread. This consists of a form of exchange in paper bills that one gets by performing services for people and that one can use to have other services done. In some cases, actual products are "sold" this way as well. The establishment of a local currency clearly points up the difference between imported and native energy and economy. Berg believes people must begin to restore a sense of a local economy that is good for the people who live there and different from the globalist economy that may seem to be good for them but mostly benefits external, multinational, financial concerns.

Presently, Planet Drum is working with a small city in Ecuador that after a series of natural disasters has proclaimed itself an eco-city and intends to rebuild itself in a way that is genuinely harmonious with the natural systems around it. No city has completely done this so far. The city is requiring itself to do this, by law, so Berg is very eager to help with this effort and assist local people and local activities. During its third visit in early January 2000, Planet Drum initiated a reforestation project for the hillsides around the city so that it does not have mudslides like those that in the past have killed as many as sixteen people in one small barrio. The project was completed in mid-summer with more than 10,000 native plants of various species which will prevent erosion and create an urban "wild corridor."

Judy Goldhaft created, directed and performed Reinhabitory Theater to promote the bioregional concept. She currently presents her performance piece, *Water Web*, at bioregional gatherings and other ecological events.

Berg and Goldhaft believe that their mission is to promote the idea of bioregions and to carry out activities that are useful to bioregionalism. As they explain, although bioregionalism is very specific to local areas, local people often do not have the wherewithal or information to begin their own group. Planet Drum has assisted in the formation of hundreds of bioregional groups. Planet Drum has also managed to spread its concepts beyond the United States. It is now a North American phenomenon, with many bioregional groups in Canada and Mexico, and the concept has become international. There are also groups in Australia, Japan, Italy, France, Spain, and South America. Berg and Goldhaft believe that the two most important proactive directions for ecological action in the future will be (1) ecosystem restoration of native plants, animals, and habitats, the recreation of forests and wild rivers, and the removal of dams, and (2) urban sustainability, the greening of cities to make them harmonious with the natural systems of their bioregions.

BIBLIOGRAPHY

Berg, P. (1978). *Reinhabiting a Separate Country. A Bioregional Anthology of Northern California.* San Francisco: Planet Drum Foundation.
Berg, P. (1995). *Discovering Your Life-Place: A First Bioregional Workbook.* San Francisco: Planet Drum Foundation.
Berg, P. (1999). *Dispatches from Ecuador.* San Francisco: Planet Drum Foundation.
Berg, P., Magilavy, B., & Zuckerman, S. (1990). *A Green City Program for the San Francisco Bay Area and Beyond.* San Francisco: Planet Drum Foundation and Wingbow Press.
Berg, P., & Robertson, R. (1998/1999, Winter). *A Deep (Ecology) Breath Before 2000, Raise the Stakes Anthology I. Raise the Stakes [The Planet Drum Review].* No. 29. San Francisco: Planet Drum Foundation.
Berg, P., & Wong, A. (1999/2000). *Eco-Realism about Water, Food, Cities, Raise the Stakes Anthology II. Raise the Stakes [The Planet Drum Review].* No. 30. San Francisco: Planet Drum Foundation.
Goldhaft, J. (1990). *Water Web.* San Francisco: Planet Drum Foundation.

—Peter Berg and Judy Goldhaft

MURRAY BOOKCHIN
(1921–   )

> Nearly all the nonhuman life-forms that exist today are, like it or not, to some degree in human custody, and whether they are preserved in their wild lifeways depends largely on human attitudes and behavior.
> —Murray Bookchin

Murray Bookchin is an anarchist, writer, and educator who has been an influential voice in the radical ecology movement for more than forty years. Bookchin regards the emergence of humans' attempts to dominate nature as being primarily due to the domination of humans by other humans, particularly within oppressive social and political structures. His views, which were unique in the 1950s and early 1960s, have since entered the general consciousness of our time as a result of the subsequent writings of others, including many ecofeminists. By advancing these ideas when he did, Bookchin exercised a strong and steady influence on the international development of radical ecological thought. From the late 1970s onward, he has been an important stimulus in the developing green (radical ecology) movements throughout the world, and he has written many works dealing with the nature and future of green politics.

Bookchin makes a distinction between "radical ecology" and "environmentalism." In his view, environmentalism addresses the human habitat as a passive habitat that people use, an assemblage of things called "natural resources" and "urban resources." He says, "Ecology, by contrast . . . is an

outlook that interprets all interdependencies . . . non hierarchically. . . .
Moreover, it affirms that diversity and spontaneous development are ends
in themselves, to be respected in their own right. . . . [T]his means that each
form of life has a unique place in the balance of nature and its removal
from the ecosystem could imperil the stability of the whole" (Biehl). In
Bookchin's view, human inequality is at the root of the biospheric degradation of today.

Bookchin was born in New York City in 1921 to Nathan and Rose
Bookchin, Russian Jewish immigrants who had been active in the 1905
Russian revolutionary movement and raised him in a tradition of opposition to authority. At the same time, he also absorbed libertarian ideas from
his maternal grandmother, who had been a member of the Socialist Revolutionaries, a populist movement in czarist Russia. Enthusiasm for the
success of the Bolshevik Revolution pervaded almost all sectors of the international Left to the degree that the ideals of socialism came to be associated with the Communist movement. In the depression era, young New
York intellectuals were raised on this version of socialism. In the early
1930s, Bookchin entered the Communist youth movement, speaking at
meetings and helping to organize unemployed workers. He eventually ran
the educational program for his branch of the Young Communist League.
Because of his family's poverty, he went to work in heavy industry directly
after high school.

By the late 1930s, Bookchin had become disillusioned with the authoritarian character of international communism, believing that it committed
abuses in the name of socialism. After breaking with the movement, he
turned to Trotskyism and later to libertarian socialism. Bookchin recalls
that in his younger days, working people were class conscious; they saw
"the bosses" as their opponents in the battles to gain union recognition,
and the atmosphere was one of conflict. As a foundry worker in New
Jersey, he became active in union organizing in heavily industrialized northern New Jersey through the militant Congress of Industrial Organizations
(CIO).

After returning from the U.S. Army during the 1940s, Bookchin found
employment as an autoworker and became deeply involved in the United
Auto Workers (UAW), a highly libertarian union at the time. When a major
General Motors strike ended with his coworkers acceptance of company
pension plans and unemployment benefits, Bookchin's disillusionment with
the workers' movement as a revolutionary force was final, and his years as
a union activist came to an end.

In 1950s America, the anarchist movement was small, fragmented, and
apparently on the decline. Yet in Bookchin's mind, anarchism's libertarian
ideal of "a stateless, decentralized society, based on the communal ownership of the means of production" seemed to be the basis for a viable
alternative (Biehl). Bookchin worked closely with German exiles in New

York who were dissident Marxists evolving toward a libertarian perspective. Many of his early 1950s articles were published in the German periodical *Dinge der Zeit*, as well as in its English-language counterpart, *Contemporary Issues*, under the pen names of M. S. Shiloh, Lewis Herber, Robert Keller, and Harry Ludd.

As Bookchin began to move toward an anarchist outlook, the American economy of the early 1950s was expanding greatly, and even industrial workers were becoming part of the middle class. New technologies were being promoted as cures for all social problems of the time and as the path to an entirely new society. Bookchin noted the increasing penetration of the capitalist market into everyday life and became increasingly concerned with the broader issue of social change. From 1952 on, he worked at formulating his anarchistic theory that the domination of nature stemmed from a capitalist economy. It was around this time that Bookchin suggested that an ecological crisis lay on the horizon. "Within recent years," he wrote in a 1952 essay, "the rise of little known and even unknown infectious diseases, the increase of degenerative illnesses and finally the high incidence of cancer suggests some connection between the growing use of chemicals in food and human diseases" (Herber 1952). He believed that the new economic and technological boom, despite all its promises, would also have harmful environmental consequences.

Little environmentalist writing existed in the United States in those years. Although there was an active conservation movement, it worked primarily for the preservation of wilderness and demonstrated little interest in social or political analysis. The existing literature on chemical pollution was mute on the role of modern capitalism in the development and use of chemicals. Bookchin not only rooted environmental problems in capitalism, he proposed a new limit to capitalist expansion: environmental destruction. He advanced, in basic terms, an anarchist solution to the problems, calling for the decentralization of society to avoid the impending ecological crisis.

In the 1960s, Bookchin was deeply involved in new countercultural and leftist movements almost from their inception, and he pioneered the foundations of social ecology in the United States. One of social ecology's most important contributions to modern ecological philosophy is the view that the basic problems that position society against nature emerge from within the process of social development by itself, not from the interaction between society and nature. Social ecology explains the ecological crisis as the result of authoritarian social structures. It strives to show how nature phases into society: it acknowledges both the differences between society and nature and the extent to which they merge with each other.

Maintaining that humans are "nature, rendered self-conscious," Bookchin and other social ecologists call for the establishment of small-scale, egalitarian societies that recognize the fact that human well-being is inextricably connected to the well-being of the natural world (Institute for Social Ecology). "Nearly all the nonhuman life-forms that exist today are, like it or not, to some degree in human custody," he says, "and whether they are preserved in their wild lifeways depends largely on human attitudes and behavior" (Bookchin 1990). Bookchin suggests that humans can construct an alternative future, "reharmonizing" people's relationship to the natural world by reharmonizing their relationships with one another. In response to the challenge of creating an ecological society, social ecology provides a critical analysis and suggests a process for building sustainable community structures through the integration of theory and practice.

For almost half a century, this assertion of the social causes of ecological problems and the insistence on their solution by a decentralization of society have remained consistent in Bookchin's writings. His first American book, *Our Synthetic Environment* (written under the pseudonym Lewis Herber) was published in 1962. A comprehensive overview of ecological degradation, it anticipated the environmentalist movement, preceding Rachel Carson's *Silent Spring* by nearly six months. Bookchin's book was the first major effort to combine ecological awareness with the need for fundamental social change. His 1964 essay "Ecology and Revolutionary Thought," the first manifesto of radical ecology, advanced a combination of anarchism and ecology to create an ecological society that would also be humane, decentralized, and cooperative.

In the early 1960s, renewable energy sources were subjects of some research and experimentation, but they essentially were ignored as practical alternatives to fossil and nuclear fuels. Not only did Bookchin show their relevance to the solution of ecological problems, he stood alone in demonstrating their integral importance to the creation of an ecological society. He also warned of the potential for global warming and the disastrous results that could be anticipated. In 1964, he wrote that increases in the "blanket of carbon dioxide" from combustion of fossil fuels "will lead to more destructive storm patterns and eventually to melting of polar ice caps, rising sea levels, and the inundation of vast land areas" (Bookchin 1989a).

Bookchin spent much of the 1960s traveling around North America, spreading his message about ecology and its revolutionary significance. Late in the decade, he taught at the Alternative University in New York, one of the largest "free universities" in the United States, and then at the City University of New York in Staten Island. In 1974, he cofounded and became the director of the Institute for Social Ecology at Goddard College in Plainfield, Vermont. The same year, Bookchin also began to teach at Ramapo College of New Jersey, where he was a full professor of social theory.

In the late 1980s, Bookchin spoke of a major dispute between segments of the ecology movement: "It is a dispute between social ecology and 'deep ecology'—the first, a body of ideas that asks that we deal with human beings primarily as social beings who differ profoundly as to their status . . . the second, that sees human beings as a mere 'species' . . . as mammals and, to some people like the 'Earth First!' leaders, as 'vicious' creatures—who are subject almost entirely to the 'forces of nature' and are essentially interchangeable with lemmings, grizzly bears . . . or, for that matter, with insects, bacteria, and viruses" (Bookchin 1988). Bookchin believes that deep ecology essentially overlooks the profound social differences that divide human from human, and that it "zoologizes" people into a biological lump called "humanity."

Since its founding, the Institute for Social Ecology has acquired an international reputation for its advanced courses in ecophilosophy, social theory, and alternative technologies. The goal of the institute's programs is the preparation of students to work effectively as participants in the process of ecological reconstruction. Bookchin says, "It [takes] a high degree of sensitivity and reflection . . . to gain an understanding of what one ultimately needs and does not need to be a truly fulfilled person. Without such people in sufficient numbers to challenge the destruction of the planet, the environmental movement will be as superficial in the future as it is ineffectual today" (Bookchin 1989b). The institute's educational programs have provided a forum for dialogue among various factions in the ecology movement and a meeting ground for activists from around the world. "In fact," Bookchin says, "real growth occurs exactly when people have different views and confront each other in order to creatively arrive at more advanced levels of truth—not adopt a low common denominator of ideas that is 'acceptable' to everyone but actually satisfies no one in the long run."

Now in his late seventies, Bookchin lives in semiretirement in Burlington, Vermont. For health reasons, his activities are increasingly restricted, but each summer he still teaches two core courses at the institute, where he is director emeritus, and he occasionally gives lectures in North America and Europe. With his life companion, Janet Biehl, and others, he publishes the theoretical newsletter *Left Green Perspectives* (formerly *Green Perspectives*). Among Bookchin's many additional books are *Remaking Society: Pathways to a Green Future* (1989), *The Philosophy of Social Ecology* (1990; revised 1995), and *The Politics of Social Ecology: Libertarian Municipalism* (1997, cowritten with Janet Biehl).

BIBLIOGRAPHY

Anarchy Archives online. Murray Bookchin biography. *http://dwardmac.pitzer.edu/ anarchist_archives/bookchin/bio1.html*
Biehl, J. (Ed.). (1997). *The Murray Bookchin Reader*. London: Cassell Academic.
Biehl, J., & Bookchin, M. (1997). *The Politics of Social Ecology: Libertarian Municipalism*. Montreal: Black Rose Books.
Black, B. (1997). *Anarchy after Leftism*. Columbia, MO: C.A.L. Press/Paleo Editions.

Bookchin, M. (1964). Ecology and revolutionary thought. In Roelofs, R. T., Crowley, J. N., and Hardestry, D. L. (eds.) (1974), *Environment & Society: A Book of Readings on Environmental Policy, Attitudes & Values*. Englewood Cliffs, NJ: Prentice-Hall.

Bookchin, M. (1988, May). The crisis in the ecology movement. *Green Perspectives*, 6, Available: *http://www.gn.apc.org/pmhp/dc/activism/e-crisis.htm*

Bookchin, M. (1989a, August). Death of a small planet. *Progressive*, pp. 19–23.

Bookchin, M. (1989b). *Remaking Society: Pathways to a Green Future*. Toronto: University of Toronto Press.

Bookchin, M. (1990). A philosophical naturalism. In *The Philosophy of Social Ecology: Essays on Dialectical Naturalism* (2nd ed., revised 1995). Montreal: Black Rose Books.

Bookchin, M. (n.d.a). Society and ecology. Available: *http://www.tao.ca/~ise/library/bookchin/society.html*

Bookchin, M. (n.d.b). Will ecology become the dismal science? Available: *http://www.tao.ca/~ise/library/bookchin/society.html*

Crute, C. (n.d.). Murray Bookchin in London. Available: *http://www.ecn.org/freedom/crute.html*

Herber, L. (pseud. of M. Bookchin). (1952, June–August). The problem of chemicals in food. *Contemporary Issues 3* (12), pp. 235–247.

Herber, L. (1962). *Our Synthetic Environment*. New York: Knopf.

Institute for Social Ecology online. Murray Bookchin biography. *http://www.tao.ca/~ise/faculty.html*

—*Bibi Booth*

LEE BOTTS
(1928–)

Lee Botts has been an environmental activist in her personal and professional adult life for more than forty years. She started out as an activist, as most do, when she became a neighborhood volunteer in Chicago. Later she worked internationally, nationally, regionally, and locally on behalf of the Great Lakes. Along the way, she has organized and managed not-for-profit organizations, worked in government at several levels, and held academic positions. In semiretirement since 1990, she still is a full-time activist, mainly as a volunteer who does occasional consulting on projects related to environmental and social issues.

Botts believes that growing up in the Dust Bowl period of the United States made her an environmentalist. She was born in 1928 in a tiny town in northwestern Oklahoma near the 100th meridian that marks the beginning of the dry Great Plains. When she was growing up, she loved to spend time at the wheat farm that her mother's parents had acquired as a homestead around the turn of the century, where the freedom to wander made

Photo © Cynthia Howe 2000.

her aware of relationships and changes in the land. She still has a little book with pressed wildflowers she collected on the farm when she was eight or nine years old.

As the drought of the 1930s killed his fruit trees, Botts's grandfather explained why he planted new trees to reduce erosion in the New Deal shelter-belt program. Sent home from school in the third grade wearing a face mask in western Kansas as a three-day dust storm approached, somehow Botts already understood that people had made the devastation worse.

Arriving in Chicago in 1949 as the wife of a graduate student after majoring in English at Oklahoma State University, Botts could not believe the vast size of Lake Michigan. As she tells it, here was blue water as far as one could see without involvement of the Corps of Engineers, the builders of the reservoirs that are Oklahoma's large lakes. She learned to organize people to make things happen. She and others saved Jackson Park on the lakefront from the expansion of a scenic roadway. She wrote a column about city gardening in the local weekly newspaper that was more about urban conservation. The annual spring fair sale of plants she and her friends began is still going strong almost forty years later in the Hyde Park neighborhood near the University of Chicago.

Through the 1960s, Botts organized environmental education projects, first in the schools her four children attended and then in inner-city schools. Becoming editor of the *Hyde Park Herald* gave her a new opportunity for advocacy on local issues, but her interests were expanding. In 1968, she went to work for the Openlands Project, the first Chicago environmental group to have a professional staff. Her main assignment was to promote environmental education in schools throughout greater Chicago. In 1969, in response to growing public concern about pollution and other threats, she organized a conference of conservation activists from the four states around Lake Michigan, Wisconsin, Illinois, Indiana, and Michigan. This activity led to the establishment of the Lake Michigan Federation in 1970 as the first regional citizens' organization for the Great Lakes. The federation's motto was Citizen Action to Save a Great Lake. The actions included organizing citizen interventions in license procedures to prevent the Atomic Energy Commission (AEC) from carrying out its plan to locate between twenty and twenty-four nuclear power plants around the lake. The issues developed here led to the 1973 federal court of appeals decision that environmental impact statements were required even for nuclear plants where construction had begun before passage of the National Environmental Policy Act in 1969 and expanded safety regulation.

Botts's group helped get the first Clean Water Act passed in 1972 and then organized local participation in all four states in the review of the discharge permits that industries and sewage-treatment plants were now required to obtain. Other major activities included leading citizen participation in implementation of the Great Lakes Water Quality Agreement that the United States and Canada adopted in 1972. The group's success in convincing the U.S. Environmental Protection Agency (USEPA) that the discovery of polychlorinated biphenyls (PCBs) concentrated in tissues of Lake Michigan fish and the discovery that toxic chemicals were reaching the Great Lakes by atmospheric deposition and long-range transport required national action is the reason that new uses and manufacture of this chemical were banned in the Toxic Substances Control Act of 1976.

Botts's first major experience in Washington was in 1969 in what was said to be the first time that heads of government agencies consulted directly with citizens on environmental issues. Six activists, including Botts, were invited to discuss the growing opposition across the country to new power plants and transmission lines. The almost five-hour meeting in the Old Executive Office Building across from the White House ended in a shouting argument with the chairmen of the Federal Power Commission and the Tennessee Valley Authority and other top officials. The issue was whether the country could continue to double its use of electricity every seven years as it had been doing since the 1950s. This experience led to Botts's first congressional testimony in 1970, on a bill to require public participation in decisions on siting power plants.

In 1972, Botts became the only female and grassroots activist on the advisory committee of the Ford Foundation Energy Policy Project. The members included chief executive officers of major corporations and leading academic and policy experts. The aim was to encourage new national policies on how the United States could meet growing energy needs and protect the environment. Mike McCloskey, then executive director of the Sierra Club, was the other environmentalist. For three years, McCloskey and Botts advocated efficiency, conservation, and use of renewable sources as the best, even the only, long-term means to meet energy needs. They were opposed by such persons as William Tavaloreas, president of Mobil Oil, and Carl Kaysen, head of the Princeton Institute for Advanced Studies. In the end, the final report and the 1973 Arab oil embargo made energy conservation a respectable topic for discussion in Washington but achieved no significant policy changes. These experiences reinforced for Botts the importance of citizen involvement at the local level to challenge conventional thinking.

Botts's first position in the government was as assistant for congressional and intergovernmental affairs to the USEPA administrator for Region 5, which included Minnesota, Wisconsin, Illinois, Indiana, Ohio, and Michigan and had the lead responsibility under the Great Lakes Agreement with Canada. Even though Great Lakes problems provided warning of global problems in many ways, EPA officials in Washington did not accept agreement objectives as national goals. At Region 5, Botts helped lead a successful internal struggle to obtain support for bans on phosphates in detergents that were essential for slowing the eutrophication that threatened the lakes.

In 1978, Botts joined other environmentalists whom President Jimmy Carter appointed in his administration as chairman of the Great Lakes Basin Commission. The planning agency was responsible for coordinating state and federal policies for water-resource development among eight states and twelve federal agencies. Botts had to convince her old friends at the Corps of Engineers not to use cooling water from nuclear plants to prevent ice from forming in the St. Lawrence River and other measures that were disruptive to the environment in order to promote year-round navigation in the Great Lakes system.

After President Ronald Reagan's secretary of the interior James Watt shut the Basin Commission down and dismissed Botts in 1981, she became a research associate at Northwestern University's Center for Urban Affairs and Policy Research. Here, with a Canadian cochairman, Henry Regier, Botts organized an interuniversity seminar with faculty from twenty-four universities in both countries that led to the first joint study by the National Academy of Sciences and the Royal Society of Canada. The resulting report on threats to wildlife and humans from toxic contamination in the Great Lakes helped convince Congress of the need for new legislation such as the Resource Conservation and Recovery Act and the Superfund law. At Northwestern, Botts helped organize a successful effort to obtain a Clean Water Act amendment to recognize the Great Lakes Agreement and Great

Lakes United, a binational coalition of environmentalists in the United States and Canada. She also teamed up with a Canadian partner to write *The Great Lakes: An Environmental Atlas and Resource Book* for EPA and Environment Canada and served as a consultant for the first five-year strategy of the Great Lakes National Program Office.

Botts's last full-time professional position was as head of Chicago's environmental agency under Mayor Harold Washington in the late 1980s. In 1990, she moved to the Indiana dunes at the south end of Lake Michigan. Since then, she has taught courses at Indiana University Northwest and has served as a consultant for such agencies as the Great Lakes Protection Fund, the Indiana Department of Natural Resources, and the Northwest Indiana Regional Planning Commission.

In 1993, Botts was the editor of *The Environment of Northwest Indiana: Contrasts and Dilemmas* and received an honorary doctorate from Indiana University in the same year. In the 1990s, her belief in the importance of grassroots citizen action was reinforced by participation in exchange programs for Lake Baikal in Siberia, in Kiev, in Ukraine, and in Estonia for six Baltic countries. This is why the Botts's home overlooking Lake Michigan is always open to people of all kinds who are organizing themselves to work for the environment. It is also why she has insisted that the audience be allowed to participate in discussions of the Quality of Life Council, a forum she helped organize to bring leaders in industry and business, academia, and the community together to consider how to achieve economic, social, and environmental sustainability around the south end of Lake Michigan.

Since 1997, Botts has worked almost full-time as a volunteer to establish a residential environmental education facility in partnership with the National Park Service. Her concern with the environmental movement of which she has been a part is that too many of the participants talk mainly, or only, to each other when the future of the environment depends on understanding and action by everyone. This is why she has recruited leaders in business and industry and educators as well as community and environmental leaders to serve on the board and provide funding and other resources to the Indiana Dunes Environmental Learning Center.

Today Botts continues to serve on the boards of directors or as an advisor for several environmental groups and to government. She is now involved in working directly with people whom she would have considered enemies in her early days. These include activists, government officials, and business leaders who agree that change is essential in the way business is conducted and the way people live in order to protect the environment for generations to come.

Many awards and other recognitions have given Botts much pleasure over the years. Still, she has reached the stage in life when the greatest satisfaction is in the work and the caring for the environment of her own children, other young people whom she has helped along the way, and the many participants at the Indiana Dunes Environmental Learning Center. Their interest in applying what they learn in their own lives is what makes Botts optimistic about their future as activists and about all life on earth.

BIBLIOGRAPHY

Botts, L. (Ed.). (1993). *The Environment of Northwest Indiana: Contrasts and Dilemmas.* Valparaiso, IN: Pahls, Inc.

Botts, L., & Gannon, J. (1988). Remediation and rehabilitation of the Great Lakes. In Caldwell, L. K. (Ed.), *Perspectives on Ecosystem Management for the Great Lakes: A Reader.* Albany: State University of New York Press.

Botts, L, & Krushelnicki, B. (1988). *The Great Lakes: An Environmental Atlas and Resource Book.* Ottawa and Washington, DC: Environment Canada and U.S. Environmental Protection Agency.

Botts, L., & Muldoon, P. (1997). *The Great Lakes Water Quality Agreement: Its Past Successes and Uncertain Future.* Hanover, NH: The Institute on International Environmental Governance, Dartmouth College.

—Lee Botts

DAVID R. BROWER
(1912–2000)

I want to restore what once was, not for an old man's memories, but for a baby's smile.

—David Brower

Born in Berkeley, CA, in 1912, David Ross Brower has been called the father of the modern environmental movement, the "Archdruid," and even the "John Muir Reincarnate." His sixty-plus years as a leader in the movement to protect the Earth made him a key figure of the American twentieth century.

The comparisons to John Muir, the legendary writer and naturalist who founded the Sierra Club in 1892, are fitting in many ways. Both men developed a deep love for the natural world during their explorations of the wild mountains of the Sierra Nevada. This love turned into a life-long crusade for the protection of wild places all over. Muir's biggest fight, to protect the Hetch Hetchy Valley in Yosemite National Park from being flooded by the construction of a large dam, also turned out to be his last as he died in 1914. But Muir's failure to save Hetch Hetchy set the stage for Brower's later campaigns to protect Dinosaur National Monument and the Grand Canyon from large dams. From these national battles, the American conservation movement emerged as a force in American politics, and David Brower emerged as its greatest field general. In 1998, David Brower

Photo courtesy of Earth Island Institute.

was awarded the world's most prestigious environmental award, the Blue Planet Prize, by the Asahi Glass Foundation of Japan. The foundation acknowledged that his successful campaigns with the Sierra Club in the 1950s and 1960s had created the tools, tactics, and standards for bold political leadership that are employed today by effective environmental groups around the world.

Growing up in the 1920s, being a conservation leader and lobbyist was not a career option that existed, and young Brower seemed an unlikely candidate. A bright but awkward boy, he struggled to fit in socially in high school and worked hard to be included in activities like neighborhood football games. He preferred wandering alone and exploring the hills of Berkeley, which were quite wild beyond the boundaries of the university campus. He liked to collect butterflies and even discovered a new sub-species which is named for him.

Entering college in 1929, he had yet to reach his seventeenth birthday and again struggled to fit in with the older students. He had hoped to study insects like his beloved butterflies and become an entomologist, but by 1931 financial and social pressures led him to drop out. The country was in the midst of the Great Depression and he had to work various jobs to make ends meet. When he was not working, he became fascinated by rock climbing and his attention returned to his early love of the Sierra Nevada.

When asked how he first became interested in protecting nature, Brower often joked that he "chose his parents very carefully." Both Mary and Ross Brower loved the outdoors and took their children on many summer trips to the Sierra Nevada and Lake Tahoe, which took three days to reach on the one-lane dirt road that is now Interstate 80. In 1921, near one of the sites they had camped at along the way, Brower had his first experience of the destruction that people can cause in wild places. While exploring in the National Forest alone, he found a spring. "I was quite pleased to find clear water bubbling from dirt," he recalled in his book *Let the Mountains Talk, Let the Rivers Run*. The following summer, when the family camped in the same spot, Brower went back to find his spring, but it was gone. Gone along with the entire forest that had been clearcut around it.

An even more important event in Brower's childhood was his mother's blindness. He wrote, "Two years before that Sierra spring was clearcut, I became my mother's eyes." Mary Brower had developed an inoperable brain tumor that left her unable to experience the sights and smells of the natural world. This left it to David to describe the landscape to her on their hikes together, which had a profound impact on his own perceptions: "You don't take eyes for granted if you grew up with a mother who lost her sight when you were eight. You begin not only to look more carefully for yourself, but also to look for her, to see for her what she once saw and loved." He would return to this skill years later as he attempted to get politicians and citizens to preserve beautiful landscapes that they had never seen before.

By 1933, Brower worked only as much as necessary to finance long trips into the Sierra Nevada. When he returned from a ten-week backpacking trip during which he nearly fell to his death trying to climb a cliff face, he decided to join the Sierra Club to learn how to climb safely. At the Sierra Club, he became an expert climber and ski mountaineer and began using his descriptive talents to write pieces for the *Sierra Club Bulletin*. In 1935, Yosemite Park and Curry Company hired him to do clerical work in Yosemite Valley, but soon discovered his other skills and made him publicity manager for the park. Living in Yosemite between 1935 and 1938, he publicized the wonders of the park by producing films and written materials. He also completed many of his 70 first ascents of peaks in the High Sierra.

Brower put his film-making skills to work for the Sierra Club as well. In 1939 he created the first of his many Sierra Club nature films, *Sky Land Trails of the Kings*, and showed it all over California and in Washington, DC. The film helped convince Congress to create Kings Canyon National Park in 1940, nearly 50 years after John Muir had first suggested it. His involvement with the club increased and by 1941 he had been elected to its Board of Directors.

In 1939, he completed his most famous ascent, Shiprock, a volcanic plug that rises 1,400 feet from the floor of the New Mexico desert. Numerous others had failed to climb it and it was generally agreed to be "insurmountable." Brower disagreed and over the course of four days, he led his climbing partners to the pinnacle of Shiprock. This was also the pinnacle of his fame as a mountain climber (he was even offered an endorsement deal from Camel cigarettes, later rescinded because of his crooked front teeth), but he was far from done tackling challenges that others felt were insurmountable.

In the 1940s, World War II broke out and there was a call for expert skiers and climbers to join a special mountain fighting force, the U.S. Army's 10th Mountain Division. As part of the Mountain Troops, Brower taught over 10,000 soldiers how to climb rocks, and entered the war in Italy as a combat-intelligence officer where he participated in the successful campaigns to drive the Germans from the high mountains of Northern Italy. He left active duty as a captain and was awarded the Bronze Star. During the war, he proposed by mail to Anne Hus, with whom he had worked at the University of California Press, and settled in Berkeley.

While fighting in the mountains of Europe, Brower was struck by the degradation of the once-wild Alps and returned home with a renewed commitment to preserving the wild mountains of America. With his war-honed skills in campaign strategy, David was now ready to lead the Sierra Club into the biggest battle it had yet fought, the battle to protect Dinosaur National Monument.

In the early 1950s the U.S. Bureau of Reclamation determined to build two dams on the Green and Yampa Rivers in Echo Park, a beautiful canyon inside Dinosaur National Monument, near the Utah/Colorado border. The dam would violate and develop lands of the National Park System, which includes most national monuments. The governing bodies of the Wilderness Society and Sierra Club decided to fight the dam proposal, which was a part of the massive Colorado River Storage Project system of dams. With the club's membership grown to over 7,000 members and the Dinosaur battle shaping up, Brower was tapped to become the Sierra Club's first executive director, a paid head administrator and field general.

Brower and Howard Zahniser of the Wilderness Society then began building a national coalition of conservation and civic organizations to defeat the dam and defend the principle of the inviolability of national park lands. It was considered a nearly futile effort by many conservationists; the Bureau of Reclamation was considered to be too powerful. However, word of the threat to the National Park System was spread through brochures, films, and public meetings, and citizens across the nation were spurred to contact their members of Congress and ask them to vote against the proposed dam projects. Congress called for hearings on the Colorado River Storage Project, and a key moment came when Brower, armed with his "ninth grade arithmetic," proved that the Bureau's engineers had their numbers wrong and that they had dramatically overstated the benefits of the project. When the Colorado River Storage Project was set to come to a vote, the Sierra Club leadership instructed Brower not to oppose the bill if the dams in Dinosaur were cancelled. Brower followed these orders and the Dinosaur dams were removed from the bill and it passed. Unfortunately, the bill that passed called for damming Glen Canyon on the Colorado River, which David had never seen. Before it was flooded, he visited Glen Canyon and found it to be the most beautiful place he had ever seen. He always regretted going along with the Sierra Club compromise that left Glen Canyon unprotected, drowned beneath the waters of Lake Powell.

But there was little time for lamenting and Brower got to work with Zahniser and others on creating a Wilderness Preservation System to protect pristine areas outside the National Park System. The Wilderness Act passed in 1964, after nine years of lobbying. Brower also helped create campaigns to expand the National Park System including Pt. Reyes, Cape Cod, and Fire Island National Seashores, and North Cascades, Redwoods, Great Basin National Parks. He drove illegal logging out of Olympic National Park and stopped a huge dam project proposed for the Yukon River in Alaska. But in many circles David Brower is best known for spearheading the most dramatic conservation campaign of the 1960s: keeping dams out of the Grand Canyon.

In the Grand Canyon campaign, Brower used all of his techniques to perfection, from breath-taking films to full-page newspaper advertisements to beautiful exhibit-format photo books. The photo books were crafted to impress upon the reader how beautiful the place was, with photographs augmented by selections of prose or poetry, and to tell them about the threat to the place. Finally, these books told readers what they could do to stop it. Brower edited twenty such books in his years at the Sierra Club.

The full-page ads got the attention of the Sierra Club's friends and enemies, with provocative headlines like "Should we also flood the Sistine Chapel, so tourists can get nearer the ceiling?" An enraged congressman got the Internal Revenue Service to cancel the Sierra Club's status as a charitable organization (donations had been tax deductible) because they were lobbying Congress too aggressively. This punishment got the media's attention and got public sympathy on the side of the Sierra Club. Brower recalls, "Even people who didn't care about the Grand Canyon knew they hated the IRS." Sierra Club membership numbers soared and sentiment against the dams grew. Finally victory came when President Lyndon Johnson announced that he would veto any legislation that came across his desk that included dams in the Grand Canyon.

In spite of the club's growth under Brower's leadership (about 70,000 new members) and his many conservation victories, a rift was growing between Brower and the Sierra Club Board of Directors who did not approve of his uncompromising stances on many issues. Finally, in 1969, they asked for his resignation, and he left the staff of the Sierra Club.

Brower had clashed with the Board because he wanted the club to oppose Diablo Canyon Nuclear Power Plant, have international chapters, and continue its publishing of exhibit format books, so he founded a new organization that would do just that. In 1969, Brower founded Friends of the Earth (along with an early version of the League of Conservation Voters) which continued to protect public lands, but also aggressively publicized the dangers of nuclear power, published new exhibit format books, and established independent affiliates in other countries. Today Friends of the Earth (FOE) has offices in 68 countries around the world.

In 1982, at the age of 70, Brower founded Earth Island Institute. This group became the focus of his efforts when he split with Friends of the Earth in the mid-1980s after conflicts on the Board of Directors. At Earth Island he helped create a project-focused umbrella organization where independent environmental campaigns could function without being slowed by Board interference, which had been a problem with the Sierra Club and FOE. In the early 1990s, Brower led several Earth Island delegations to Russia to help build the Russian environmental movement and to secure protection for Lake Baikal (the world's oldest, deepest lake) and the Siberian wilderness. These delegations resulted in the formation of the Baikal Watch project at Earth Island and the EcoJuris Environmental Law Network in Russia, both of which have won significant protections for the Siberian Frontier. Lake Baikal also achieved United Nations recognition as a World Heritage Site of outstanding world-wide significance. David served as Chairman of Earth Island Institute.

Even on into his late eighties, David Brower continued to speak publicly all over the country, trying to inspire the next generation of Earth protectors. His speeches called for Global CPR, resuscitating the Earth and its life-support system that we all depend on with Conservation, Preservation, and Restoration. He believed Restoration must be the theme of the twenty-first century. "Too much of our work," he opined, "merely slows the rate at which things are getting worse, we must invest ourselves and our time towards making things better." Brower feared that human society is being eroded by its addiction to growth, living off natural resource capital, not just the interest. Despite many discouraging signs, Brower was ever the optimist, intrigued by human potential. He felt that we have a long way to go but we're getting somewhere.

In addition to winning the Blue Planet Prize and ten honorary degrees, David Brower was nominated three times for the Nobel Peace Prize (1978, 1979, and 1998—jointly with Paul Ehrlich). But he refused to rest on his laurels, despite heart problems and an ongoing battle with bladder cancer. He appreciated that even as his mobility decreased, his increased visibility allowed him to participate in forming high-profile and often unlikely alliances. In his final years, he campaigned to restore Glen Canyon and co-founded the Alliance for Sustainable Jobs and the Environment with leaders from the United Steelworkers of America to promote policies that protect both the natural world and quality jobs. He also served as an advisor to several large corporations.

Comments about his efforts have ranged widely. John McPhee's *Encounters with the Archdruid* is about David Brower and three of his natural enemies—and is in its twenty-seventh printing. Brower especially liked what Russell Train, an early head of the Environmental Protection Agency, said: "Thank God for David Brower; he makes it so easy for the rest of us to be reasonable." Brower believed that if he had been reasonable, we would have dams in the Grand Canyon and far fewer wildlands for today's youth to enjoy. Said President Jimmy Carter of Brower, "He has been for many years a steady force of nature, drawing us to see the natural world as nurturer, teacher, inspirer and partner. He has been a pathbreaker, not given to easy answers or ruinous compromises and we are the richer for it. . . . [He] is a man of great insight who cares very deeply for his world."

Brower hoped that others will follow his example by heeding the words of Johann Wolfgang von Goethe: "Anything you can do, or dream you can, begin it; Boldness has genius, power, and magic in it." He learned this lesson as a mountaineer when he climbed the insurmountable Shiprock. "Some of the environmental problems may at times seem insurmountable," he often reminded us, "But we old mountaineers don't believe in insurmountability, and you don't have to either."

—*Mikhail Davis*

LESTER R. BROWN
(1934–   )

We do not have generations, we only have years, in which to turn things around. Still, I'm not by nature a pessimist. You don't work on environmental issues for 20 years unless you think you can change things.

—Lester R. Brown

Lester R. Brown is the president and senior researcher of the Worldwatch Institute, a nonprofit research organization based in Washington, D.C., and devoted to the analysis of global environmental issues. The institute endeavors to act, in Brown's words, as a "global early warning system." Brown is considered by many environmentalists to be one of the world's most influential thinkers today.

Lester Brown was born in 1934 in Bridgeton, New Jersey, the son of farmers Calvin and Delia Smith Brown. In the late 1940s, at the age of fourteen, Brown began a part-time career as a farmer with his older brother. The two boys were "real hustlers," according to Brown. They bought a used tractor for $200, rented some nearby agricultural fields, and established themselves as tomato growers, working their fields after school each day. Eventually, despite being only teenagers, they found themselves among the top 1 percent of tomato producers in the United States. This achievement more than qualified them for membership in the Ten Ton Tomato Club, the upper echelon of tomato growers, which was open only to those who were harvesting at least that quantity per acre.

In 1955, shortly after earning his bachelor's degree in agricultural science from Rutgers University, Brown spent six months living in a rural village in India as part of the International Farm Youth Exchange Program (now the International 4-H Youth Exchange). After deciding to devote his career to world food issues, Brown earned his master's degree in agricultural economics from the University of Maryland in 1959 and joined the U.S. Department of Agriculture's Foreign Agricultural Service as an analyst of international food, population, and the environment. He helped India to design its Green Revolution strategy, which resulted in the doubling of that nation's wheat crop in just seven years. Brown also developed the world's first systematic projections of global food, population, land, and fertilizer to the end of the twentieth century.

In 1962, Brown received his master's degree in public administration from Harvard University. In 1964, he became an advisor to Secretary of Agriculture Orville Freeman on foreign agricultural policy. Brown foresaw the impending Indian crop failure in 1965 in time to mobilize aid to that country, and in 1966, Freeman appointed him administrator of the department's International Agricultural Development Service. Brown resigned from this position in 1969 to establish the Overseas Development Council with Jim Grant, then the executive director of UNICEF.

Such global trends as deforestation, overgrazing, and the startling depletion of grain reserves during the 1970s, together with the agricultural near catastrophe in India, prompted Brown to create the Worldwatch Institute in 1974 to help prevent calamities in the future. The institute was established just after the Rome World Food Conference, with support from the Rockefeller Brothers Fund. The institute is dedicated to fostering the evolution of an environmentally sustainable society in which human needs are met in ways that do not threaten the health of the environment or the well-being of future generations. The institute seeks to achieve this goal through its unbiased, interdisciplinary research on emerging global environmental issues, the results of which are widely disseminated throughout the world. The institute's philosophy is that information is a powerful tool of social change; its mission is to raise public awareness of global environmental threats to the point of producing effective policy responses.

Brown reflects today that his boyhood experiences with tomato farming in the 1950s greatly affected his career direction and later focused his assessments of global food-production trends. At the time, he remembers, all tomato farmers were affected when the Campbell's Soup Company, seeking to lower the cost of tomatoes, turned its sights to increasing crop productivity. Within a few years, the club of tomato "supergrowers" could only continue to represent the best by becoming the Twenty Ton Tomato Club. As it turned out, however, such acceleration could not be sustained indefinitely, and despite much research and effort over the next four decades, the growth in tomato yield per acre eventually slowed. Today, the best growers, Brown notes, get only around thirty tons per acre. This, he explains, exemplifies what is now happening to food supplies the world over. "The near tripling of the harvest by the world's farmers [from the 1950s to the 1990s] . . . was a remarkable performance, one without historical precedent," he says. "But now the world has suffered a dramatic loss of momentum" (Lean). Many experts, he says, assume that technology will always ensure plentiful food. Instead, both farmers and politicians are going to face "tough choices" as the new era unfolds. As population has risen, the amount of grain land per person has fallen by nearly half since 1950, and the Worldwatch Institute predicts that it will fall by another third by 2030. Since 1979, the amount of irrigated land per person has also been falling. Meanwhile, yield per hectare, after increasing more between 1950 and 1990 "than during the 10,000 years since agriculture began," scarcely rose at all in the 1990s. Brown believes that this "hints at a much slower future rise in cropland productivity" (Lean).

In 1984, Brown launched the report series *State of the World*, a highly acclaimed publication that is mapping out what an environmentally sustainable society will look like. These annual assessments, translated into almost thirty languages, have become an indispensable guide for national leaders as well as environmentalists and concerned citizens everywhere. *State of the World* is widely used by governments, policymakers, business leaders, colleges, and environmentalists and has achieved semiofficial status as a result. *State of the World* is not simply updated and revised every year; each edition consists of new essays on subjects of crucial importance and on the key environmental issues of the preceding year. In 1988, Brown expanded Worldwatch's publications by launching *World Watch*, a bimonthly magazine featuring articles on the institute's research.

In 1991, the institute inaugurated the Environmental Alert book series with a book Brown coauthored with Christopher Flavin and Sandra Postel, entitled *Saving the Planet: How to Shape an Environmentally Sustainable Global Economy*. The book describes how a less polluting economy based on solar energy would function, and how already-existing technologies could make it a reality. The book suggests new ways of using and reusing

materials, discusses less resource-intensive means of growing food, and proposes strategies for preserving forests. In 1992, Brown launched and coauthored a new annual publication, *Vital Signs: The Trends That Are Shaping
Our Future*. In 1995, he wrote *Who Will Feed China? Wake-up Call for
a Small Planet*, which challenged the Chinese government and led to hundreds of conferences, seminars, and new assessments of China's food pros-
pects.

In Brown's view, worldwide food security is going to be the defining
issue of the new millennium as rising demand hits global ecological limits.
"Rising food prices," he predicts, "will be the first major economic indicator to show that the world economy is on an environmentally unsustainable path." On the positive side, food scarcity could "rouse us from our
sleepwalk through history, convincing us to take the steps needed to create
a sustainable balance between ourselves and our natural food support systems" (Lean). The two most important of these steps, he argues, are sta-
bilizing population and climate. The bottom line, says Brown, is that
human population growth is destined to slow, one way or the other. Developing societies will either recognize problems on the horizon and act to
encourage smaller families, or unchecked births will have their price in
rising death rates. "We do not have generations, we only have years, in
which to turn things around," Brown says. "Still, I'm not by nature a
pessimist. You don't work on environmental issues for 20 years unless you
think you can change things" (Environmental group sees . . .).

Brown has been awarded more than twenty honorary degrees from both
American and foreign institutions. He was also a MacArthur fellow (1986–
1991) and has been the recipient of countless prizes and awards, including
the 1987 United Nations Environment Prize, the 1989 World Wide Fund
for Nature Gold Medal, and the 1994 Blue Planet Prize for his "exceptional
contributions to solving global environmental problems." He also received
the 1966 National Wildlife Federation Special Conservation Award and
the 1982 United Nations Environment Programme Environmental Leadership Medal. He was elected one of the "100 Who Made a Difference"
by *The Earth Times* in 1995 and was included as one of the National
Audubon Society's "100 Champions of Conservation" in 1998. He also
serves on the Board of Directors of the Environmental and Energy Study,
as well as the advisory boards of the United Nations Foundation, Zero
Population Growth, and the International Fund for Agricultural Research,
among many other organizations. He is a member of the International
Council for Earth Day 2000. Additional books by Brown include *By Bread
Alone* (1974), *The Twenty-Ninth Day* (1978), and *Full House: Reassessing
the Earth's Population Carrying Capacity* (with Hal Kane) (1994). His latest book is *Beyond Malthus: Nineteen Dimensions of the Population Challenge* (1999).

BIBLIOGRAPHY

Audubon Magazine. (1998). Champions of Conservation. Available: *http:// magazine.audubon.org/century/champion.html*

Barnes and Noble online. Biographical information for Lester Russell Brown. *http:// www.bn.com*

Biography online. Brown, Lester. *http://www.biography.com/cgi-bin/biomain.cgi*

Brown, L. R. (1978). *The Twenty-Ninth Day*. New York: W. W. Norton.

Brown, L. R. (1995). *Who Will Feed China? Wake-up Call for a Small Planet*. New York: W. W. Norton.

Brown, L. R. (1999, May 27). *The Worldwatch Report*: Shifting views of population. Environmental News Network. Available: *http://www.enn.com/ features/1999/05/052799/worldwatch_3421.asp*

Brown, L. R., with Eckholm, E. P. (1974). *By Bread Alone*. New York: Praeger.

Brown, L. R., Flavin, C., & Postel, S. (1991). *Saving the Planet: How to Shape an Environmentally Sustainable Global Economy*. New York: W. W. Norton.

Brown, L. R., Gardner, G., & Halweil, B. (1999). *Beyond Malthus: Nineteen Dimensions of the Population Challenge*. New York: W. W. Norton.

Brown, L. R., with Kane, H. (1994). *Full House: Reassessing the Earth's Population Carrying Capacity*. New York: W. W. Norton.

Environmental group sees new stress on planet. (1996, January 13). Reuter Information Service. *The Nando Times*. Available: *http://www4.nando.net/ newsroom/ntn/health/011396/health26_20766.html*

Facing the Future: People and the Planet. (n.d.). How many people can the earth support? The carrying capacity debate. Available: *http://www.facingthefuture. org/howmany/howmany-2.htm*

Hager, M. (1998, November 2). How "demographic fatigue" will defuse the population bomb. *Newsweek*. Available: *http://newsweek.washingtonpost.com/ nw-srv/printed/us/dept/ml/ml0118_1.htm*

Lawrence, M. (1996, May/June). The coming food crunch. The Foundation for Global Community, *Timeline*. Available: *http://www.globalcommunity.org/ cgpub/27/27articl.htm*

Lean, G. (1996, November). The era of scarcity is upon us: Lester Brown tells Geoffrey Lean that food security is going to be the defining issue of our time, as rising demand hits the world's ecological limits. United Nations Environment Programme, *Our Planet*. Available: *http://www.ourplanet.com/ imgversn/84/brown.html*

Perdue, M. (1999, March 30). Modernization interrupted: Unprecedented population problems proliferate. Scripps Howard News Service. Available: *http:// abcnews.go.com/sections/science/DailyNews/population990330.html*

Review of "State of the World 1997". Earthscan Publications online. *http:// www.earthscan.co.uk/books/427_0.html*

Worldwatch Institute online. Lester R. Brown, President, Worldwatch Institute. *http://www.worldwatch.org/bios/brown.html*

W. W. Norton and Company online. Lester R. Brown et al. State of the World 1995. *http://www.wwnorton.com/blurbs/031261.htm*

—Bibi Booth

Photo courtesy of Robert Bullard and the Environmental Justice Resource Center, Clark Atlanta University.

Robert Bullard
(1946–)

No community, rich or poor, urban or suburban, black, brown, red, white, or yellow should be allowed to become an environmental "sacrifice zone."

—Robert Bullard (*People of Color Environmental Groups Directory*, p. 12)

Robert Bullard is the founder of the Environmental Justice Resource Center at Clark Atlanta University in Atlanta, Georgia. The center was established in 1994 after Bullard moved from California, where he had served as a full professor of sociology at the University of California, Riverside. In a short period of time, he has built the center into a nationally recognized powerhouse for environmental policy analysis, community-driven research, education, and training. He established a world-class media archive of videos, photographs, slides, tapes, and other multimedia materials that document the history and accomplishments of the environmental justice movement. He supervised the development of a comprehensive Web site that serves as an information clearinghouse, database, reference, and curriculum resource guidebook on environmental justice, environmental racism, healthy communities, transportation equity, and suburban sprawl.

Bullard's environmental justice career began in 1979 when his wife, attorney Linda McKeever Bullard, represented a group of African-American Houston homeowners opposed to a plan that would locate a municipal landfill in the middle of their backyards. Only a couple of years out of graduate school at Iowa State University, where he received his Ph.D. in sociology, Bullard was asked to conduct a study to document the spatial location of municipal waste-disposal facilities in Houston. He was also asked to serve as an expert witness on the case. At the time, he was an untenured assistant professor at Texas Southern University, a predominately black state-run institution in Houston. The request was part of a class-action lawsuit his wife filed against the city of Houston, the state of Texas, and the locally headquartered Browning Ferris Industries, the nation's second-largest waste-disposal company. The 1979 lawsuit, *Bean v. Southwestern Waste Management, Inc.*, was the first of its kind in the United States that charged environmental discrimination in waste-facility siting under the civil rights laws. Houston's middle-class, suburban Northwood Manor neighborhood was an unlikely location for a garbage dump except that it was over 82 percent black.

In order to obtain the history of waste-disposal-facility siting in Houston—the only major U.S. city that does not have zoning—government records (city, county, and state documents) had to be manually retrieved because the files were not yet computerized. On-site visits, windshield surveys, and informal interviews, done in a sort of "researcher as detective" role, were conducted as a reliability check. Bullard completed the Houston garbage study in 1979. It was later published as an article in a 1983 issue of *Sociological Inquiry*. Bullard's research eventually documented Houston's history of placing waste facilities in predominately African-American, though not always low-income, communities. All five city-owned garbage dumps, six of the eight city-owned garbage incinerators, and three of the four privately owned landfills were sited in black neighborhoods. In the case of landfills in Houston, the task was made easier because of the city's flat terrain. Whenever a "mountain" was encountered—and quite a few were scattered across the urban landscape—Bullard and his students suspected an old dump site.

After poring over sixty years' worth of "musty files and archives," collecting the data for *Bean v. Southwestern Waste Management*, and interviewing citizens from other African-American neighborhoods, Bullard realized that the siting of local waste facilities was not random. He also discovered that this was not a problem of the chicken or the egg (which came first?). In all cases, the residential character of the neighborhoods had been established long before the waste facilities invaded the neighborhoods. Many residents came to understand the research he was conducting and to recognize the noble profession of sociology as a field in which grandiose theories are developed, hypotheses formulated, and data collected that result in the verification of the obvious: Most residents of Houston's segregated neighborhoods knew not only on which days the garbage was collected but also the addresses of the existing and abandoned landfills and incinerators. Many of these same residents had spent much of their lives escaping from waste sites only to find waste-facility disputes following them to their new neighborhoods.

Bullard was curious to know whether the Houston case was typical of African-American communities in the South, a region where over half of all African Americans reside. He extended his research focus to include four additional African-American communities. He decided to explore the thesis that African-American communities in the South—the nation's Third World—because of their economic and political vulnerabilities, have been routinely targeted for the siting of noxious facilities, locally unwanted land uses (LULUs), and environmental hazards. The subjects were drawn from an array of mostly black areas, including neighborhoods in Houston and Dallas, Texas, and communities of Alsen, Louisiana; Institute, West Virginia; and Emelle, Alabama. He hypothesized that residents in these communities, in turn, are likely to suffer greater environmental and health risks than in the general population.

In 1990, Bullard authored *Dumping in Dixie: Race, Class, and Environmental Quality*. The same year, the National Wildlife Federation presented Bullard with its Conservation Achievement Award in science for this groundbreaking work. In the South, African Americans make up the region's largest racial minority group. Bullard's analysis could have easily focused on Latino Americans in the Southwest or Native Americans in the West. People of color in all regions of the country bear a disproportionate share of the nation's environmental problems. Racism knows no geographic bounds. However, Bullard sought to identify the major economic, social, and psychological impacts associated with the siting of noxious facilities (municipal landfills, hazardous-waste facilities, lead smelters, chemical plants) and their significance in mobilizing the African-American communities in the South, the region where he then resided.

Bullard's *Dumping in Dixie* demonstrated that limited housing and residential options, combined with discriminatory facility-siting practices, contributed to the imposition of all types of hazards on African-American communities. This included garbage dumps, hazardous-waste landfills, incinerators, lead smelters, paper mills, chemical plants, and a host of other polluting industries. These industries have generally followed the path of least resistance, which has been to locate in economically poor and politically powerless African-American communities.

Poor African-American communities are not the only victims of environmental discrimination, however. Middle-income African-American communities are confronted with many of the same land-use disputes and environmental threats as their lower-income counterparts. Increased income has enabled few African Americans to escape the threat of unwanted land uses and potentially harmful environmental pollutants. In the real world, racial segregation is the dominant residential pattern, and racial discrimination is the leading cause of segregated housing in America.

Bullard's work made it clear that middle-class people of color are not immune to environmental racism. For those making environmental and industrial decisions, African-American communities—regardless of their income or class status—were considered to be "throw-away" communities where residential land use is compatible with garbage dumps, transfer stations, incinerators, and other waste-disposal facilities.

African-American grassroots activists and those of other people of color have challenged public policies and industrial practices that threaten the residential integrity of their neighborhoods. Activists began to demand environmental justice and equal protection. The demands are reminiscent of those voiced during the civil rights era for an end to discrimination in housing, education, employment, and the political arena. Many exhibited a growing militancy against industrial polluters and government regulatory agencies that provided these companies with permits and licenses to pollute. Bullard was the lead author of the position paper that created the Washington Office on Environmental Justice, a think tank of people of color based in the nation's capital.

Bullard's work also shattered the myth that poor people and people of color are not concerned about the environment or involved in environmental issues. In 1990, he began with a list of about 30 groups of people of color he knew of who were working on environmental problems. With a small grant from the Charles Stewart Mott Foundation, he and his students called the grassroots leaders and asked them about their issues and tactics and whether they knew of other grassroots environmental justice groups. His initial list grew to over 300 groups of people of color. This list formed the basis for the invitation roster to the First National People of Color Environmental Summit, held in October 1991. The expanded list later became the first edition of the *People of Color Environmental Groups Directory*, published in 1992 by the Charles Stewart Mott Foundation. In 1994 and 2000, Bullard expanded the directory to include groups in the United States, Puerto Rico, Canada, and Mexico. Because of its popularity, the Mott Foundation distributed over 10,000 copies of the directory free of charge to grassroots groups, researchers, students, academic institutions, and governmental officials. The directory is also available via the Internet.

Bullard has spent more than two decades of intense study and targeted research documenting the intersection of race, class, gender, power, and environmental protection. He has attended and participated in hundreds of public hearings and community strategy meetings. He has prepared written testimony, testified, and served as an expert witness on dozens of environmental racism cases. It was Bullard's environmental justice analysis and expert testimony that provided the margin of victory in the *Citizens against Nuclear Trash (CANT) v. Louisiana Energy Services (LES)* lawsuit, the first major environmental justice lawsuit that was favorably decided by a federal court. The Nuclear Regulatory Commission Licensing Board in October 1997 ended a nine-year struggle by CANT members to block the construction of the nation's first privately owned uranium-enrichment plant, which had been planned for the mostly black communities of Forest Grove and Center Springs, Louisiana. In April 1998, the NRC upheld its decision to deny the permit after an appeal by the company.

Bullard has worked with grassroots community groups from West Dallas to West Harlem and from Southside Chicago to South Central Los Angeles. He makes it crystal clear that he does not speak for anyone, but offers his expertise to disenfranchised groups who are demanding an end to environmental and economic injustice at home and around the world. For him, environmental justice does not stop at the U.S. borders.

Bullard was a key organizer of a delegation of environmental leaders that traveled to the Earth Summit in Rio de Janeiro in 1992. He was a cosponsor of an Environmental Justice Exchange Study Tour that took twelve environmental justice leaders to South Africa in 1996. He was a member of the Washington Office on Environmental Justice delegation that participated in the 1996 United Nations Habitat II Summit in Istanbul, Turkey. In 1999, he assisted in preparation of the environmental racism documents that were presented by environmental justice leaders at the United Nations Human Rights Commission in Geneva, Switzerland.

The decade of the 1990s was a different era from the late 1970s. Some progress was made in mainstreaming environmental protection as a civil rights and social justice issue. Bullard has taken the environmental justice message to numerous radio, television, and news programs. He has given keynote addresses at major international environmental conferences in the United States, the United Kingdom, Australia, Chile, Brazil, Mexico, Canada, and South Africa. When he started his work in the late 1970s, few environmentalists, civil rights advocates, or government policymakers understood or were willing to challenge the regressive and disparate impact of this country's environmental and industrial policies. In the end, lower-income communities and those of people of color paid a heavy price in terms of their health, lowered property values, and overall diminished quality of life.

Today, groups like the NAACP Legal Defense and Education Fund, the

Earthjustice Legal Defense Fund, the Lawyers Committee for Civil Rights under the Law, the Center for Constitutional Rights, the National Lawyers Guild's Sugar Law Center, the American Civil Liberties Union, and the Legal Aid Society are teaming up on environmental justice and health issues that differentially affect poor people and people of color. Panels on environmental racism and environmental justice have become hot topics at conferences sponsored by law schools, bar associations, public health groups, scientific societies, and social science meetings.

Bullard's work has had a profound impact on public policy, industry practices, national conferences, funding from private foundations, and academic research. Environmental justice courses and curricula can be found at nearly every university in the country. Bullard has proven that it is now possible to build an academic career—and get tenure, promotion, and merit raises—studying environmental justice issues. A dozen environmental justice centers and legal clinics have sprung up across the nation. Four of these centers are located at historically black colleges and universities (HBCUs): the Environmental Justice Resource Center (Clark Atlanta University, Atlanta, Georgia), the Deep South Center for Environmental Justice (Xavier University of Louisiana, New Orleans, Louisiana), the Thurgood Marshall Environmental Justice Legal Clinic (Texas Southern University, Houston, Texas), and the Environmental Justice and Equity Institute (Florida A&M University, Tallahassee, Florida).

Environmental justice groups are beginning to sway administrative decisions their way. They have even won a few important court victories. Groups have been successful in blocking numerous permits for new polluting facilities. They have forced the Environmental Protection Agency (EPA) to permanently relocate an African-American community in Pensacola, Florida, from a toxic-waste dump that was tagged "Mount Dioxin."

Environmental justice has trickled up to the federal government and the White House. Bullard and a handful of environmental justice advocates convinced the U.S. Environmental Protection Agency (under the Bush administration) to create an Office on Environmental Equity. The Reverend Benjamin Chavis (who at the time was executive director of the United Church of Christ Commission for Racial Justice) and Bullard were selected to work on President Bill Clinton's transition team in the natural-resources cluster (the EPA and the Departments of Energy, the Interior, and Agriculture).

Bullard was part of a national coalition of environmental justice leaders who developed a position paper on environmental justice that was later submitted to the Clinton-Gore transition team. This position paper was instrumental in getting the Clinton administration to establish a National Environmental Justice Advisory Council (NEJAC) to advise EPA, an expanded Office of Environmental Justice at EPA, and an Environmental Justice Executive Order. On February 11, 1994, President Clinton signed the Environmental Justice Executive Order 12898.

Bullard currently holds an endowed chair as Ware Professor of Sociology and directs the Environmental Justice Resource Center at Clark Atlanta University. Prior to joining the faculty at Clark Atlanta University in 1994, he served as a professor of sociology at the University of California, Riverside, and visiting professor in the Center for African-American Studies at UCLA. He has worked on and conducted research in the areas of urban land use, housing, transportation, urban sprawl, economic development, siting of industrial facilities, environmental quality, and community health. His scholarship has made him a leading expert and highly sought-after speaker on environmental justice.

Bullard served on the U.S. EPA National Environmental Justice Advisory Council (NEJAC), where he chaired the Health and Research Subcommittee. He also served on the U.S. EPA's National Advisory Council for Environmental Policy and Technology (NACEPT) Title VI Implementation Advisory Committee. He is the author of numerous articles, monographs, and scholarly papers and nine books that address environmental and economic justice concerns. His *People of Color Environmental Groups Directory* (1994) won a Silver Medal for publications from U.S. foundations. His *Dumping in Dixie* (2000) has become a standard text in the environmental justice field. Some of his other books include *In Search of the New South* (1989), *Confronting Environmental Racism: Voices from the Grassroots* (1993), and *Unequal Protection: Environmental Justice and Communities of Color* (1994). He coedited with J. Eugene Grigsby and Charles Lee *Residential Apartheid: The American Legacy* (1994). Bullard also coedited with Glenn S. Johnson *Just Transportation: Dismantling Race and Class Barriers to Mobility* (1997). His most recent book, coedited with Angel O. Torres and Glenn S. Johnson, *Sprawl City: Race, Politics, and Planning in Atlanta* (2000), addresses the impact of urban sprawl on communities of color in the widely sprawled Atlanta metropolitan region.

No other single individual, activist or academician, has done more to make the terms "environmental justice," "environmental racism," and "environmental equity" household words.

BIBLIOGRAPHY

Bullard, R. D. (1983). Solid waste sites and the black Houston community. *Sociological Inquiry 53*, pp. 273–228.

Bullard, R. D. (1987). *Invisible Houston: The Black Experience in Boom and Bust.* College Station: Texas A&M University Press.

Bullard, R. D. (1989). *In Search of the New South: The Black Urban Experience in the 1970s and 1980s.* Tuscaloosa: University of Alabama Press.

Bullard, R. D. (1993). *Confronting Environmental Racism: Voices from the Grassroots.* Boston: South End Press.

Bullard, R. D. (1994). *Unequal Protection: Environmental Justice and Communities of Color*. San Francisco: Sierra Club.

Bullard, R. D. (2000a). [1990]. *Dumping in Dixie: Race, Class, and Environmental Quality*. Boulder, CO: Westview Press.

Bullard, R. D. (2000b). *People of Color Environmental Groups Directory*. Flint, MI: Charles Stewart Mott Foundation.

Bullard, R. D., Grigsby, J. E., III, & Lee, C. (1994). *Residential Apartheid: The American Legacy*. Los Angeles: Center for Afro-American Studies.

Bullard, R. D., & Johnson, G. S. (1997). *Just Transportation: Dismantling Race and Class Barriers to Mobility*. Gabriola Island, BC: New Society Publishers.

Bullard, R. D., & Johnson, G. S. (1998). Environmental and economic justice: Implications for public policy. *Journal of Public Management and Social Policy* 4 (4), pp. 137–148.

Bullard, R. D., Johnson, G. S., & Torres, A. O. (1999, Fall). Atlanta: Mega sprawl. *Forum: For Applied Research and Public Policy* 14 (3), pp. 17–23.

Bullard, R. D., Johnson, G. S., & Torres, A. O. (2000, February/March). Dismantling transportation apartheid through environmental justice. *Progress: Surface Transportation Policy Project* 10 (1), pp. 4–5.

Bullard, R. D., Johnson, G. S., & Torres, A. O. (2000). *Sprawl City: Race, Politics, and Planning in Atlanta*. Washington, DC: Island Press.

Bullard, R. D., Johnson, G. S., & Wright, B. H. (1997). Confronting environmental injustice: It's the right thing to do. Environmentalism and Race, Gender, Class Issues. *Race, Gender and Class* 5 (1), pp. 63–79.

Environmental Justice Resource Center at Clark Atlanta University. *http://www.ejrc.cau.edu*

—Glenn S. Johnson

ARCHIBALD FAIRLY CARR
(1909–1987)

Archibald ("Archie") Carr, a noted taxonomist, evolutionary biologist, and conservationist, was born in Mobile, Alabama, on June 16, 1909. He was the son of Archibald Fairly Carr, a Presbyterian minister, and Louise Gordon Deaderick, a piano teacher. The Carrs lived in Fort Worth, Texas, and later in Georgia, where Carr entered Davidson College in 1928 as an English major. After deciding to study biology, Carr attended the University of Florida at Gainesville, where he received the university's first doctorate in biology in 1937.

In his early career, Carr was an assistant biologist at the University of Florida in the 1930s. He obtained a fellowship to work at Harvard University's Museum of Comparative Zoology from 1937 to 1943 and continued as a biology instructor at the University of Florida from 1938 to 1940, becoming a full professor in 1949.

During the 1940s and 1950s, Carr was professor of biology at Escuela

Agricola Panamerica in Honduras and a technical advisor to the faculty of the University of Costa Rica. In addition, he served in research positions for the United Fruit Company in Honduras and for the American Museum of Natural History, beginning in 1949. Carr served as technical director and later as executive vice president of the Caribbean Conservation Corporation in the 1950s and 1960s.

Carr coauthored a monograph, *Antillian Terrapins*, that described new turtles classified in Cuba, Haiti, and the Bahamas. Another publication, *Handbook of Turtles* (1952), classified seventy-nine turtle species and subspecies in the United States, Canada, and Baja California. As a principal investigator for the National Science Foundation, Carr participated in a project to study marine turtle migration, making expeditions to the Caribbean, Central America, South America, Portugal, Africa, and Madagascar. The Brotherhood of the Green Turtle, the organization that funded Carr's projects, later became the Caribbean Conservation Corporation.

Carr received research assistance from the Office of Naval Research, which was interested in the navigational abilities of sea turtles. In 1978, he was an assistant in the Green Turtle Expedition of the research vessel *Alpha Helix* project in Costa Rica. Carr invented the "five-dollar tag," a device for tagging turtles.

During his lifetime, Carr, the "Turtle Man," was recognized as the foremost authority on turtles. He helped dispel the myths and folklore about turtles. His extensive studies of the migrations and habits of turtles enabled him to locate the optimal areas for turtles to live and breed. His consistent efforts for the conservation of turtles have helped to increase their population throughout the world. Carr maintained a systematic, organized standard when researching turtles throughout the world. He aggressively pursued the opportunity to educate people about turtles and his studies. His peaceful nature and sense of humor helped him to engage everyone in his efforts of research and conservation.

Carr published hundreds of articles and contributed to the Time-Life Books series on the natural history of Africa. From 1937 to 1986, he authored *Handbook of Turtles* (1952), *High Jungles and Low* (1953), *The Windward Road* (1956), *Amphibians and Freshwater Fishes of Florida* (1956), *Guideposts of Animal Navigation* (1962), *The Reptiles* (1963), *Ulendo* (1964), *African Wildlife* (1964), *So Excellent a Fishe: A Natural History of Sea Turtles* (1967), and *The Everglades* (1973).

Carr received many awards for his work, including decoration as an officer of the Order of the Golden Ark (the Netherlands), the Daniel Giraud Elliot Medal of the National Academy of Sciences, the Smithsonian Institution's Edward H. Browning Award for outstanding achievement in conservation, the Florida Audubon Society's John Burroughs Medal for exemplary nature writing, and the World Wildlife Fund's Gold Medal. He was a fellow of the Linnaean Society of London. For the "Black Beach" chapter of his book *The Windward Road,* he won the O. Henry Award in 1956. In 1979, Florida State Museum established the Archie F. Carr Medal in his honor. Carr died in Micanopy, Florida, in 1987.

BIBLIOGRAPHY

Carr, A. (1952). *Handbook of Turtles: The Turtles of the United States, Canada, and Baja California.* Ithaca, NY: Cornell University Press.

Carr, A. (1953). *High Jungles and Low.* Gainesville: University Press of Florida.

Carr, A. (1956). *The Windward Road: Adventures of a Naturalist on Remote Caribbean Shores.* New York: Knopf.

Carr, A. (1962). *Guideposts of Animal Navigation.* Boston: Heath.

Carr, A. (1964). *Ulendo: Travels of a Naturalist in and out of Africa.* New York: Knopf.

Carr, A. (1967). *So Excellent a Fishe: A Natural History of Sea Turtles.* Garden City, NY: Natural History Press.

Carr, A. (1994). *A Naturalist in Florida: A Celebration of Eden.* New Haven, CT: Yale University Press.

Carr, A., & the editors of *Life.* (1963). *The Reptiles.* New York: Time.

Carr, A., & the editors of Time-Life Books. (1973). *The Everglades.* New York: Time-Life Books.

Sterling, K. B., Harmond, R. P., Cevasco, G. A., & Hammond, L. F. (Eds.). (1997). *Biographical Dictionary of American and Canadian Naturalists and Environmentalists.* Westport, CT: Greenwood Press.

—*John Mongillo*

RACHEL L. CARSON
(1907–1964)

> The more I learned about the use of pesticides, the more appalled I became. . . . What I discovered was that everything which meant most to me as a naturalist was being threatened.
>
> —Rachel L. Carson

Rachel Carson was a writer and marine biologist whose 1962 book *Silent Spring,* a groundbreaking study of the dangers of DDT and other insecticides, helped to create a moral foundation for the modern environmental movement. "Can anyone believe it is possible to lay down such a barrage of poisons on the surface of the earth without making it unfit for all life?" she asked (Simon). In the estimation of many historians, *Silent Spring* was the most revolutionary book since Harriet Beecher Stowe's antislavery novel *Uncle Tom's Cabin.* Carson was middle-aged before she became famous, and she died before she could observe the growing power of her influence. As *Time* magazine subtitled her listing in 1999 as one of the "*Time* 100," "Before there was an environmental movement, there was

Photo by Bob Hines. Used by permission of Rachel Carson History Project.

one brave woman and her very brave book." *Silent Spring* did much to spark widespread public support for more stringent federal regulation of the use of agricultural chemicals. Long after her death, Carson's life continues to inspire other biologists, writers, and environmentalists. In his foreword to the most recent edition of *Silent Spring*, Vice President Al Gore credits Carson with helping him and millions of others to develop an environmental consciousness.

Rachel Carson was born in 1907 to a relatively poor family in the Allegheny Valley town of Springdale, Pennsylvania. Her early love for nature came from her mother, who encouraged Carson to investigate anything that piqued her curiosity. As a child, Carson spent her free time roaming the sixty-five acres of woods and farmland that surrounded her family's house. The landscape offered Carson an intimate view of what she later called "that intricate fabric of life" (Russell). Here Carson also got her first glimpse of the effects of pollution. She watched the beautiful little town of Springdale, on the banks of a formerly clean river, become dirty and dreary when two enormous utility companies began to spew smoke into the air and release wastes into local waters.

At age eleven, Carson published her first essay in *St. Nicholas Magazine*, a literary magazine for children. A shy loner, Carson continued to write throughout her adolescence. She chose to major in English at Pennsylvania College for Women (later known as Chatham College) in Pittsburgh and frequently submitted poetry to periodicals. In her junior year, a biology course taught by a unique professor, Mary Scott Skinker, reawakened her interest in the natural world; she quickly switched her major to zoology, not yet aware that her literary and scientific passions might complement each other. Carson's science sprang from a passionate love of all natural things and her sense that nature held the answers to all questions of the spirit.

Carson graduated magna cum laude in 1929 and, on full scholarship, enrolled in graduate school at Johns Hopkins University in Baltimore. She received her master's degree in zoology, but family responsibilities caused her to abandon her hope for a doctorate. In 1931, she joined the zoology faculty of the University of Maryland, where she taught for five years; she continued her studies during summers at the Marine Biological Laboratory in Woods Hole, Massachusetts. It was there, while she was in her early twenties, that she first saw and became intrigued by the sea.

When Carson's father died suddenly in 1935, she took part-time work writing broadcast scripts for the Bureau of Fisheries (now the U.S. Fish and Wildlife Service), a job that led to her appointment as a full-time junior aquatic biologist in 1936. To supplement her income, she contributed feature articles to the *Baltimore Sun*, most of them related to marine zoology. Though her poetry was never to be published, the lyricism of her prose style was already evident. One of her government publication pieces, "Undersea," struck the editor as so elegant and unusual that he urged her to submit it to the *Atlantic Monthly*, which published it in its September 1937 issue. The article formed the basis for Carson's 1941 book *Under the Sea Wind*—her own favorite among her four books—which was published to little notice.

As early as 1945, Carson had become alarmed by government abuse of new chemical pesticides such as DDT; federal programs to control predators and pests, in particular, were broadcasting poisons with little regard for the welfare of nontarget species. That same year, Carson offered *Reader's Digest* an article about ongoing insecticide experiments at Patuxent, Maryland, not far from her home in Silver Spring, to determine the effects of DDT on wildlife in affected areas. The *Digest* declined to publish the piece, and Carson returned her full attention to her government job, where her editorial duties were increasing. In 1946, she was promoted to information specialist, and in 1949, she became editor-in-chief of publications, including guides to wildlife refuges.

In 1951, the editor of the *New Yorker* magazine broke format by serializing fourteen chapters of a Carson work in progress as "A Profile of the

Sea," the first nonhuman profile in the publication's history. The piece was a rousing international success. In July 1951, the entire manuscript was published in book form as *The Sea around Us*, which Carson described as "a biography of the sea." It won the 1952 John Burroughs Medal and the National Book Award for nonfiction and within a year sold more than 200,000 copies in hard cover. In 1951, a Guggenheim Fellowship enabled Carson to begin work on her third book, *The Edge of the Sea*, which was published in 1955.

Success permitted Carson to retire from the Fish and Wildlife Service in 1952 to write full-time. That summer, she bought land and built a cottage on the Sheepscot River near West Southport, Maine, where she and her mother had vacationed since 1946. Her new celebrity also gave her the opportunity to speak out about her concerns.

In the late 1950s, there was much controversy over the fire ant in the southern states, which the U.S. Department of Agriculture was avidly—and unsuccessfully—attempting to exterminate. Agency scientists were pouring tons of pesticides on the southern landscape, to no avail. Carson began to look into the matter and was hooked by what she discovered: that scientists did not even understand the life cycle of the very insect that they were trying to eliminate. "The more I learned about the use of pesticides, the more appalled I became," Carson recalled. "I realized that here was the material for a book. What I discovered was that everything which meant most to me as a naturalist was being threatened, and that nothing I could do would be more important" (Matthiessen). Though others had also been warning of pesticide dangers, it was Carson who struck upon the metaphor that would draw all these dire warnings to a point. "There was once a town in the heart of America where all life seemed to live in harmony with its surroundings. . . . Then a strange blight crept over the area and everything began to change. . . . There was a strange stillness. . . . The few birds seen anywhere were moribund; they trembled violently and could not fly. It was a spring without voices" (Carson 1994).

Carson's *Silent Spring* was serialized in the *New Yorker* in June 1962. Even before the book's publication, Carson was violently threatened with corporate lawsuits and was derided as a "hysterical woman" who was unqualified to write such a book. Though Carson's accessible prose was not what the public expected of a "serious" scientist, underneath her lyricism lay solid, meticulous research; her carefully assembled scientific facts spoke for themselves. A huge counterattack was organized by the chemical industry and supported by the Department of Agriculture as well as some in the media. With the support of her peers, Carson remained remarkably serene in the face of her assailants. For its part, the *New Yorker* received more letters from concerned readers than it had ever received on any topic before.

By the end of 1962, both *Audubon* and *National Parks* magazines had published additional excerpts from the book, and Carson's attackers were rapidly retreating. With their self-serving campaign against a courageous scientist's protest, the chemical interests had only increased public awareness of the dangers of their products. *Silent Spring* became a runaway best-seller, with international reverberations. "Man is a part of nature, and his war against nature is inevitably a war against himself," Carson said in a 1963 television interview. Testifying before the U.S. Congress in 1963, Carson called for new policies to protect human health and the environment.

By writing passionately about a specific danger, Rachel Carson helped people to recognize that there are many unknown and misunderstood dangers caused by humans' impact on the earth. *Silent Spring* did much to spark widespread public support for federal regulations on the use of agricultural chemicals and was the book that defined the modern ecological movement, putting the environment on the political agenda for the first time. Nearly forty years after its initial publication, *Silent Spring* is still regarded as the cornerstone of modern environmentalism. The book's conclusions were officially endorsed by the report of President John F. Kennedy's specially appointed Science Advisory Committee panel. Carson won the National Wildlife Federation's 1963 Conservationist of the Year award.

At fifty-two, and at the height of her fame, influence, and reputation, Carson fell ill with metastatic breast cancer. From 1960 on, she was never completely well again. Nearing the end of her life and in great pain, Carson still drove herself to speak in public in defense of her book, with little support from established environmental groups. In April 1964, she died at the age of fifty-six at her home in Silver Spring. She left a substantial bequest to the Sierra Club that included royalties from sales of *Silent Spring*, given with the hope that the club would carry on her crusade to protect the earth. That gift continues to help support the Sierra Club's conservation campaigns today. Her legacy also led the club to create the Rachel Carson Society, established to honor individuals who include the Sierra Club in their estate plans. A National Audubon Society fund begun in Carson's name shortly after her death also led to the establishment of the independent Environmental Defense Fund in 1967 to finance lawsuits aimed at stopping pesticide contamination and solving other environmental problems, with its first goal a federal ban on the use of DDT.

In 1999, Carson was chosen as one of *Time* magazine's "Top 20 Most Influential Scientists and Thinkers of the 20th Century" and one of only five featured on its March 29, 1999, cover. Carson was also chosen as one of "The 20 Greatest Innovators of the Century" by *American Heritage* magazine.

The Pennsylvania Department of Environmental Protection created the Rachel Carson Forum to honor her spirit of inquiry and communication and to showcase new ideas and opinions on environmental issues facing Pennsylvania. The Rachel Carson National Wildlife Refuge in Maine was established in 1966 and dedicated in her honor in 1970. The refuge will ultimately consist of more than 7,000 acres of salt marsh and adjacent freshwater habitat for migratory waterfowl between Kittery and Cape Elizabeth, Maine.

"The beauty of the living world I was trying to save," Carson wrote in a 1962 letter to a friend, "has always been uppermost in my mind—that, and anger at the senseless, brutish things that were being done. I have felt bound by a solemn obligation to do what I could—if I didn't at least try I could never be happy again in nature. But now I can believe that I have at least helped a little. It would be unrealistic to believe one book could bring a complete change" (Matthiessen).

BIBLIOGRAPHY

Brooks, P. (1991). Carson, Rachel. In E. Foner and J. A. Garraty (Eds.), *The Reader's Companion to American History* online. Boston: Houghton Mifflin. Available: *http://www.elibrary.com*

Carson, R. L. (1937, September). Undersea. *Atlantic Monthly*, pp. 322–325.

Carson, R. L. (1989). [1951]. *The Sea Around Us*. New York: Oxford University Press.

Carson, R. L. (1994). *Silent Spring*. Boston: Houghton Mifflin.

Carson, R. L. (1996). [1941]. *Under the Sea Wind: A Naturalist's Picture of Ocean Life*. New York: Viking Press.

Carson, R. L. (1998). [1955]. *The Edge of the Sea*. Boston: Houghton Mifflin.

Earth Explorer online. (1995). Rachel Carson (1907–64): Sounding the alarm on pesticides. Available: *http://www.elibrary.com*

Ecology Hall of Fame [online]. Rachel Carson. *http://www.ecotopia.org/ehof/carson/index.html*

The global environment: a natural woman. (1998). *Economist, 347*. Available: *http://www.elibrary.com*

Gregory, D. (1998, April 19). Earth mother. *Independent on Sunday*, p. 29.

Janson-Smith, D. (1998, March 29). Mother of the greens: Deirdre Janson-Smith on the woman who inspired the environmental movement. *Sunday Telegraph*, p. 14.

Lean, G. (1998). [Review of *Silent Spring*]. *New Statesman, 127*, 46.

Lear, L. (1997). *Rachel Carson: Witness for Nature*. New York: Henry Holt and Company.

Linden, E. (1997, October 6). Poet of the tide pools: Unveiling the life and longings of Rachel Carson. *Time*, p. 97.

Mardon, M. (1994). A gift for the future: Sierra Club establishes the Rachel Carson Society. *Sierra, 79*, 84.

Matthiessen, P. (1999, March 29). *Time* 100: The most important people of the century: Rachel Carson. *Time*, p. 187.

McHenry, R. (1995). Carson, Rachel Louise. *Her Heritage: A Biographical Encyclopedia of Famous American Women* [online]. *http://www.plgrm.com/Her_Heritage.HTM*

Mt. Holyoke College, Women in Science and Medicine online. Rachel Carson (1907–1964). *http://www.mtholyoke.edu/proj/cel/carson.htm*

National Women's Hall of Fame online. Rachel Carson. *http://www.greatwomen.
    org/carson.htm*
Pennsylvania Department of Environmental Protection online. A biography: Rachel
    Carson. *http://www.dep.state.pa.us/dep/Rachel_Carson/Carson.htm*
Rachel Carson Web. *http://www.rachelcarson.org/rcbio.html*
Revkin, A. (1994). Passing the baton: Rachel Carson and the start of Environmental
    Defense Fund online. *http://www.edf.org/pubs/EDF-Letter/1994/May/
    n_earlyedf.html*
Roorbach, B. (1995, April 16). Enduring friendship. *Newsday*, p. 33.
Russell, D. (1998). Where the land meets the sea: The delicate fabric of the world's
    coastal regions is being torn apart. *E Magazine, 9*, 36.
Simon, S. (1997, December 13). Rachel Carson remembered [radio interview with
    Linda Lear]. National Public Radio Weekend Saturday, Washington, DC.
                                                                —*Bibi Booth*

GEORGE CATLIN
(1796–1872)

> The history and customs of [American Indians], preserved by pictorial
> illustrations, are themes worthy of the lifetime of one man, and nothing
> short of the loss of my life shall prevent me from visiting their country
> and becoming their historian.
>
>   —George Catlin

George Catlin, an artist and ethnographer, was born in Wilkes-Barre, Pennsylvania, one of fourteen children of Polly Sutton and Putnam Catlin, a
lawyer and gentleman farmer. Even in his early years, Catlin's life was
strongly influenced by Native Americans; his mother had been temporarily
held captive by Iroquois at the age of seven and told her son about her
experiences with the tribe. Catlin was educated at home and collected Indian relics. Though he was later trained as a lawyer, Catlin gave up that
profession early in his life to devote himself to painting Indians in their
native lands. He spent the rest of his life championing the cause of Native
American cultural preservation.

Motivated by anxiety over the effects of expanding white civilization on
indigenous peoples, Catlin vowed to "become the historian and limner of
the aborigines of the vast continent of North America" and to help "a
dying nation, who have no historians and biographers of their own . . .
thus snatching from approaching oblivion what could be saved for the
benefit of posterity" (Yale University Wildernet). Like some other idealists
of his time, Catlin believed that nature was holy, and that primitive man
was nature's nobleman. Indians, he wrote, were "[t]he finest models in all
Nature, unmasked and moving in all their grace and beauty" (Yale Uni-

versity Wildernet). To sustain this belief, it was necessary to view Indians as existing in a timeless, unchanging, and pristine state. Catlin and his contemporaries did not see the connections that bound the Native American to the European. Instead, the Indian, from their perspective, was an innocent primitive, untouched by the corruption of white society, who was proclaimed to be "the chief exponent of natural law, the one splendid example who remained close to God's image of mankind" (Compton's Encyclopedia). In reality, by the time Catlin began to record his images, Native Americans were already becoming a peripheral part of a global economic system that had its center in Europe.

In 1823, after teaching himself to draw and paint, Catlin gave up a budding law practice, sold his law books, and moved to Philadelphia, where he opened a portrait studio. From the beginning, he was more successful at painting portraits of the socially prominent than he had been at winning lawsuits. In February 1824, he enrolled at the Pennsylvania Academy of the Fine Arts. While he was a student there, he encountered a delegation of tribal chieftains passing through Philadelphia on their way to Washington, D.C., to visit President Andrew Jackson. Seeing the Indians sparked memories of Catlin's youthful interest in them. Suddenly, he realized what his life's goal should be. Catlin recalled the fervor with which he proposed to fulfill his mission, writing in his notes, "The history and customs of such people, preserved by pictorial illustrations, are themes worthy of the lifetime of one man, and nothing short of the loss of my life shall prevent me from visiting their country and becoming their historian" (Jewett).

In the spring of 1830, Catlin launched his great western painting crusade. Leaving his young wife, Clara, behind in the East, he settled in St. Louis, where New York governor DeWitt Clinton had arranged a meeting for him with General William Clark, the renowned mapmaker and explorer. Clark, along with Meriwether Lewis, had surveyed the Louisiana Purchase twenty-five years earlier and was now superintendent of Indian affairs for the western tribes. There was immediate friendship between Clark and Catlin, who had in common a natural sympathy for, and understanding of, the Indians' plight. Clark became Catlin's patron, taking the artist with him to several negotiating and ceremonial sessions with Indians, whom Catlin studied and painted. Clark invited Catlin to accompany him on a treaty-making trip to Prairie du Chien and Fort Crawford, Wisconsin; on this journey, Catlin visited the Sioux and the Iowa, among other tribes, and painted portraits of such leaders as Mew-Hew-She-Kaw, chief of the Iowa. Catlin also sketched and painted Indians who visited Clark at his St. Louis office.

In 1831, Catlin joined Major Jean Dougherty on a trip up Nebraska's Platte River. For the next six years, he traveled the immense area between the Mississippi River and the Rocky Mountains, and from North Dakota to Oklahoma, capturing the images that would later form the core of his famous "Indian Gallery." One of his most extensive expeditions was up the Missouri River to North Dakota on the supply and trading steamboat *Yellowstone*. His tentative plans were to reach Fort Union, Montana, at the mouth of the Yellowstone River. For Catlin, it was the chance of a lifetime; beyond Fort Union lay uncharted wilderness—virgin territory as far as artists were concerned. No artist had ever painted Indians in this region, and Catlin intended to be the first.

Catlin spent a year on this particular journey and made over one hundred paintings, detailing specific information on all aspects of Indian life. He became one with his subjects, living with the Indians, hunting with them, and sharing their ceremonies. Between 1832 and 1837—traveling by boat, on horseback, and on foot—he visited forty-eight different Indian tribes. He painted portraits of the major tribal leaders, sketched and painted scenes of village life, and took precise notes. Catlin worked in the classic European tradition of heroic painting, using its techniques to pay tribute to Native Americans. But his approach was not reflective of a naïve faith in the nobility of the savage; his conclusions resulted from personal experience with the people he painted and were moderated by an even-handed judgment of the Indians' virtues and vices.

Catlin visited the Mandan people in 1832. The Mandan had first been contacted by the French in 1738 and had become familiar with European goods through trade with the Assiniboin tribe. Even before Lewis and Clark passed through their villages, they had already been greatly affected by epidemics, primarily smallpox; but by 1837, only five years after Catlin's visit, the tribe had been completely wiped out by disease. When fellow painter John James Audubon traced Catlin's path up the Missouri River, he found the situation so completely changed that he accused Catlin of misrepresentation.

In 1834, Catlin joined a diplomatic mission into Comanche country under the command of General Henry Leavenworth. Until this time, the Comanches had had little contact with white men. The first days into the journey were happy, Catlin recorded in his journal. But the idyllic existence did not last long. Malaria, combined with illnesses due to bad food and contaminated water, struck the expedition, sickening over half of the men. Catlin was among those who were afflicted, but his sickness did not deflect him from his goal. He struggled to keep up with the rest of the party, painting and making notes about every scene he felt had value. He was particularly impressed with the Comanches' horsemanship. Catlin ultimately befriended most of the Comanche chiefs and won the respect of many braves. He returned from this expedition confined to a stretcher in a baggage train. More than a third of those who had started the journey ended up dead, including General Leavenworth.

A year later, Catlin traveled to Minnesota, where he painted Ojibwe and Santee Sioux portraits. He also began to search for a legendary quarry where the Indians of many tribes obtained a soft, easily carved stone, said to be sacred, which they used in crafting smoking pipes. Catlin discovered the quarry in 1836 in what is now Pipestone, Minnesota; no white man had ever seen it before. He painted a large panorama of the quarry and collected samples of the rocks. A leading geologist who later examined the rocks declared them to be an unknown specimen, which was later called "catlinite" in Catlin's honor.

On his journeys, Catlin also took meticulous notes on Indians and the varying geography and wildlife he encountered. Of the great prairie, he wrote, "There are many meadows on the Missouri of many miles in breadth, which are perfectly level, with a waving grass so high, that we are obliged to stand erect in our stirrups in order to look over its waving tops as we are riding through it" (Kansas City Museum). His written work recorded a world that was rapidly dying; in 1842, he published *Letters and Notes on the Manners, Customs, and Conditions of the North American Indians*.

Catlin used impressive showmanship as he tried to market the products of his life's work. First, he opened his "Indian Gallery" in New York in 1837. The gallery featured portraits, landscapes, and other paintings and also included Catlin's extensive collection of artifacts. Catlin then presented his paintings and artifact collection to the U.S. government in 1838, only to have Congress refuse to appropriate funds for their purchase, despite the urging of many prominent citizens.

Rejected in his own country, Catlin, accompanied by his wife and children, took his Indian Gallery to Europe in 1839. The gallery opened on December 30, 1839, in London, and he became an instant celebrity. He also employed costumed actors to perform his versions of Indian dances and songs. Later, to insert an additional element of realism, a number of Indians who had been brought overseas by early promoters joined the living portion of Catlin's gallery. After running in London for six years, the show began to languish, and Catlin's funds began to dwindle. He took his production to Paris in April 1845, and it opened to a warm reception in June. However, during the next month, Catlin's wife and son died of pneumonia, and his circumstances continued to deteriorate afterwards. He was unable to market the contents of his gallery in Europe, and in 1853 he again failed to sell his collection of art to the U.S. government: the proposal to purchase Catlin's artwork was defeated by one vote in the Senate. Hopelessly in debt, Catlin went bankrupt and was forced to sell his entire collection at auction.

From 1852 to 1858, Catlin continued his wanderings. He traveled to Brazil in an unsuccessful search for gold and stayed to paint the native peoples. Catlin returned to the United States in 1870 and died a pauper in Jersey City, New Jersey, on December 23, 1872. After his death, many of Catlin's hundreds of paintings were donated to and installed in several important American museums, including the Smithsonian Institution's National Museum of American Art.

As one of America's first documentary artists, Catlin sought to capture on canvas the history and now-vanished lifestyle of the American Indian. Catlin's work has been particularly noted for the "roughness and energy" of its subject matter, which left an impression on art critics not only in America, but also in Europe (Vivian Kiechel). As one British newspaper critic wrote, "The puny process of the fox chase sinks into insignificance when compared with . . . the grappling of a bear or the butting of a bison" (Vivian Kiechel). Perhaps most important, Catlin introduced the world to previously undepicted tribes, such as the Blackfoot, Crow, Plains Cree, and Yanktonai Sioux Indians. Though many of Catlin's tribes are now extinct, and all of the chiefs and warriors whose portraits he painted have long since died, through his art, Catlin ensured their place in history and American culture.

BIBLIOGRAPHY

Catlin, G. (1973). [1842]. *Letters and Notes on the Manners, Customs, and Conditions of the North American Indians.* Phoenix, AZ: Dover Publications.

The Columbia Encyclopedia (5th ed.). (1993). Catlin, George. New York: Columbia University Press.

Compton's Encyclopedia online. Catlin, George. *http://www.comptons.com*

Daniel, J. (1997, July 6). Cowboys and Indians: Art from 2 views. *St. Louis Post-Dispatch*, p. 4F.

Diener, R. E. (1996). The magnificent park of George Catlin. *http://rumford.com/DienerCatlin.html*

Encarta online. Catlin, George. *http://encarta.msn.com*

Haley & Steele online. *http://www.haleysteele.com/exhibition/west/artists/catlin_toc.html*

Jewett, P. L. (1997). Artist George Catlin roamed the early West, driven by a desire to capture on canvas its natural wonders and the people who lived there. Available: *http://www.thehistorynet.com/WildWest/articles/1997/0497_text.htm*

Kansas City Museum online. *http://www.kcmuseum.com/explor02.html*

Moorman, M. (1994, September 23). A poignant, historic portrait of Indian life. *Newsday*, p. B21.

Tuleja, T. (1994). Catlin, George. The New York Public Library Book of Popular Americana online. Available: *http://www.elibrary.com*

University of Nebraska Museum of Nebraska Art online. *http://monet.unk.edu/mona/artexplr/catlin/catlin.html*

Vivian Kiechel Fine Art online. *http://www.kiechelart.com/artists/catlin/about georgecatlin.htm*

Wortham, J. (n.d.). George Catlin. University of South Dakota online. *http://www.usd.edu/~jwortham/crow/catlin.html*

Yale University Wildernet online. *http://pantheon.cis.yale.edu/~thomast/art/catlin.html*

—*Bibi Booth*

BARRY COMMONER
(1917–   )

> My entry into the environmental arena was through the issue that so dramatically—and destructively—demonstrates the link between science and social action: nuclear weapons.
>
> —Barry Commoner

Barry Commoner is an American biologist, educator, and environmental activist who, since the early 1950s, has warned the public about the dangerous effects of modern technology on the environment. Many people believe that Commoner's ideas helped prepare the world for the grassroots environmentalism and citizen activism that emerged in the 1960s and 1970s, and he is credited with being one of the founders and leaders of the modern environmental movement. He became a powerful voice for protection of the environment years before most people ever heard the words "environmentalist" or "green." Put simply, Commoner's "law" of pollution prevention is this: if you don't put something into the environment, it isn't there. "The environmental crisis arises from a fundamental fault: our systems of production—in industry, agriculture, energy and transportation—essential as they are, make people sick and die" (Hall). In 1995, he said, "[T]he conflict between the environment and the economy is not an ecological given, but an outcome of the failed attempt to improve the environment by relying on the strategy of control. . . . Pollution prevention . . . eliminates this conflict" (Commoner 1995).

Commoner was born in 1917 in Brooklyn, New York, the son of a Russian immigrant tailor. In 1933, he entered Columbia University and then attended Harvard University, where he earned a doctorate in cellular biology in 1941. During World War II, he served in the Naval Air Force. In 1947, he accepted a faculty position with Washington University in St. Louis, Missouri, where he did pioneering work on interactions of viruses and plant cells and helped establish the important role of free radicals (molecules with unpaired electrons) in cell metabolism.

In the early 1950s, an event occurred that caught Commoner's attention and turned his interest to larger questions. On the morning of April 25, 1953, a nuclear bomb was exploded as part of a series of tests in Nevada. The tests were done in secret, their occurrence marked only by Atomic

Energy Commission (AEC) announcements that the emitted radiation was confined to the test area and was, in any case, "harmless." This conclusion reflected the agency's assumption that the resultant radioactive debris would remain aloft in the stratosphere for years, allowing time for much of its radiation to decay.

A day and a half later, there was an intense rainstorm in the city of Troy, New York, 2,300 miles away from the Nevada test site. Radiation monitors at nearby Rensselaer Polytechnic Institute began to record atomic fallout at levels that were three times as high as natural background radiation. The radioactive fallout in Troy became important news, and suddenly citizens began to suspect that they did not have to live near the Nevada test site to be affected by tests of atomic bombs. Commoner's training as a biologist convinced him that such human intrusions into the global biosphere would have unforeseen consequences. He commented, "It is prudent to regard any addition of a potentially toxic substance to the biosphere as capable of producing a total biological effect which is roughly proportional to its concentration in the biosphere" (Commoner 1964).

When Commoner tried to learn more about atomic fallout in 1953, he found that much of the information had been classified as secret by the federal government. This made it impossible for academic scientists to examine the data, which was a violation of the guiding principles of scientific inquiry. As Commoner expressed it in 1958, "[T]he development of a scientific truth is a direct outcome of the degree of communication which normally exists in science. . . . What we call a scientific truth emerges from investigators' insistence on free publication of their own observations. This permits the rest of the scientific community to check the data and evaluate the interpretations, so that eventually a commonly held body of facts and ideas comes into being. Any failure to communicate information to the entire scientific community hampers the attainment of a common understanding" (Commoner 1958). After 1954, when some of the secret reports were declassified, independent scientists were able to analyze the fallout data that AEC scientists had developed but had not understood. The new analyses confirmed that they had grossly underestimated the dangers.

The issue of atomic fallout occupied Commoner for twelve years. While he was studying it, he derived many of the principles of environmental protection that formed the basis for the emergence of grassroots environmentalism. For example, he asserted that in a democracy, scientists have no more right to make decisions than anyone else, that is, decisions with major social consequences must not be left to "experts," who have no special qualifications in matters of moral judgment. On the contrary, Commoner said, experts have an obligation to inform the public about the scientific facts and then let the public decide. "Anyone who attempts to determine whether or not the biological hazards of world-wide fallout can be justified by necessity must somehow weigh a number of human lives against deliberate action to achieve a desired military or political advantage. Such decisions have been made before . . . but never in the history of humanity has such a judgment involved literally every individual now living and expected for some generations to live on the earth" (Commoner 1958).

Commoner believed that a scientist was just one more citizen making a moral judgment; his or her scientific expertise did not enter into the moral equation. Indeed, it was "self-evident," Commoner argued in 1958, that "the public must be given enough information about the need for testing and the hazards of fallout to permit every citizen to decide for himself whether nuclear tests should go on or be stopped" (Commoner 1958). As he said in 1997, "My entry into the environmental arena was through the issue that so dramatically—and destructively—demonstrates the link between science and social action: nuclear weapons" (Hall).

Commoner put his ideas into practice: he helped organize scientists and citizens into the St. Louis Committee for Nuclear Information (CNI). The group started a newsletter called *Nuclear Information* that evolved into the magazine *Scientist and Citizen*. Commoner and his fellow scientists at CNI in St. Louis formed a working alliance with local citizens. "Our task was to explain to the public—first in St. Louis and then nationally—how splitting a few pounds of atoms could turn something as mild as milk into a devastating global poison" (Hall).

Commoner's studies of atomic fallout had convinced him that it represented a biological hazard to humans. However, the federal government was still insisting that fallout was benign. For example, in 1956, President Dwight Eisenhower said, "The continuance of the present rate of H-bomb testing, by the most sober and responsible scientific judgment . . . does not imperil the health of humanity" (Commoner 1969). The Committee for Nuclear Information began collecting baby teeth and sending them to a lab for testing of their radioactivity in order to show that strontium 90, one of the main components of fallout from testing of atomic bombs, was indeed building up in humans. They succeeded in their task. They then drafted a petition, eventually signed by thousands of scientists worldwide, that is credited with persuading President John Kennedy to propose the 1963 Nuclear Test Ban Treaty, which would phase out above-ground testing of nuclear weapons worldwide. It was a triumph of citizen action, with scientists helping bring critical facts to light. Commoner says, "The Nuclear Test Ban Treaty victory was an early indication of the collaborative strength of science and social action" (Hall). This conclusion eventually led CNI to become the Committee for Environmental Information and to extend its mission to the environmental crisis as a whole.

Commoner often acknowledged the important role of an active citizenry in the victory over nuclear testing: "[T]he collaboration between scientist and citizen [is not] a one-way street. Citizens have contributed significantly to what scientists now know about fallout. Through the St. Louis Baby

Tooth Survey, the children of that city have contributed . . . some 150,000 teeth to the cause of scientific knowledge about fallout. . . . By such means, and through hard work and financial support many citizens have become partners in the scientific effort to elucidate the fallout problem" (Commoner 1964).

In 1965, Commoner and his colleagues warned that industry's supposedly scientific "risk assessments" were actually political in nature, and that many industry scientists overstepped the bounds of scientific legitimacy when they tried to impose their employers' political views on the public, using science as a screen. Therefore, when scientists and scientific committees stated that the risk from a particular hazard was "negligible," "acceptable," or "unacceptable," this was actually not a scientific conclusion, but was rather a social judgment.

In 1966, Commoner established the Center for the Biology of Natural Systems at Washington University to study humans' relationship with the environment, including environmental, energy, and resource problems and their economic implications. On the first Earth Day in 1970, Commoner said, "We are in a period of grace, we have the time—perhaps a generation—in which to save the environment from the final effects of the violence we have already done to it" (Soulutions).

In 1981, Commoner moved his Center for the Biology of Natural Systems from Washington University to New York's Queens College, where he joined the biology faculty. There the research team he directs has continued to make major advances in environmental science: investigating the origin of dioxin in trash incinerators, developing alternatives to incinerators, and developing a computer model that tracks long-range transport of air pollutants from their sources through the food chain and into the human body. The model is being used to evaluate dioxin contamination of milk on dairy farms in Wisconsin and Vermont. "What is needed now is a transformation of the major systems of production more profound than even the sweeping post–World War II changes in production technology. Restoring environmental quality means substituting solar sources of energy for fossil and nuclear fuels; substituting electric motors for the internal-combustion engine; substituting organic farming for chemical agriculture; expanding the use of durable, renewable and recyclable materials—metals, glass, wood, paper—in place of the petrochemical products that have massively displaced them" (Hall). At the same time, however, "There are powerful reasons why environmental advocates should favor economic development and growth, as long as they are based on ecologically benign technologies of production. . . . The most cogent reason is that the massive transformation of our major systems of production—which is essential to environmental quality—cannot achieve this goal if it is pursued only in developed countries. The environmental crisis is a global problem, and only global action will resolve it. . . . This is the rational, logical outcome of the environmental experience" (Hall).

Commoner is the author of nine books, including the highly influential *Science and Survival* (1966), *The Closing Circle* (1971), and *Making Peace with the Planet* (1990). He was the 1980 presidential candidate of the Citizens' party, of which he was a founder. In its February 2, 1970, cover story, *Time* magazine called Barry Commoner "The Paul Revere of Ecology" and said that he was "endowed with a rare combination of political savvy, scientific soundness and the ability to excite people with his ideas" (Queens College). Today, Commoner's views are deeply ingrained in American society. From kindergarten through high school, classes incorporate environmental concepts across the curriculum, and many schools provide teacher training in environmental topics. In 1995, *The Earth* hailed Commoner as one of the "100 Who Made a Difference" worldwide in the past fifty years and called Commoner "the dean of the environmental movement, who has influenced two generations" (Queens College).

BIBLIOGRAPHY

Commoner, B. (1958 May 2). The fallout problem. *Science* 127 (3505), pp. 1023–1026.

Commoner, B. (1964, December). Fallout and water pollution—parallel cases. *Scientist and Citizen* 7 (2), pp. 2–7.

Commoner, B. (1966). *Science and Survival*. New York: Viking Press.

Commoner, B. (1969, March). The myth of omnipotence. *Environment* 11(2), pp. 8–13, 26–28.

Commoner, B. (1971). *The Closing Circle: Man, Nature and Technology*. New York: Alfred Knopf.

Commoner, B. (1975, August/September). How poverty breeds overpopulation (and not the other way around). *Ramparts*, pp. 21–25, 58–59.

Commoner, B. (1992). *Making Peace with the Planet*. New York: The New Press.

Commoner, B. (1995, April 25). Prevention is the best pollution Rx. *Newsday*, p. A31.

Hall, A. (1997). Barry Commoner: A leading environmentalist reviews his long, contentious past and sets new directions for the future [interview]. Available: *http://www.sciam.com/interview/commoner/062397commoner.html*

Mills, M. (1990, May 21). New Life for environmental humbug: With saving the earth suddenly in the mainstream, the movement's erstwhile leaders are again in demand; they're as wrong as ever. *Fortune*, p. 159.

O'Riordan, T., Clark, W. C., Kates, R. W., & McGowan, A. (1995). The legacy of Earth Day: Reflections at a turning point. *Environment*, 37 (3), pp. 7–15.

Queens College, Center for Biology of Natural Systems online. The environmental community celebrates one of its founders. *http//qcunix1.acc.qc.edu/CBNS/release.html*

Rachel's Environmental and Health Weekly online. Origins of environmental ideas. *http://garynull.com/Documents/erf/origins_of_environmental_ideas.htm*

St. Louis Walk of Fame online. St. Louis walk of fame: Barry Commoner. *http://www.stlouiswalkoffame.org/inductees/barry-commoner.htm*

Soulutions online. The Soulutions *Now* Environmental Directory. *http://www.soulutions.com/soulu_env.html*

Stelzer, C. D. (1994, August 3). Commoner cause. *Riverfront Times* (St. Louis, MO). Available: *http://home.stlnet.com/~cdstelzer/diex7.html*

—*Bibi Booth*

MAX CORDOVA
(1946–   )

> I am proud of my Spanish heritage, my pueblo Native American heritage and of being an American. I have visited many areas but Northern New Mexico is unique, it is a land of not much money but rich in traditional knowledge acquired and passed down from generation to generation. We only take what we need to survive.
>
> —Max Cordova

Max Cordova lives in Truchas, New Mexico, an area that contains the Black Mesa, the sacred mountain of San Ildefonso, and Truchas Peak, a mountain that has a cap of snow on top even during the early summer months. Truchas is Spanish for "trout," and there are several trout streams in the community that Cordova has fished since he was a boy. One can also see juniper and pinion-pine forests across the mountains north of Santa Fe.

Cordova is a historian. He can talk about his people's history going back several hundred years. He is an environmental activist, the president of the Truchas Land Grant, and the founder and director of the Los Siete Project. Today, Los Siete is a cooperative for about seventy-four artists and is dedicated to helping Truchas and surrounding villages establish economic viability and regain cultural pride. Cordova says that the Los Siete Project is making a difference in his community because the economy is growing due to the artworks and the marketable woven goods that are sold. There is also a renewed sense of pride in the communities' rich cultural heritage.

The Los Siete Project began in the late 1980s when Cordova was seeking a grant from the Self-Development People Fund of the Presbyterian church. His goals were to develop a woodworking business, an arts and crafts co-op, and a large building to help his community be self-sufficient and more productive. He got his grant, and for two years, Cordova, a licensed contractor, and hired local youths built the Los Siete Project. It was completed in 1991. The Los Siete Project trains people and sells local art products made by the artists.

Max Cordova was born in 1946 and was raised in Truchas, New Mex-

Photo courtesy of Lillian Trujillo Studios.

ico, one of ten children of Albert and Cordelia Cordova. While the Spanish European side of his family settled in the area in 1750, the residency of his Native American ancestors dates to the eleventh century. His mother was the center of their lives, the bond that held them together as a family. His father was a good provider but at times a heavy drinker. His mother taught him the values he holds dear today; she laid the foundation as he matured. Grandparents and parents were equally in charge of teaching a child and nurturing him and his needs. Cordova tended to chores at his home and also at his grandparents' homes. The chores included chopping wood, feeding animals, and household chores such as washing dishes and sweeping floors. As he grew older, his father taught him the value of working hard and to be dedicated to what one starts.

Cordova recalls one of his early school experiences. When he was attending mission schools, every time the children spoke other than in English, they had to put their hands on top of the desk and were hit with a yardstick. The grandparents of the children strongly objected to this treatment. Their philosophy was that the more a person knows, the better off he is, and this includes language. Grandparents told the teachers that it was good for the children to learn English, but it was also important for them to retain their native languages.

Cordova married Lillian Trujillo de Rio Chiquito in 1968. They have three children, David, Bonnie, and Max. Lillian has worked for Espanola Public Schools for thirty-two years. Lillian's family includes eight generations of weavers, while Max's family has three generations. Lillian taught him the fine art of weaving, which he now does professionally to generate an income. They make fine Chimayó blankets, vests, and other weavings. All of their children also weave. David was a master weaver at fifteen, graduated from the University of New Mexico with a degree in economics and business administration, and is currently the director of economic development at Eight Northern Indian Pueblos in San Juan. Bonnie works in Albuquerque. Max is an eighth-grader at McCurdy School.

Lillian has been very supportive of Cordova's work through thick and thin. She has provided counsel and advice. She handles all of the work of an assistant; she does most of the typing and spell checking for his letters. At times she has carried the financial load while Cordova spent his time as an activist. She has always encouraged him and has never discouraged him from the work he wanted to do. His family is also very supportive.

Cordova is the president of the Truchas Land Grant, a forested area whose 290 member families depend on it for their survival. The Truchas Land Grant is one of the few surviving Spanish land grants of northern New Mexico. Cordova's ancestors came to the area of northern New Mexico in 1721. The government of Spain established two land grants, La Merced de Quemado Land Grant and Nuestra Senora del Rosario de Las Truchas Land Grant. These two land grants were buffer communities used to protect the established Spanish cities in northern New Mexico from marauding Apache and Comanche tribes. The area contained about 21,000 acres. The land was given to the communities by the government of Spain in 1754 and later the government of Mexico in 1821.

However, the government of the United States in 1892 through the U.S. court of private land claims took most of the land away, leaving Cordova's people with 7,000 acres of what Cordova states is inhospitable and resource-depleted land. Today, the federal government manages 70 percent of the land, 15 percent is in private hands, and 15 percent belongs to Native American tribes. Cordova's opinion is that most of the good land is held by national forests. Cordova, his father, and his grandfather all held positions as officers of the land grant.

Cordova's community is dependent on forests. The forests provide wood for heat, cooking, and building materials, fish and game, pinion-pine nuts, herbs, and water for irrigation. At times, Cordova has had his conflicts with government officials and other environmental groups who do not consider the forest needs of his community in their environmental programs. He has spoken up and written letters to the Forest Service and even to large environmental organizations. As one example, one environmental group wanted to stop all commercial logging on public lands. Although Cordova sympathized with the group's stewardship of forests, he stated that the forests were overgrown and needed to be thinned out. The wood could be used by his community, which depended on these resources.

Cordova is a member of the National Network of Forest Practitioners and the Land Grant Forum, a New Mexico association of Spanish land grants. He also sits on the board of the Rio Arriba County Extension Service. Cordova is chairman of a nonprofit community-based wood-products business called Montana de Truchas. Currently it employs twenty-eight contractors and three managers.

Cordova's work with the Forest Service earned him in 1997 the Al Gore Hammer Award for cutting government red tape and making government more efficient. In 1988, he was on the team that won the Innovation in American Government Award of the John F. Kennedy School of Government at Harvard University for a project called Collaborative Stewardship.

Cordova has a strong belief in divine intervention. He believes that he could not have done his work without God's help to open doors for him and give him the courage to meet the challenges of today's world. He also has a number of good friends and organizations who have supported him, including Gie Laksman, Hunimon Foundation Santa Fe, Wayne Mueller, and Bread for the Journey. Cordova believes that there is a goodness in all of us, even radical special-interest groups or people. All we need to do is sit at the table and try to work out our differences—try to educate and communicate with each other to break down barriers that separate people.

—*John Mongillo*

JAY NORWOOD DARLING
(1876–1962)

Land, water and vegetation are just that dependent on one another. Without these three primary elements in natural balance, we can have neither fish nor game, wild flowers nor trees, labor nor capital, nor sustaining habitat for humans.

—Jay Norwood Darling

Jay Norwood Darling was a two-time Pulitzer Prize–winning political cartoonist, artist, active conservationist, leading environmentalist, chief of the U.S. Biological Survey, and founder of the National Wildlife Federation. Darling was born in Norwood, Michigan, October 21, 1876. His family settled in Sioux City, Iowa, when he was about ten years old. He spent much of his youth in Sioux City. He graduated from the local high school there and went to Yankton (South Dakota) College in 1894. He attended Beloit (Wisconsin) College from 1895 to 1898 and worked for a year before

Photo courtesy of J. N. "Ding" Darling Foundation.

returning to graduate with a bachelor's degree in 1900. One of Darling's favorite subject areas was the field of biology with an ecological approach. In the course, his professor stressed how human life shares the earth's natural resources of soil, water, and air with plants and animals.

After college, Darling joined the *Sioux City Journal*, where he worked as a photographer and news reporter. One day at the newspaper, he was assigned a story and needed to get a photo. Unable to get the photo, Darling drew a sketch instead and submitted it to the editor, who liked it. Darling's career as a cartoonist was launched.

In 1906, Darling left Sioux City and joined the *Des Moines Register and Leader*, where he worked with Gardner Cowles, Sr., in rebuilding the faltering newspapers. Darling's work as a cartoonist became extremely popular. He began signing his cartoons with the nickname Ding as his signature, a contraction of his last name. "Ding" soon became a household word.

Ever since Darling was a youngster, he had been concerned about environmental issues such as overhunting, particularly of migratory birds. He observed overgrazing, soil erosion on farms, and little or no management of natural resources. He reflected these themes in his political and conservation cartoons with strong messages and humor. He drew his first conservation cartoon during the term of President Theodore Roosevelt, a man Darling admired. When Roosevelt died January 6, 1919, Darling sketched a farewell cartoon of Roosevelt as a Rough Rider titled the "Long, Long Trail." The cartoon became so popular that it has been displayed in more public places than any other Darling cartoon (see Roosevelt entry).

In 1911, Darling accepted an offer from the *New York Globe*, in part because the paper offered to syndicate his cartoons. He disliked living and working in New York, however, and returned to the *Register*, newly renamed the *Tribune* in Des Moines in 1913. In 1916, his work was syndicated nationally by the *New York Herald Tribune*. Darling's cartoons were so popular that they were syndicated in 130 daily newspapers in the United States. In 1924, he received his first Pulitzer Prize for editorial cartooning. He received his second for a cartoon that appeared in 1942. In words and pictures, Darling was able to explain complex issues in politics and in the environment in terms that lay people could easily understand. Many of his cartoons appear in contemporary texts and cartoon anthologies.

Darling enjoyed cartooning very much, but it was hard work. Sometimes it took hours and hours to simplify a picture. He was constantly asked for advice by young cartoonists. In David L. Lendt's book *Ding: The Life of Jay Norwood Darling*, Darling wrote of one hopeful lad, "If he is not willing to sacrifice . . . everything in order to devote himself to the development of his drawing ability then he better not try."

In 1932, as a member of the Iowa Fish and Game Commission, Darling proposed the establishment of the nation's first Cooperative Wildlife Research Unit at Iowa State University, using $9,000 of his own money for start-up costs. Darling and others believed there was a need to do a biological survey of the state's marshes, rivers, and other water resources. They emphasized that the work should be done only by unbiased professionals with strong backgrounds in scientific research. As a result, and under the leadership of the noted ecologist Aldo Leopold, a complete biological survey of Iowa's resources was developed and a twenty-five-year conservation plan was established, the first of its kind in the nation. Darling also established the system of Cooperative Wildlife Research Units still present on more than fifty university campuses for education in conservation-related fields.

In July 1934, President Franklin Delano Roosevelt asked Darling to be the chief of the U.S. Biological Survey, the predecessor of the U.S. Fish and Wildlife Service. In this capacity, Darling worked hard for greater national attention and expenditures for conservation. Darling was responsible for securing $17 million for restoration of wildlife habitats. He established the Migratory Bird Conservation Commission and made great strides toward bringing hunters and conservationists together. He also pioneered leadership in the field of proper game management. In 1935, he issued strict hunting restrictions to stem the decline of duck populations and other migratory birds. The hunting season was cut down to thirty consecutive days, duck quotas were dropped to ten ducks and four geese, live bait and live decoys were banned, and shotguns holding more than three shells were made illegal. He cracked down on violators of game laws as well.

Darling initiated the federal Duck Stamp Program, officially known as the federal Migratory Bird Hunting and Conservation Stamp. Proceeds from the sale of Duck Stamps are used to acquire and protect waterfowl habitats. Darling drew the first Duck Stamp, *Mallards Alighting*, whose image today is used to mark every federal wildlife refuge. It is probably one of the most widely recognized works of wildlife art ever produced. Since its inception in 1934, the federal Duck Stamp Program has raised more than $500 million for the purchase of more than 5 million acres to be set aside for wetland habitats. More than 123 million stamps have been sold. While Darling was chief, he greatly expanded the national wildlife refuge system. Largely responsible for the establishment of the network of game refuges in the United States today, Darling was called the best friend ducks ever had. He liked to remind overzealous developers that ducks can't lay eggs on picket fences.

In 1935, Darling, resigned as chief of the Biological Survey to concentrate on creating a federation of conservation organizations to help combat abuse and waste of soil, wildlife, and water. As a result of his efforts, the National Wildlife Federation was founded, and he was appointed the first president of the new organization, which has become the largest of its kind in the world.

Darling led the drive that culminated in the creation on December 1, 1945, of the Sanibel National Wildlife Refuge in Florida. Since 1935, Darling and his family had had a winter home on Captiva Island near Sanibel Island, off the west coast of Florida. At that time, these islands were remote places. There were no businesses, only a few cottages, and no electricity. But they were a major bird habitat with an abundance of brown pelicans, ring-necked and scaup ducks, greater and lesser white herons, egrets, black-necked stilts, and willets. However, in the early 1940s, Darling learned that the state of Florida was getting ready to sell the 2,400 acres of land that it owned to developers for twenty-five cents an acre. The land had previously been designated as a State of Florida Wildlife Refuge. At the time, Darling was president of a local group that called itself the Sanibel and Captiva Conservation Association. (This is not to be confused with the present Sanibel-Captiva Conservation Foundation, which was founded twenty-five years later.) Operating under the auspices of the Sanibel and Captiva Conservation Association, Darling immediately put together a bid of $1.50 per acre for the land. The real purpose of the bid, however, was to delay any potential sale while Darling could muster his forces. Darling had considerable forces at his disposal. They included Ira Gabrielson, who succeeded him as chief of the U.S. Biological Survey, now the U.S. Fish and Wildlife Service; J. Clark Salyer, who had become the service's assistant director for national wildlife refuges; and Spessard L. Holland, governor of the state of Florida. When it became clear that Darling was putting together a transaction that would preempt the developers, those developers raised their bid to $2.50, ten times their original offer. A public auction was then announced that would sell the land to the highest bidder. Just eight days before the auction, Governor Holland announced an agreement to lease the land to the U.S. Fish and Wildlife Service. The U.S. Fish and Wildlife Service in turn announced that it would establish a National Wildlife Refuge encompassing the state lands and an additional 1,000 acres that the federal government already owned. Darling had succeeded. On February 4, 1978, the refuge was officially dedicated as the J. N. "Ding" Darling National Wildlife Refuge in his memory.

By the time he retired in 1949, Darling had fashioned approximately 15,000 cartoons for the *Des Moines Register and Tribune*, the *New York Globe*, the *New York Herald Tribune* and its syndicate, *Collier's*, the *Saturday Evening Post, Better Homes and Gardens*, and other publications. Anthologies of his work were printed by the *Register and Leader* and the *Register and Tribune*. He wrote and illustrated several books, including *Ding Goes to Russia* (1932), *The Cruise of the Bouncing Betsy* (1937), and *It Seems Like Only Yesterday* (1960). With John Henry, he compiled *Ding's Half Century* (1962).

Darling died in Des Moines, Iowa, February 12, 1962. The national wildlife refuge system is perhaps his greatest legacy. The J. N. "Ding" Darling National Wildlife Refuge is one of more than 500 U.S. Fish and Wildlife Service refuges managed within the United States. The refuge of more than 6,300 acres of wetlands, islands, and mangrove forests, is located on Sanibel Island, Florida, and on islands in the adjacent Pine Island Sound. The purpose of the J. N. "Ding" Darling National Wildlife Refuge is to conserve and protect native and migratory birds and resident amphibians, reptiles, and mammals. It is not a "park" with the expected tourist attractions and exhibits, but rather a sanctuary where a visitor can see large numbers of birds and reptiles in their natural habitat. During the year, large numbers of migratory birds arrive either to stop and rest during their journey or to winter there before starting the long trip back to their breeding grounds. There is a large resident population of native long-legged wading birds as well as many alligators and even a resident American crocodile. Mammals that may be seen in the refuge include the river otter, the bobcat, and the ubiquitous raccoon.

BIBLIOGRAPHY

Darling, J. N. (1932). *Ding Goes to Russia*. New York: Whittlesey House.

Darling, J. N. (1937). *The Cruise of the Bouncing Betsy*. New York: Frederick A. Stokes.

Darling, J. N. (1962). *Ding's Half Century*. Edited by J. M. Henry. New York: Duell, Sloan and Pearce.

Laycock, G. (1973). *The Sign of the Flying Goose: The Story of Our National Wildlife Refuges*. Garden City, NY: Anchor Press.

Lendt, D. L. (1979) *Ding: The Life of Jay Norwood Darling*. Ames: Iowa State University Press.

Mills, G. (1977). *Harvey Ingham and Gardner Cowles, Sr.: Things Don't Just Happen*. Ames: Iowa State University Press.

Trefethen, J. B. (1961). *Crusade for Wildlife: Highlights in Conservation Progress*. Harrisburg, PA: Stackpole.

—*John Mongillo*

WILL DILG
(1867–1927)

Will H. Dilg was a publicist, advertising salesperson, fisher, nature writer, and conservationist. He was the founder and first president (1922–1926) of the Izaak Walton League of America, the first mainstream conservation organization with mass appeal. Dilg was born in Milwaukee, Wisconsin, in 1867 and spent his career in advertising and public relations.

Dilg and about fifty people who were active in fishing and hunting established the Izaak Walton League in Chicago in January 1922. They started the league to battle the depletion of fish stocks in their area and to save the Mississippi River. The league was formed in the name of Izaak Walton, a well-known English angler and author of *The Compleat Angler*, one of the most famous books in the English language. Walton's book, published in 1653, discusses his love of the out-of-doors, fishing adventures, brooks and streams, and his exemplification of true sportsmanship. The book, containing dialogue, poetry, and songs, describes Walton's observations of weather conditions, fishing equipment such as lines and rods, tips on how to catch trout, and several ways to prepare fish to cook. A later edition, written with the English poet Charles Cotton, discussed fly fishing. Walton's love of angling was revered by fishers throughout the world. Dilg knew that Walton's name would add importance to the movement to conserve resources.

In time, Dilg's league spread throughout the country, and several state divisions were formed. "Ding" Darling (profiled in this book), the cartoonist and conservationist, played a key role in organizing the Iowa division of the Izaak Walton League of America. The Iowa branch was one

Izaak Walton. Courtesy of Izaak Walton League
of America.

of the most progressive and industrious of the league's chapters. Under
Dilg's leadership, the organization grew to 100,000 members by 1925. Dilg
also wrote and edited the *Izaak Walton League Monthly,* a periodical later
called *Outdoor America.*

In 1924, Dilg and his organization influenced and supported U.S. congressional approval for the establishment of the Upper Mississippi River
Wild Life and Fish Refuge. This was the first major such refuge to be
funded by Congress and to be established under the supervision of the
federal government. The refuge was located on the Mississippi River near
Lansing, Iowa, and was also the country's first sanctuary for fish. Many
believe that without the league's influence, this large marsh area would have
been drained for farmland. Dilg also set up the Izaak Walton League Fund
to save starving elk in Jackson Hole, Wyoming. In his later years, Dilg also
spent much time in Washington, D.C., encouraging lawmakers to set up a
special conservation department in the U.S. president's cabinet. Dilg died
in March 1927 in Washington, D.C. Today, the Izaak Walton League of
America is still active, with a membership of about 50,000 and 400 local
chapters dedicated to the protection of soil, streams and rivers, watersheds,
and wildlife. The Save Our Streams (SOS) Program of the Izaak Walton

League of America has greatly enhanced the quality of surface water across the United States and has increased the number of citizen watershed stewards. The program received the 1990 Renew America award as the nation's best project for the protection of surface water. It also helps communities achieve sustainability of their water resources through education and technical support.

BIBLIOGRAPHY

The Izaak Walton League of America. 707 Conservation Lane, Gaithersburg, Maryland.
James, F. (1990, Spring). Izaak Walton: His life and times. *Outdoor America*.
Karas, N. (1993, July). Izaak Walton, the father of fishing. *Outdoor Life*.
New York Times. (1927, 29 March). Obituary.
A 60-year battle to preserve the outdoors. (1982, March/April). *Outdoor America, 47*.
Sterling, K. B., Harmond, R. P., Cevasco, G. A., & Hammond, L. F. (Eds.). (1997). *Biographical Dictionary of American and Canadian Naturalists and Environmentalists*. Westport, CT: Greenwood Press.
Voigt, W., Jr. (1992). *Born with Fists Doubled: Defending Outdoor America*.
—*John Mongillo*

MARJORY STONEMAN DOUGLAS
(1890–1998)

Marjory Stoneman Douglas is often referred to as the "mother of the Everglades." Douglas was an environmentalist, journalist, playwright, activist, feminist, and independent thinker. Her name is synonymous with the Everglades for her grassroots efforts to protect and preserve this watery region, which many considered to be a large, useless, and worthless swamp.

Douglas was born on April 7, 1890, in Minneapolis, Minnesota. She was the daughter of Frank Bryant Stoneman of Minneapolis and Florence Lillian Trefethen of Providence, Rhode Island. Her ancestors included her great-great-uncle Levi Coffin, who had headed the Underground Railroad. Her family later moved to Providence, Rhode Island. Due to the breakup of her parents' marriage, Douglas was raised by her grandparents and an aunt. She attended local schools, went on to college, and graduated from Wellesley College in 1912. After college, she worked in a department store helping young women in sales. A short time later, she met Kenneth Douglas, her future husband.

After an unsuccessful marriage, Douglas was reunited with her father, who was then the editor and founder of Miami's first morning paper, the *News Record*. The newspaper later evolved into the *Miami Herald*. She went to work as a society writer and editor. She also wrote about the Everglades, describing its geography and animal life.

The Everglades is an unusual, vast ecosystem. It stretches from a watershed north of Lake Okeechobee to the southern tip of Florida. Now a national park, it is the largest remaining subtropical wilderness in the continental United States and has extensive freshwater and saltwater areas, open prairies, and mangrove forests. Abundant wildlife includes rare and colorful birds, and this is the only large place in the world where alligators and crocodiles exist side by side. The park is 1,506,539 acres (606,688 hectares) in size. The great floral variety of the Everglades is one of the key resources of the park. Among its more prominent and colorful plants are bromeliads and epiphytic orchids. As many as 25 varieties of orchids are known to occur in the park, in addition to over 1,000 other kinds of seed-bearing plants and 120 species of trees. Over 36 threatened or endangered animal species reside in Everglades National Park, such as the American alligator and crocodile, the Florida panther, the West Indian manatee, and the Cape Sableseaside sparrow. Over 300 species of birds have been recorded, 7 of which are rare or endangered. The Everglades is a refuge for 13 threatened or endangered animal species.

In the 1920s, Douglas worked on a committee to establish the Everglades as a national park, and she continued her interest in the Everglades as she started a new career, writing for magazines. From 1926 into the early 1940s, she wrote mainly short stories for the *Saturday Evening Post, Ladies Home Journal,* and other periodicals. In 1942, an editor asked Douglas to write a book about the Miami River. At the time, the editor was editing a series of books on America's great rivers. However, Douglas suggested another idea, a book about the Everglades that would include the Miami River. She remarked to the editor that she had been thinking about this kind of book for a long time, ever since she had been on the committee to establish the Everglades as a national park. It was "an idea that would consume me for the rest of my life," Douglas later wrote. The editor liked her idea and told her to go ahead.

Douglas knew that it would take much research and some time to write the book. Even though she was an excellent researcher, she sought the help of an out-of-state geologist and expert on Florida groundwater. The geologist explained to Douglas that while the Everglades was swamplike, it also featured an almost imperceptible flow of water from north to south. The geologist's description prompted Douglas to conceptualize the image of the Everglades as a "River of Grass." Taking nearly five years to research and write, *The Everglades: River of Grass* went on sale in November 1947 and sold out in a few months. It became a best-selling book. Douglas was now an established writer and spokeswoman and defender for a vast swamp that few but Native Americans and hunters knew anything about. Her book describes the beauty of the Everglades and its diversity of wildlife.

She discussed the history of the Native Americans and the hunters and explorers who traveled in and near the Everglades. Douglas also discussed the impact of human population on this fragile ecosystem. The same year her book was published, the Everglades National Park was created. The Everglades National Park became a World Heritage Site on October 26, 1979. It was designated a wetland site on the Ramsar List of Wetlands of International Importance on June 4, 1987.

In Douglas's 1987 autobiography, *Marjory Stoneman Douglas: Voice of the River*, written with John Rothchild, the environmentalist discusses the Everglades' role as an important watershed for southern Florida:

Much of the rainfall on which South Florida depends comes from evaporation in the Everglades. The Everglades evaporate, the moisture goes up into the clouds, the clouds are blown to the north, and the rain comes down over the Kissimmee River and Lake Okeechobee. Lake Okeechobee, especially, is fed by these rains. When the lake gets filled, some of the excess drains down the Caloosahatchee River into the Gulf of Mexico, or through the St. Lucie River and into the Atlantic Ocean. The rest of the excess—the most useful part—spills over the southern rim of the lake into the great arc of the Everglades.

In 1970, Douglas founded the 6,000-member Friends of the Everglades, a group that successfully defeated a proposal to build a jetport in the fragile wetlands. In 1983, Douglas helped persuade the Florida governor to begin the Save the Everglades campaign that led to the Everglades Forever Act in 1994 and the federal Water Resources Development Act of 1996.

Her interest in environmental issues was only natural, Douglas told *Time* magazine in 1983. "It's women's business to be interested in the environment. It is an extended form of housekeeping" (May 25, 1998, p. 33). In 1993, President Bill Clinton invited Marjory Douglas to the White House and presented her with the Medal of Honor.

"I want to do more than I'm doing now," Douglas said when she was 96. "I get tired physically, but I don't tire of my beliefs. Ideas aren't tiring, they're what keep you going. There's no end to what can happen, and there's no end to the fight (for the Everglades)." Marjory Stoneman Douglas died in 1998 in Coconut Grove, her home since 1926. She was 108 years old.

BIBLIOGRAPHY

Douglas, M. S. (1997). *The Everglades: River of Grass* (50th anniversary ed.). Sarasota, FL: Pineapple Press.

Douglas, M. S., & Rothchild, J. (1987). *Marjory Stoneman Douglas: Voice of the River* [autobiography]. Sarasota, FL: Pineapple Press.

Everglades National Park. *Everglades History, A Closer Look: Marjory Stoneman Douglas: Visionary of the Everglades. http://www.americanparknetwork. com/parkinfo/er/history/msd.htm*

Friends of the Everglades. Marjory Stoneman Douglas, 1890–1998. Available: *http://www.everglades.org/youngfriends.html*

—*John Mongillo*

Photo courtesy of James A. Brown.

NEVADA DOVE
(1980–   )

Nevada Renee Dove was born in Los Angeles, California, on September 9, 1980. She was raised by her mother and father, Melodie and Reginald Dove. She is the second child of four children and has one sister and two brothers. She grew up in South Central Los Angeles. Her fondest memories of childhood were of time spent playing school or taking ballet lessons. Dove attended McKinley Avenue Elementary School, where she was classified as gifted. Throughout elementary school, she was never absent from school. Her most memorable award was given to her at her sixth-grade graduation for perfect attendance, seven years straight. Since an early age, dedication has been one of her strongest suits, and she has displayed leadership skills. When she was in the seventh grade at Horace Mann Middle School, Mrs. Atchison, her English teacher, gave her a piece of advice that would never be forgotten. She sat Dove down and told her, "With your leadership skills and big mouth, you could really make a difference." At that age, as a class clown, she did not pay much attention to this statement; however, as she looks back, it has proven to be true.

Dove's mother, Melodie Dove, who is an environmental organizer for Concerned Citizens of South Central Los Angeles, initially sparked her interest in community organizing and fighting environmental racism. When she was about thirteen years old, she began to develop a reoccurring curiosity about what exactly an environmental organizer does. This curiosity reached its peak after she attended several meetings regarding the New Jefferson Middle School. The New Jefferson Middle School is a newly constructed school in the community that was built on a contaminated site. This site is also directly adjacent to a federal Superfund site. When Dove realized that African-American and Latino children were being endangered, she felt an instant passion and realized that this was a cause worth fighting.

In May 1998, Dove was hired by Concerned Citizens of South Central Los Angeles as an intern on a youth organizing project that was later named POWER (People Organizing for Workplace and Environmental Rights). She has spearheaded several campaigns that directly affect the community in which she lives, especially youth in the community. Currently, youth are seen as troublemakers. Many people feel that investing in today's youth is a waste of time, but Dove believes that children can make a difference.

Dove has gone head-to-head with many public officials and members of regulatory agencies who have tried to deter her. She has always stood up for what she believes in and is not easily intimidated. She is seen by many people as the voice that youth in South Central Los Angeles needs. She organized her first community meeting on August 4, 1998. This meeting served to educate the community on the various issues surrounding the New Jefferson Middle School. The meeting brought all the key players to the table. Present in this meeting were the Los Angeles Unified School District's (LAUSD) Environmental Health and Safety Branch, South Coast Air Quality Management District officials, Department of Toxic Substance Control scientists, Southern California Regional Water Quality Control Board analysts, and concerned community members. Nevada Dove, Maria Perez, and Fabiola Tostado were the only youth on the project at the time, and they led the meeting. During this meeting, these young activists became enraged at the lies that were being told to their community members.

Prior to this meeting, Dove spent countless hours after school and on weekends reading health risk assessments (HRAs) and environmental impact reports (EIRs). From these reports, she learned that the site was contaminated with chromium, which is a carcinogenic chemical, and that anyone attending this school was at risk of being exposed to and developing cancer. During this meeting, the blame was being shifted from agency to agency, and lies were being told in bulk. The girls were angry because Bill Piazza of LAUSD's Environmental Health and Safety Team delivered testimony that was unbelievable and false, and Nevada felt that he should be exposed. He stated, "I'm an expert, I have credentials and degrees, so I know what I'm saying. The school is safe and no one is at risk." At that moment, Dove stood up and began to confront him with the support of

Perez and Tostado. Once they exposed him, he appeared angry and embarrassed by the fact that he had been challenged by minority teenagers. Consequently, he stormed out of the meeting. Soon after that, he lost his position and was demoted to an unknown position within the LAUSD.

This was Dove's first victory. Finally, what Mrs. Atchison had said made sense. Dove stood up to an adult, and she says that she has to admit that "once he became upset and left the meeting, I knew there was a reason for his actions and I am determined to figure it out." Her elders have always told her that the truth hurts, and this had proven to be true. The fight was just beginning.

From this experience, Dove learned that the words "adult" and "expertise" are not synonymous. People often assume that having a degree or reaching a certain age is synonymous with credibility. Dove learned early on, however, that there was little truth in this.

Dove has faced numerous obstacles, many stemming from her growing up in the inner city as a female youth of color. One obstacle has been the media. In this day and age, young people are seen on the news daily, usually for high-profile negative behavior. Here she is an activist for environmental justice at the age of seventeen, and the media would rather report teen violence.

This was not enough to grasp the attention of the media. Whenever members of the media were asked to attend events she organized or helped organize, they would interview Melodie Dove. Therefore, Melodie often ended up receiving credit for the work Nevada had done. Even when Melodie specified that a particular event was planned by youth, the media still did not pay attention. After four years of her work toward justice for the community, CNN and *Times Magazine* developed an interest in her and her coworkers' accomplishments. This media coverage was not just a story on the evening news but also a milestone in her career as one of the first young, African-American, female environmentalists.

Although Dove's advocacy has consumed much of her time, she has still found time to be a teenager. She enjoys shopping, gossiping with her girlfriends, and just being a teenager. Throughout her youth, she has spent much time babysitting her younger cousins and siblings. The young activist has also spent many Saturdays going door to door doing outreach in the community to ask their opinion and to gain support on certain issues. To her surprise, adults are very receptive to her efforts and are glad to help. She feels that the organizing she does is greatly needed in her community. In South Central Los Angeles, one bears witness to things that one would never encounter in Beverly Hills or Westwood. The air quality in the zip code in which she resides has been labeled the worst air quality in the county of Los Angeles and the United States. Children in her community are faced daily with environmental hazards that are the result of poor planning and mixed-use zoning. Most of the houses in the community are contaminated with lead because they were built prior to 1950 and utilized lead-based paint. There are several homes in close proximity to industries that house harmful carcinogens. These are just a few of the problems that exist in South Central Los Angeles.

Dove's mother has always taught her that she cannot always wait for an outside force to come into her community to change things. Changes are the responsibility of community residents. Individuals within the community must step up to the plate and take the initiative to help bring about change. Dove has worked with groups such as YUCA (Youth United for Community Action), CBE (Citizens for a Better Environment), SCYEA (South Central Youth Empowered through Action), WLCAC (Watts Labor Community Action Committee), and KIWA (Korean Immigrant Workers' Association) to help organize youth-driven conferences that served to educate, empower, and mobilize youth to get involved in the environmental justice movement. Many changes need to be made to environmental laws, and she has been instrumental in doing what it takes to ensure that these changes take place. Along with California state senator Tom Hayden, she has worked to draft legislation to amend laws that are supposed to protect children from environmental hazards. Dove has been called upon by Senator Hayden to give testimony at Senate hearings that provided support for improving environmental justice.

Dove's role model throughout her life has been her mother Melodie. Melodie Dove has been a community activist for twenty years, seven of which have been spent with Concerned Citizens of South Central Los Angeles. Nevada Dove feels that she wouldn't be where she is today if it were not for her mother. "It feels good to have such a positive role model in the same line of work, who shares the same last name," says Dove.

After working as organizers, Dove, Perez, and Tostado were named "toxic crusaders." This name came from Dove's older brother Earnest Dove, Sr., who continuously made light of the fact that she constantly worked for community improvement and changes did not come immediately. Her brother says the name came about because she is able to spot a toxic site from a mile away, and after a while, the name caught on.

Dove plans to continue her college education and obtain a law degree to pursue a career as an environmental attorney. She also wants to establish her own youth-based nonprofit organization in South Central Los Angeles that will continue the work of CCSCLA with more of a youth focus. She is currently residing in South Central Los Angeles. At nineteen years of age, she is still an active organizer in her community, where she plans to stay. "It would be so easy for me to move to a community without so many environmental hazards, but how would I sleep at night knowing the problems still exist? My mother never quit, even when times got tough, and I was raised the same way. If I quit now, not only would I let my community down but myself as well. If I let myself down, then I've let my mother down, and that is not something I plan to do."

Many youth, for various reasons, feel that their voice will not be heard. Dove feels that youth need to combat this notion and break the cycle. For many years, adults have made decisions for youth, and one thing youth should keep in mind is, "If we are the future, when are we, as youth, going to be allowed to start acting accordingly? The time is now!" Dove's accomplishments can be achieved by anyone. All it takes is hard work, dedication, and motivation. Sometimes, as an activist, it may feel as if you are working for nothing. When changes do not come suddenly, just remember that slow and steady wins the race.

—Nevada Dove

MARK DUBOIS
(1949–   )

Mark Dubois was international coordinator for Earth Day 1990, an event which involved 200 million people in 141 countries—more people than any other peace event in history. He again became part of the Earth Day team as international coordinator in 1998 to prepare for celebration of Earth Day 2000, which involved 184 countries. Dubois helped found Friends of the River (1973), and co-founded International Rivers Network in 1984, which are today the leading proponents of river preservation and restoration.

Dubois was born in Sacramento, California. In his teens and twenties, he rafted and played in the Stanislaus River in central California and explored the nearby caves. He was dimly aware, as many were, that federal engineers and planners had already begun building a dam which would drown the river and its ancient canyon full of life and mystery. The river's fate gradually became known to those who either depended upon it, enjoyed it, or defended it in principle. These people made up a movement opposed to the building and operating of the New Melones Dam and Reservoir, a project managed by the U.S. Army Corps of Engineers.

Dubois is the fellow who anchored himself to the river canyon bedrock during a showdown with the Corps in 1979. The dam was near completion, and it had effectively relocated or drowned all the life and beauty of the canyon. But the Army intended to illegally extend their human-made lake miles farther up the canyon, where Dubois positioned himself to either be drowned or saved along with the canyon. The Corps was compelled by one plain-speaking human to reconsider its plan.

This widely reported story was the culmination of a campaign which included river rafters, environmentalists, and politicians, who had fought unsuccessfully to save the Stanislaus. While colleagues continued the day-to-day legislative work, fellow paddlers manned the publicity machine, and Dubois quietly reached beyond his final heart-to-heart plea with the project's commanding Colonel. Chained closely to a rock which would soon become part of the reservoir's bottom, he told the Army he would be very difficult to locate. Amid the publicity, the river and Dubois gained the support of the governor. The flooding was halted.

It was only a three-year reprieve. The wild and ancient river had been lost to dam builders, but the struggle stirred deep feelings. With widespread press coverage the dismay felt by Dubois and team over the loss of the Stanislaus helped catalyze the growing movement to protect rivers.

Dubois observed that dams were being sold as third-world "development," while planners seemed blind to the devastating environmental and social impact. He spent the next ten years leading grassroots efforts to lobby directors of the World Bank and the International Monetary Fund (IMF) in a campaign of reform and enlightenment. Dubois became fascinated with the possible effectiveness of an evolving idea of "heart politics" to reach decision makers. Recalling Gandhi's insistent appeal to humankind's better nature, Dubois maintains that individuals do respond to encouragement and appeals to conscience. When their hearts are touched, he says, they and we *will* do the right thing.

Dubois likes to share the magic and the message of the river. It is an amazing natural world, and it transcends physics and biology to become a deep metaphor for the beautiful calm and breathtaking turbulence of free-flowing humanity. He recounts that he broke seven oars when he first "confronted" the rapids. Eventually he learned to "dance" with the powerful forces of the river, rather than struggle against them. The most effective and direct way to create long-lasting or real change, he says, is to use the power of our hearts, and the power of truth.

In struggling for Earth's needs and for overcoming anger, the best path, Dubois believes, is to follow your heart. He points out that more long-term solutions may be found with this approach. It also helps, he says, to have fun.

Mark's strong tendency toward activism began when he fell in love with the river. Falling in love, he believes, may be the key to all necessary environmental and social work, the best motivation for working hard to preserve and protect—whether it be a river, a forest, people, or anything else.

Mark has been a good example of living lightly, practicing much of his preaching about recycling, reducing consumption, and living simply. He has led hundreds of river trips (often in an old, recycled, patchwork canoe). He has spoken on water issues, international development bank reform, sustainable development, heart politics and communication.

Based for most of his life in Sacramento, California, he now lives on Bainbridge Island near Seattle with his wife, Sharon Negri, who leads efforts to protect wildlife and their ecosystems, and their son Tevon, who is nine years old.

—Mark Dubois

SYLVIA EARLE
(1935–   )

Far and away the greatest threat to the sea and to the future of mankind is ignorance. But with knowing comes caring, and with caring the hope that maybe we'll . . . achieve an enduring place for humankind on a planet that got along without us for billions of years, and no doubt could do so again.

—Sylvia Earle

Sylvia Earle is an internationally renowned marine biologist, author, lecturer, and scientific consultant who has devoted her life to studying the world's oceans. Earle has spent more than 6,000 hours underwater, has led over fifty expeditions, and has authored more than one hundred scientific, technical, and popular publications. She has become one of the world's most eloquent spokespersons for the protection of the fragile and endangered marine ecosystem and has presented scientific, technical, and general-interest lectures in more than sixty countries. Her oceanographic research has made her a truly exceptional figure in the world of conservation and has led the way to increased understanding and enhanced protection of the earth's oceans.

Earle was born in Gibbstown, New Jersey, and was raised with her two brothers on a small farm near Camden. From the time she was very small, Earle loved exploring the woods near her home and was fascinated by the creatures and plants she found in the wild. Her mother and father were a practical nurse and a self-trained engineer, respectively. Though neither had received a college education, they both loved nature and taught their young daughter to respect wild creatures and not to be afraid of the unknown. Her fascination with the sea began on the beaches of New Jersey during family vacations. According to Earle, "We didn't live near the beach. . . . I think in my earliest years, the ocean became particularly special because it wasn't there all the time, I never took it for granted. It was a very special treat. . . . The most important thing, to me, aside from the freedom and the power of it, was the kind of creatures that you could see along the beach that you can't find anywhere else" (American Academy of Achievement).

When Earle was twelve, her family moved to Clearwater, Florida, and the Gulf of Mexico became her backyard. Permitted the freedom to explore on her own, Earle was soon learning all she could about the wildlife of the Gulf waters and its coastal environment. She developed a wonder and respect for nature that would eventually inspire her to explore the ocean's depths and to bear witness to a beauty and variety of life that few people ever see. As she said in a 1991 interview, "I grew up more or less fearless with respect to all sorts of things—spiders, squirrels, birds, mammals— because of the gentleness that both my father and my mother and my family in general expressed toward our fellow citizens on the planet. That empathy for living things became naturally expanded as I grew up into a study of living things. I became a biologist just following my heart, I suppose" (American Academy of Achievement).

Earle's parents could not afford to send her to college without financial assistance, but she was an exceptional student and won academic scholarships to Florida State University. Throughout her school years, Earle supported herself by working as an assistant in college laboratories. There she first learned scuba diving and became determined to use this new technology to study marine life firsthand. Fascinated by all aspects of the ocean and marine life, Earle nevertheless decided to specialize in botany. "By knowing the plants, you get some feel for how the whole system works," Earle has said. "Plants provide shelter, whether it's underwater or above. They provide food. They provide the energy that supplies a whole interacting system. So if you know the plants, you get a good cornerstone on how all the rest of the system works. . . . When I was still an undergraduate student at Florida State University, I began my lifetime project with that as a starting point. I am still working on it, and expect to continue working on it" (American Academy of Achievement).

A gifted student, Earle graduated from Florida State University at nineteen. After earning her master's degree at Duke University at the age of twenty, Earle married a fellow Duke student and started a family. She still remained active in marine exploration, however. In 1964, when her children were only two and four, respectively, she left home for six weeks to join a National Science Foundation expedition in the Indian Ocean. Throughout the mid-1960s, she struggled to balance the demands of her family with contributions to scientific expeditions that took her all over the world. Her first marriage ultimately ended in divorce.

In 1966, Earle received her doctorate from Duke University. Her groundbreaking postgraduate research on the aquatic plant life of the Gulf of Mexico had earned her great respect within the oceanographic community, but Earle soon found an interested audience in the general public as well. In 1970, Earle participated in the Tektite Project, jointly sponsored by the U.S. Navy, the U.S. Department of the Interior, and the National Aeronautics and Space Administration. At Tektite II, Mission 6, an expedition led by Earle herself, she and four other female aquanauts spent two weeks living in an enclosed habitat under 50 feet of water on the ocean floor, 600 feet off the island of St. John's in the Virgin Islands. "[W]e were just one of ten [Tektite] teams in all. The others were all-male teams. We constituted the only all-women's team" (American Academy of Achievement). Upon their return, the scientists, whom the media had dubbed "aquababes," were honored with a ticker-tape parade and a reception at the White House. This much-publicized experiment made Earle a celebrity outside the scientific community. She was increasingly in demand as a public speaker and found her voice as an outspoken advocate of undersea research. At the same time, she began to write for *National Geographic* and to produce books and films. Besides trying to arouse greater public interest in the sea, she hoped to raise public awareness of the damage being done to our hydrosphere by pollution and environmental degradation.

In the 1970s, scientific missions took Earle to the Galapagos Islands, Panama, China, the Bahamas, and the Indian Ocean. During this period, she began a productive collaboration with undersea photographers Al Giddings and Chuck Nicklin. Together they investigated the battleship graveyard in the Caroline Islands of the South Pacific. In 1977, they made their first voyage in pursuit of humpback whales. In a series of expeditions, they followed the whales from Hawaii to New Zealand, Australia, South Africa, Bermuda, and Alaska. Their journeys were recorded in the 1980 documentary film *Gentle Giants of the Pacific*.

In 1979, Earle walked untethered on the sea floor at a lower depth than any living human being before or since. In the "Jim Suit," a cumbersome, pressurized one-atmosphere (i.e., surface-pressure) diving suit, she was transported by a small submarine to a depth of 1,250 feet below the ocean's surface off the island of Oahu, Hawaii. At the bottom, she detached from the vessel and explored the area for two and a half hours with only a communication line connecting her to the submersible vehicle, and nothing whatsoever connecting her to the world above. She described this adventure in her 1980 book for the National Geographic Society, *Exploring the Deep Frontier*.

From 1980 to 1984, Earle served on the President's Advisory Committee on Oceans and Atmosphere. Earle was curator of phycology at the California Academy of Sciences from 1979 to 1986 and worked as a research associate at the University of California, Berkeley, from 1969 to 1981. She was also a research fellow and associate at Harvard University from 1967 to 1981. In 1990, Earle was appointed as chief scientist of the National Oceanic and Atmospheric Administration (NOAA), where she served until 1992. There, among other duties, Earle was responsible for monitoring the health of the nation's waters. In this capacity, she also reported on the environmental damage wrought by Iraq's burning of the Kuwaiti oil fields.

In 1981, Earle cofounded Deep Ocean Technology and Deep Ocean Engineering, now DOER Marine Operations, a leading design company for underwater equipment that permits scientists to maneuver at unprecedented depths. Earle has designed, operated, and consulted on manned and robotic subsea systems, including the Deep Rover, a spherical diving system launched in 1984 that can descend more than 3,000 feet. The company has also worked on Deep Flight, a positively buoyant deep-water system that can move rapidly underwater, much as an aircraft flies through the air. Currently the chairperson of DOER and several other marine research and conservation organizations, Earle is still most often recognized as the former NOAA chief scientist and by her affectionate NOAA sobriquet, the U.S. "sturgeon general."

Earle's research has focused on the ecology of marine ecosystems, particularly marine plants, and on the development of technology for access and research in the deep sea. Over years of experience, Earle has witnessed how disturbing environmental developments, such as pollution and overfishing, have continued to alter the character of formerly pristine waters. She has intimately observed the effects of oil spills, agricultural runoff, solid waste, sewage, whaling, and commercial fishing and their devastating impacts on water quality and wildlife. As a result, Earle has become a passionate advocate for the protection of marine ecosystems and has worked ardently to educate people about growing threats to the health of our oceans and our planet.

Earle notes that mankind often appears more interested in exploring the mysteries of outer space than the vast and unknown depths of our own planet, and she believes that this lack of knowledge and interest in the sea is contributing to its degradation. "Far and away the greatest threat to the sea and to the future of mankind is ignorance," Earle wrote in her 1996 book *Sea Change*. "But with knowing comes caring, and with caring, the hope that maybe we'll find the Holy Grail of understanding, strike a balance with the natural systems that sustain us, and thus achieve an enduring place for humankind on a planet that got along without us for billions of years, and no doubt could do so again" (National Wildlife Federation).

Earle is currently a research associate in botany at the Smithsonian Institution and was the 1998 explorer in residence at the National Geographic Society. She has been the recipient of the Prince of the Netherlands Order of the Golden Ark, the Department of the Interior's Conservation Award, the Geographers' Gold Medal Award, the American Academy's Golden Plate Award, and the National Wildlife Federation's Conservationist of the Year Award. Along the way, she also has had a sea urchin (*Diadema sylvie*) and a marine plant (*Pilina earli*) named in her honor.

BIBLIOGRAPHY

American Academy of Achievement online. *http://www.achievement.org/autodoc/page/ear0bio-1*
The birth of an expedition. (1999, February 2). *USA Today*, p. 5D.

Discover Magazine, "Discovering Great Minds in Science" Series. (1995). Ocean-
    ography: Dr. Sylvia Earle. Discover Magazine online. *http://www.discover.
    com/educators/index.html*
Divernet. *http://www.divernet.co.uk/profs/earl1196.htm*
Earle, S. A., & Giddings, A. (1980). *Exploring the Deep Frontier: The Adventure
    of Man in the Sea.* Washington, DC: National Geographic Society.
Earle, S. (1996). *Sea Change: A Message of the Oceans.* New York: Ballantine
    Books.
Earle, S. (1997, Spring). Ocean advocate. *Amicus Journal.* Available: *http://www.
    nrdc.org/eamicus/clip01/sylviaea.html*
Environmental News Network. (1998, May 29). Sylvia Earle wins U.N. Environ-
    ment Award. Available: *http://www.enn.com/enn-news-archive/1998/05/
    052998/earle.asp*
Friend, T. (1999, February 2). Hearing the seas' siren call: In a one-seat sub, ex-
    plorer Sylvia Earle charts a course toward ocean conservation. *USA Today*,
    p. 1D.
National Geographic Society. (1998, April 23). America's fragile sea sanctuaries to
    be explored and documented. Available: *http://www.nationalgeographic.
    com/events/releases/pr980423.html*
National Science Teachers Association online. *http://www.nsta.org/conv/areabios/
    earle.htm*
National Wildlife Federation. (1998). 1998 National Conservation Achievement
    Awards. Available: *http://nwf.org/nwf/about/annmtg63/connies/earle.html*
Rosenblatt, R. (1998). Call of the sea: Her deepness welcomes us into her world of
    wonders. *Time* Magazine Heroes for the Planet, Heroes Gallery. Available:
    *http://www.pathfinder.com/time/reports/environment/heroes/heroesgallery
    /0,2967,earle,00.html*
Stover, D. (1995, April 1). Queen of the deep: Marine scientist Sylvia Earle cam-
    paigns for conservation of marine resources. *Popular Science*, pp. 67–71.
Van Putten, M. (1999, April/May). Honoring conservation heroes. *National Wild-
    life.* Available: *http://www.nationalwildlife.org/nwf/about/annmtg63/connies
    /earle.html*
Wexler, M. (1999, April/May). Sylvia Earle's excellent adventure. *National Wild-
    life.* Available: *http://www.nationalwildlife.org/nwf/natlwild/1999/earleam9.
    html*

—*Bibi Booth*

MABEL ROSALIE EDGE
(1877–1962)

Mabel Rosalie Edge was a noted conservationist who was born in New
York City on November 3, 1877. She was the daughter of John Wylie
Barrow (a first cousin of Charles Dickens) and Harriet Bowen Barrow.
Mabel was given a private-school education. She married Charles Noel
Edge, a mining engineer, in 1909 and had two children. Following her

marriage, Edge lived in Europe for several years. She became involved in the women's suffrage movement. Edge developed an interest in birds about 1915 and became a conservation activist.

Edge was the creator and chair of the Emergency Conservation Committee (ECC) from 1929 to 1962. In this capacity, she attacked the inertia of the National Audubon Society Board, its finances, and its association with hunting interests at the 1929 meeting. Ultimately, in 1934, she helped bring about the reorganization of the Audubon Society and the rejuvenation of its conservation ethos. The small ECC was sustained principally by Edge's dwindling fortune and modest contributions. Through ECC, she continued her advocacy of conservation causes on a limited basis in her later years. By 1947, the ECC budget was less than $3,000, and membership was about 900.

Among ECC's priorities was the Hawk Mountain Sanctuary Association, of which Edge was organizer, director, and president from 1936 to 1962. She purchased an option on Hawk Mountain, near Drehersville, Pennsylvania, and turned it into a refuge for hawks, which previously had been targets for hunters. Beginning in 1931, Edge, a dedicated preservationist, inveighed against the U.S. Biological Survey for its practice of trapping and poisoning predators and small mammals to protect the interests of ranchers and livestock farmers. Terming it "the Bureau of Destruction and Extermination," she unsuccessfully sought to overturn the Biological Survey's policy. While her efforts to end the Biological Survey's poisoning and trapping practices in the 1930s did not prevail, later reformers brought about such changes in federal wildlife management procedures in the 1960s and 1970s.

In the West, Edge overcame bitter opposition of lumber companies and brought about the founding of Olympic National Park (1938). She also played a major role in the creation of Kings Canyon National Park (1940). She spearheaded the effort to place 6,000 additional acres of old-growth sugar pines within Yosemite National Park in 1937. The defeat of the Echo Canyon Park Dam project in the mid-1950s was Edge's last great success, her voice having been among the chorus of opposition to the project. As a persistent and dedicated advocate of conservation causes throughout the mid-twentieth century, Edge was described by a reform colleague as "the only honest, unselfish, indomitable hellcat in the history of conservation" (Sterling et al.).

Edge was the author of *White Pelicans of Great Salt Lake* (1935), *Finishing the Mammals* (1936), *Migratory Bird Treaty with Mexico* (1936), and *Roads and More Roads in the National Parks and National Forests* (1936). Her poem "Motor Power" appeared in the December 1940 issue of *Nature Magazine*. Wagner College (Staten Island, New York) awarded Edge an honorary doctorate of letters in 1948.

Edge's death in New York City on November 30, 1962, marked the demise of the ECC. An edition of the typescript of her autobiography was released under the title *An Implacable Widow* in 1978.

BIBLIOGRAPHY

Edge, R. (1936). *Roads and More Roads in the National Parks and National Forests.* New York: Emergency Conservation Committee.

Sterling, K. B., Harmond, R. P., Cevasco, G. A., & Hammond, L.F. (Eds.). (1997). *Biographical Dictionary of American and Candian Naturalists and Environmentalists.* Westport, CT: Greenwood Press.

—*John Mongillo*

PAUL EHRLICH AND ANNE HOWLAND EHRLICH
(1932– and 1933–   )

If you're concerned about dying and your children and grandchildren having a miserable life, and you try to do something about it, and try to get the people who, in essence, are destroying the world to stop—if that's radical, then I sure as hell am a radical.

—Paul Ehrlich

Paul Ehrlich and Anne Howland Ehrlich are biologists, educators, and, for the past thirty years, America's best-known activists for human population control. In 1968, Paul Ehrlich published *The Population Bomb*, which argued that unchecked human reproduction was positioning the planet for catastrophic human overcrowding, widespread food shortages, and mass starvation. The book had an enormous impact and may have ultimately led to a voluntary downsizing of the American family. It became the best-selling environmental book in history, selling three million copies in ten years, and it catapulted Paul Ehrlich into prominence as one of America's leading prophets of ecological danger.

Paul Ehrlich was born in 1932 in Philadelphia and was raised in the New Jersey suburbs. His own awareness of humans' influence on the natural world developed when he started collecting butterflies at age twelve. The caterpillars that Ehrlich tried to raise into butterflies on local plant cuttings kept dying. The culprit turned out to be the insecticide DDT, which was being sprayed on wild foliage to keep mosquitoes away from new housing developments. "Almost any plant you brought in had DDT on it," he says (Rodriguez).

Ehrlich received early inspiration to study ecology when in high school he read William Vogt's 1948 *Road to Survival*, an early study of the problem of rapid population growth and food production. "I got interested in these issues when I was a freshman in college. I was reading . . . and talking in a lot of bull sessions. I went to college sort of with the last group of Second World War veterans, and we had a lot of discussions of this kind of issue, and my interest in it has persisted, basically" (Altenberg). He received his bachelor's degree in zoology from the University of Pennsylvania in 1953. He later donated his collection of butterflies, numbering in the thousands, to the American Museum of Natural History, where Ehrlich worked with the curator of entomology, Charles Michener, mounting collections in exchange for duplicate specimens.

Ehrlich followed Michener to the University of Kansas, where he met Anne Howland. They married in 1954. "Anne and I would go to a drive-in theater and they'd haul around a blower and just blow DDT through the car window on you" (Allen 1993b). As a graduate student, he studied evolutionary biology and the process of natural selection by which insects, specifically fruit flies, developed resistance to DDT. "I was [also] very heavily involved in the civil rights movement in Lawrence, Kansas. Another faculty member and I organized the desegregation of the restaurants . . . when a visitor who was coming to visit our group turned out to be Black, arrived on a Saturday and couldn't get fed in town till he came and found us on Monday, and we had a very exciting time; we independently invented sit-ins" (Altenberg).

Ehrlich received his doctorate in 1957 and worked as an entomologist at the University of Kansas from 1953 to 1959. From 1957 to 1958, he was also a research associate at the Chicago Academy of Science. Ehrlich's fieldwork included an insect survey in the Canadian Arctic to which he credits his interest in human cultures. In 1959, he joined the faculty of Stanford University, where he began his thirty-five-year study of local populations of checkerspot butterflies, the most thorough research ever done on their ecology and evolution. He became a full professor of biology in 1966.

The Ehrlichs have worked together since the 1950s, beginning their scientific collaboration on butterfly research to answer key questions about biological classification, ecology, and evolution. It was while the Ehrlichs were working with butterflies, reef fish, and birds in the 1950s that they first began to think about human population's impact on ecosystems. Paul Ehrlich's studies of the population ecology of butterflies led to the development (with Peter Raven) of the powerful concept of coevolution, an important understanding of the dynamics of animal populations. These ecological and evolutionary principles were later applied by the Ehrlichs to help assess the impact of human populations on the environment. Though much of their early research was done in the field of entomology, the Ehrlichs' overriding concern became unchecked population growth.

Ehrlich became a crusader for human conservation of natural resources after a 1966 trip to India made him aware of the ecological effects of poverty and overpopulation. In 1968, he distilled many of his articles and lectures on the subject into his first book, *The Population Bomb*. Controversial and hotly debated, it has been referred to as "a wake-up call for an entire generation" (Rodriguez). Its publication brought the Ehrlichs instant celebrity.

In 1968, the Ehrlichs also cofounded a group called Zero Population Growth, a grassroots organization concerned with the impacts of rapid population growth and wasteful consumption. According to the Ehrlichs, "You can reduce the impacts by reducing the number of people, reducing the level of affluence, or by improving the efficiency of the technology that one uses to achieve that affluence" (Altenberg). When it comes to the environment, say the Ehrlichs, the number of people is the factor in the equation that counts the most. (By choice, they themselves have only one child.)

In the 1970s, Paul Ehrlich appeared on television's *Tonight Show* twenty-five times to discuss the consequences of human overpopulation. At that time, the word "ecology" was new to the American vocabulary, and Ehrlich says that he was a "babe in the woods when it came to being an environmental spokesperson" (Rodriguez). When he warned that the world's expanding population would have catastrophic effects on our dwindling natural resources, nobody wanted to believe it.

In 1980, Julian Simon, a population economist and business professor at the University of Maryland, very publicly offered to take anyone on in a wager. His opponent would buy an imaginary bundle of natural resources for $1,000 at 1980 prices. Ten years later, they would check the new prices of the materials, and if the bundle's total price had increased—indicating greater scarcity—Simon would pay his taker the difference. If the bundle's price had decreased—indicating greater abundance—Simon would receive the difference. Paul Ehrlich took him up on the bet, writing that it would be "like taking candy from a baby," and that he was eager to get the money before everyone else cashed in. Under the terms of the bet, Ehrlich was permitted to select any natural resources, as long as their prices were not government controlled. He chose five industrial metals: copper, tungsten, nickel, chrome, and tin. Ten years later, in 1990, Ehrlich wrote a check to Julian Simon for $576.80. All of the metals Ehrlich chosen had declined in price. For his part, Ehrlich contended that prices do not reflect the essentials of the environmental crisis.

In 1990, the Ehrlichs published a follow-up book to *The Population Bomb*. Entitled *The Population Explosion*, it contended that "the population bomb had already exploded." The effects of that explosion, the Ehrlichs wrote, included global warming; destruction of rainforests; garbage overload; air, soil, and water pollution; acid rain; and chronic famine. Certain environmental trends that did not exist when *The Population Bomb* was written have reinforced the Ehrlichs' concerns. "For the first time in the history of our planet," he reminds us, "one species is altering the composition of the atmosphere, changing the basic weather mechanism of the planet, allowing flux of poisonous ultraviolet light to get through"

(Public Broadcasting Service). In 1990, Paul Ehrlich won Sweden's Crafoord Prize in Population Biology and the Conservation of Biological Diversity and also became a MacArthur Fellow.

The Ehrlichs have been forceful advocates for broadening the agenda of the environmental movement to include such issues as biodiversity, poverty, carrying capacity, energy supplies, global warming, environmental ethics, and sustainable development. Long dedicated to responsible science and public education on environmental issues, the Ehrlichs say that humans often have difficulty recognizing environmental destruction. Our nervous systems, Paul Ehrlich says, were originally designed to ignore what appeared to be minor environmental changes. "It's much more important to spot the saber-toothed tiger coming at you" (Rodriguez).

Today, Paul Ehrlich says, "[P]opulation has become more a part of the standard discourse, but one of the things people still don't understand is how big a connection there is between environmental problems and population size. There is . . . almost no concern about consumption control in the United States. We have a dual problem: the third-largest population in the world and this incredibly high level of per-capita consumption, serviced very often by sloppy technologies. I'm afraid that most Americans who are aware of population problems tend to think of them in terms of poor countries" (Motavalli).

Some label Paul Ehrlich an environmental naysayer, but he freely admits improvements humans have made as caretakers for the environment, and his conversations focus as much on human caring as human carelessness. The Ehrlichs' critics point out that the worst of their doomsday predictions have not come to pass, but they believe that the weight of scientific evidence clearly supports their point of view: that humanity has only briefly postponed a catastrophic collision with the consequences of runaway population growth. This perspective, uncommon among scientists, has sometimes made them the target of strident criticism, which they accept as the price of forthrightness. Paul Ehrlich says, "If you're concerned about dying and your children and grandchildren having a miserable life, and you try to do something about it, and try to get the people who, in essence, are destroying the world to stop—if that's radical, then I sure as hell am a radical. . . . The biggest hope is that dramatic changes are possible" (Allen 1993b).

In a 1996 interview, the Ehrlichs said, "The carrying capacity of the Earth depends on the behavior of the individuals. At current behavior we're clearly above the carrying capacity because we're reducing the capacity of the planet to support people in the future" (Motavalli). They are also concerned about loss of biodiversity and human-caused species extinctions. "People do not seem to understand that their fates are imminently intertwined with the other organisms of the planet. That is, they are working parts of the life support system, the ecosystems that supply our economy with absolutely irreplaceable services. In other words, if we lose most biodiversity, we will also lose our industrial civilization" (Motavalli).

Anne H. Ehrlich is currently a senior research associate in the Department of Biological Sciences and associate director of the Center for Conservation Biology at Stanford University, which the Ehrlichs founded. Paul Ehrlich is Bing Professor of Population Studies and professor of biological sciences at Stanford University. Through the Center for Conservation Biology, they work with an international team of scholars to use science to help conserve humanity's "biological capital."

Paul Ehrlich is the author of thirty-seven books and more than 600 technical and popular articles on subjects ranging from overpopulation and consumption in the United States to the population dynamics of checkerspot butterflies in central California. Among his books are *The Stork and the Plow* (with Anne Ehrlich and Gretchen Daily); *The Science of Ecology* (with Jonathan Roughgarden); *Healing the Planet* (with Anne Ehrlich); *Ecoscience: Population, Resources, Environment* (with Anne Ehrlich and John P. Holdren); and *Betrayal of Science and Reason: How Anti-Environmental Rhetoric Threatens Our Future* (with Anne Ehrlich). The Ehrlichs are currently publishing a series of newsletters entitled *Ecofables/Ecoscience* that use science to shatter myths about humans' relationship to the environment.

Among the awards and honors the Ehrlichs have received is the 1998 Tyler Prize for Environmental Achievement. Paul Ehrlich also received the 1993 World Ecology Medal from the International Center for Tropical Ecology and the 1994 United Nations Environment Programme Sasakawa Environment Prize. He is also a member of the Board of Directors of the Audubon Society; a fellow of the American Association for the Advancement of Science, the American Academy of Arts and Sciences, and the American Philosophical Society; and a member of the National Academy of Sciences. Anne Ehrlich serves on advisory boards for several organizations, including the National Parks Association and the Winslow Foundation, and is a fellow of the American Academy of Arts and Sciences.

BIBLIOGRAPHY

Aldrich, R. (1998, April 23). Population ecologist Paul Ehrlich to speak at Dodd Center. University of Connecticut News Release.

Allen, W. (1993a, October 16). "Population bomb" ticking, expert says. *St. Louis Post-Dispatch*, p. 1B.

Allen, W. (1993b, December 13). Bombshell revisited [interview with Paul Ehrlich]. *St. Louis Post-Dispatch*, p. 1D.

Altenberg, L. (1983, April 1). From the bedroom to the bomb: An interview with Paul Ehrlich. *Stanford Daily*. Available: *http://dynamics.org/altenber/papers/EHRLICH*

Biography online. Ehrlich, Paul R. (Ralph). *http://www.biography.com/cgi-main/ biomain.cgi*

Business Wire. (1998, March 18). 1998 Tyler Prize for Environmental Achievement awarded to Anne and Paul Ehrlich of Stanford University. Available: *http:// www.elibrary.com*

Carter, T. (1999, February 7). Will world be too crowded to manage? Global meeting to revive debate on population explosion. *Washington Times*, p. A1.

Countering the brownlash: Sound science and the environment. (1996, October). Island Press *Eco-Compass*. Available: *http://www.islandpress.com/ecocompass/ BeSciRes/BSRec.html*

Daily, G. C., Ehrlich, A. H., and Ehrlich, P. R. (1994, July). Optimum human population size. *Population and Environment, 15* (6), pp. 469–475.

A doomsday message that won't die [editorial]. (1999, November 23). *Tampa Tribune*, p. 14.

Earth Explorer. (1995). The Ehrlichs: Population prophets. Available: *http://www. elibrary.com*

Ehrlich, P. (1976). *The Population Bomb*. Rivercity, MA: Rivercity Press.

Ehrlich, P. R., & Ehrlich, A. H. (1990). *The Population Explosion*. New York: Simon and Schuster.

Ehrlich, P. R., & Ehrlich, A. H. (1991). *Healing the Planet: Strategies for Resolving the Environmental Crisis*. New York: Addison-Wesley, Inc.

Ehrlich, P. R., & Ehrlich, A. H. (1996). *Betrayal of Science and Reason: How Anti-Environmental Rhetoric Threatens Our Future*. Washington, DC: Island Press.

Ehrlich, P. R., Holdren, J. P., & Ehrlich, A. H. (1986). *Ecoscience: Population, Resources, Environment*. New York: W. H. Freeman and Co.

Ehrlich, P. R., & Roughgarden, J. (1987). *The Science of Ecology*. New York: MacMillan.

Encyclopaedia Britannica online. Ehrich, Paul (Ralph). *http://www.britannica.com*

Envirolink Environmental Ethics online. *http://www.envirolink.org/enviroethics/ ehrlichindex.html*

Foster, J. B. (1998). The scale of our ecological crisis. *Monthly Review, 49* (11), pp. 5–16.

Green plus red-tape makes brown [book review]. (1998, March 14). *The Economist*. Available: *http://www.britannica.com*

Gupta, A. (1999, October 20). 6 billion people—too many or too few? University Wire. Available: *http://www.elibrary.com*

The Heinz Awards online. *http://www.awards.heinz.org/ehrlich.html*

Hewson, G. (1998, May). Dr. Paul Ehrlich wins major environmental award. Duke University, Organization for Tropical Studies, *Amigos Newsletter, 49*. Available: *http://www.ots.duke.edu/en/amigos/49/ehrlich.htm*

Humphrey, J. (1999, November 15). Stanford expert speaks at U. South Florida on population, consumption. University Wire. Available: *http://www. elibrary.com*

Motavalli, J. (1996, November 21). The countdown continues on the population bomb: Paul and Anne Ehrlich [interview with population-control advocates]. *E Magazine, 7*, p. 10.

Overpopulation.com. *http://www.overpopulation.com/paul_ehrlich.html*

Public Broadcasting Service (PBS)/KQED (San Francisco, CA). (1996). Paul Ehrlich and the Population Bomb [documentary background]. Available: *http://www.pbs.org/kqed/population_bomb*

Rodriguez, B. (1996, November 11). Ehrlich's earth: One of the nation's best known ecologists launches a strike against the "anti-scientists." *St. Louis Post-Dispatch*, p. 1E.

Stanford University Center for Conservation Biology online. *http://www.stanford.edu/group/CCB/Staff/anne.htm*

Wertheimer, L. (1994, September 13). Population growth also an American issue [discussion]. National Public Radio, All Things Considered. Available: *http://www.elibrary.com*

Zaleski, R. (1997, August 25). Another person, another problem. *Capital Times* (Madison, WI), p. 1D.

—*Bibi Booth*

DAVE FOREMAN
(1946–   )

> For over twenty-five years now I've been doing the same thing. I've never changed my goal as a conservationist—and that's been to protect as much wilderness and as many intact ecosystems and native species as possible.
>
> —Dave Foreman

Dave Foreman is a conservationist, biocentrist, and deep ecologist who for almost thirty years has been at the forefront of the conservation movement, working where political activism intersects with ecological philosophy. Foreman has been a longtime champion of "deep ecology," which maintains that humankind is no more important than any other species. Although he has long mistrusted many traditional American institutions, Foreman's chronic problems with authority—"I just can't stand to be pushed around," he says—have more to do with individualism than anarchism. Foreman's conservation ethic is rooted in a fundamental conservatism. "Conservation isn't just about protecting beautiful places or protecting our outdoor recreation opportunities," he says. "It's really come down to protecting the integrity of the diversity of life and the evolutionary process." The world is losing species at "a thousand, maybe ten thousand times the overall background rate of extinction in the fossil records," says Foreman, who lays the blame on "six billion human beings and a civilization that's fatally out of touch with wildness" (Bergman 1998).

Foreman was born to Benjamin and Lorane Juanita Foreman in Albuquerque, New Mexico, but was raised in many places around the world because his father was a sergeant in the U.S. Air Force. Foreman's love of the outdoors took root when he was a child; with all the constant moving, Foreman found stability in the woods. As a teenager, he campaigned for Barry Goldwater in the 1964 presidential election and still loves the late Arizona senator's famous statement, "Extremism in the defense of liberty is no vice."

While Foreman was attending San Antonio Junior College, he established a chapter of the conservative group Young Americans for Freedom. He then went on to the University of New Mexico, where he majored in history. As chairman of Students for Victory in Vietnam, Foreman was once asked to name his hero. Foreman said later, "To my undying embarrassment, I put down 'J. Edgar Hoover' " (Looney).

Upon graduation in 1968, he enrolled in the Marine Corps officer candidate school in Quantico, Virginia. He lasted in the Corps only sixty-one days, half of which were spent in the brig for insubordination and for being away without leave. "I think I had a nervous breakdown," says Foreman (Looney). He eventually received a dishonorable discharge.

In the years that followed, Foreman worked as a schoolteacher on a Zuni Indian reservation in New Mexico and then as a horseshoer in northern New Mexico. In 1973, he began working for the Wilderness Society, a mainstream environmental group based in Washington, D.C., first as the organization's Southwest representative, and then as its lobbying coordinator. In 1977, Foreman and his first wife, Debbie Sease, moved to Washington, D.C., as full-time lobbyists. Foreman's personal breaking point came after working for a year and a half on a U.S. Forest Service survey of America's last wilderness areas. When the Wilderness Society went along with the government's recommendation to open up 65 million of the remaining 80 million acres to logging and mining, Foreman took a leave from the organization.

Foreman returned to New Mexico and separated from his wife. When he and his friends conceived Earth First! in 1980, Foreman says, it was a "rambunctious, rowdy period in my life. But when people immediately branded us as radical, I thought, 'O.K., if that's what they think, let's show 'em what environmental extremism is' " (Looney). "The people who started Earth First!, we'd all been in the mainstream groups, we were solid conservationists. But we were frustrated at making too many compromises," Foreman explains, debunking the popular myth that Earth First! sprang more or less full-blown from the plot of Edward Abbey's novel *The Monkey Wrench Gang* (Bergman 1998).

After working for mainstream environmental organizations, Earth First!'s five founders believed that they had compromised too much for the small land areas they had managed to save from development. They created Earth First! to stop the destruction of wilderness and to bring attention to the issues of biodiversity and biocentrism. Foreman reluctantly admits to "monkeywrenching" (i.e., performing ecological sabotage) in the name of

Earth First!'s objectives, but declines to be more specific. In his 1991 book *Confessions of an Eco-Warrior*, Foreman wrote: "We are warriors. Earth First! is a warrior society" (Looney). Earth First!'s monkeywrenching was meant to block projects that were destroying the earth by increasing the costs of those projects and by attracting attention to the environmental crisis. It has included actions such as pulling up survey stakes on proposed wilderness roads, tearing down billboards, sabotaging machinery, and "spiking" trees. Although the group was not extremely successful in stopping the destruction of old-growth forests, the issue of biodiversity gained public attention as a result of Earth First!'s activities.

Earth First! has been on the radical fringe of the environmental movement since the group's inception. Foreman and his friends soon made their mark with a spectacular forgery at Lake Powell. Using huge rolls of black plastic, duct tape, and nylon rope, they created what looked like a 300-foot crack down the middle of Glen Canyon Dam. "Glen Canyon Dam is probably the single most objectionable feature in the West," Foreman says. "It's symbolic of the industrial conquest of the wilderness" (Bergman 1998). The prank put Earth First! on the map and began Foreman's friendship with Abbey, who had been a guest at the event. Abbey, he says, was "a visionary, the greatest inspiration my generation of conservation activists in the West ever had" (Bergman 1998).

In 1986, while speaking in Chico, California, Foreman met Nancy Morton, a registered nurse and former Sierra Club member. The two married that summer at Earth First!'s Round River Rendezvous in Idaho. Instead of a honeymoon, they went to Yellowstone National Park to demonstrate against the park's newest development, which infringed on the habitat of the grizzly bear. Throughout the 1980s, Foreman was the classic ecowarrior; as Earth First!'s slogan stated it, he was willing to brook "no compromise in defense of Mother Earth." Still, Foreman insists, "We were not violent. We were confrontational. There's a difference" (Looney). As the decade progressed, however, the group took on a more countercultural, extreme character than Foreman was comfortable with. Foreman recalled in 1997, "I offended a bunch of the new, radical-type Earth First!ers in the mid-eighties when I said that if I had to choose between the Sierra Club and Earth First! there'd be no choice at all—I'd pick the Sierra Club without thinking twice. Because Earth First! only has meaning as part of the larger conservation movement. And that's what the Sierra Club is" (Bergman 1998).

In February 1989, the *Earth First! Journal* announced that Foreman was stepping down from his informal position as Earth First!'s spokesperson. By 1990, realizing that the group had evolved into something very different from their initial vision, the group's founders voluntarily severed their ties with Earth First! They were unaware at the time that during 1988 and 1989, the Federal Bureau of Investigation had been infiltrating Earth First!

and conducting surveillance on the group's activities, targeting Foreman in order to send a message to the entire environmental movement. In May 1989, Foreman and three others from Earth First! were arrested, and a fifth person was charged in December. In 1991, Foreman and the others—known collectively as the "Arizona Five"—were tried on charges of conspiring to sabotage nuclear power plants in California, Colorado, and Arizona. The case ended with a joint plea bargain to a lesser charge of conspiring to damage property, and with recriminations from some former supporters, who concocted conspiracy theories to explain how Foreman had dodged a jail term while his codefendants had received sentences ranging from thirty days to six years.

Foreman says that he now wants to work with mainstream environmental groups to help make them more visionary. In 1995, Foreman was elected to a three-year term as a Sierra Club director. Even many mainstream conservationists find Foreman's conversion baffling, since he had spent more than a decade sniping at the national green groups, claiming that they had been co-opted by their access to power and their appetite for expansion.

Since 1991, Foreman has been chairman of the Wildlands Project, a small group of activists dedicated to preserving and restoring "true wilderness" across North America. "We need to look at the whole landscape, at the connectivity between protected areas," he says, citing a central tenet of the emerging field of conservation biology (Bergman 1998). The Wildlands Project aims to establish wilderness corridors, buffer zones, and other forms of protected habitat for wildlife, especially for large carnivores, which cannot survive in ecologically fragile "islands" such as national parks and refuges. "We see implementation not as one fell swoop, one piece of legislation, but as a jigsaw puzzle with a lot of pieces," Foreman says, adding that it will take larger groups, such as the Sierra Club, to put all the pieces together (Bergman 1998). "If we picture life as one of those great American chestnut trees just spreading thousands and thousands of branches out—if we see that as the tree of life—then we aren't just snapping off a few twigs out on the end of branches that may grow back in other ways," he says. "What we're doing is taking a chainsaw to huge limbs of that tree of life" (Bergman 1998). The group is working with Sierra Club activists at the national level and in places from New Mexico to Vermont to establish scientifically sound wildlife reserves. Foreman's organization also publishes *Wild Earth*, the literary vehicle of the Wildlands Project, and he is a member of the Board of Directors of the New Mexico Wilderness Coalition.

In November 1996, Foreman cast his vote, along with the rest of the fifteen-member Sierra Club Board, for a resolution that called for draining Lake Powell in order to reclaim the lost resources below. As to whether the government is likely ever to consider doing so, Foreman takes the long view. "Sometimes it doesn't matter whether you have a chance of making something happen," he says. "Sometimes it's important just to raise the issue. Do we live in a natural ecosystem, or do we live in an elaborate plumbing system? Talking about draining the reservoir forces us to think about the future" (Bergman 1998).

Today, Foreman lives with his wife in his hometown of Albuquerque. "For over twenty-five years now I've been doing the same thing," he insists. "I've never changed my goal as a conservationist—and that's been to protect as much wilderness and as many intact ecosystems and native species as possible. And as times change, as a person changes, you seek new ways of doing that. I've been looking for new, effective strategies the whole time" (Bergman 1998).

Currently, Foreman's main source of income is public speaking. "People often ask me, 'What does conservation biology say we should do with this area, with this landscape?' Or they come to me with a Forest Service document and ask, 'What did the Forest Service do wrong here with their science?' Well, often we can find scientific or technical flaws. But in most cases the agencies and even many industries have good scientists and they're doing good scientific work. It's just that the goals driving their process are different from the goals we share as conservationists. Science can't really tell us where to go; values have to tell us that" (Looney). "Each of us is an animal, and a child of this earth. Each of us has responsibility to all other animals and plants and to the process of evolution that created us. All of us alive now are members of the most important generation of human beings who have ever lived, because we're determining the future, not just for a hundred years, but for a billion years. When we cut a huge limb off the tree of natural diversity, we're forever halting the evolutionary potential of that branch of life. That's what it fundamentally comes down to. Nobody has ever lived who is more important" (EcoFuture PlanetKeepers). Additional books by Foreman are *EcoDefense: A Field Guide to Monkey-wrenching* (1985) and *The Big Outside* (with Howie Wolke; 1992).

BIBLIOGRAPHY

Barnes and Noble online. Dave Foreman biography. *http://www.bn.com*

Bergman, B. J. (1996). Thinking like a mountain (Sierra Club conference on conservation biology). *Sierra, 81,* 21.

Bergman, B. J. (1998). Wild at heart: Environmentalist Dave Foreman. *Sierra, 83,* pp. 24–33.

EcoFuture PlanetKeepers online. A Dialogue with Derrick Jensen. *http://www.csn.net/~felbel/pk/pkar9510.html*

Foreman, D. (1991). *Confessions of an Eco-Warrior.* New York: Harmony Books.

Foreman, D., & Haywood, B. (1993). *Ecodefense: A Field Guide to Monkey-wrenching.* Chico, CA: Abbzug Press.

Foreman, D., & Wolke, H. (1992). *The Big Outside: A Descriptive Inventory of*

the Big Wilderness Areas of the United States. Nevada City, CA: Harmony
    Books.
Looney, D. (1991, May 27). "Some people think I'm the devil incarnate," says
    Dave Foreman. Sports Illustrated, p. 54.
Reed, S., & Benet, L. (1990, April 16). Eco-warrior Dave Foreman will do whatever
    it takes in his fight to save Mother Earth. People, p. 113.
The Wildlands Project online. Earth First! http://www.wildlandsproject.org/html/
    ef.htm

—Bibi Booth

DIAN FOSSEY
(1932–1985)

> When you realize the value of all life, you dwell less on what is past
> and concentrate on the preservation of the future.
>
> —Dian Fossey

Dian Fossey was an American zoologist and wildlife environmentalist who
conducted research of one of the most endangered mammals on earth, the
mountain gorillas of Rwanda and the Democratic Republic of the Congo
in central Africa. Fossey was born in California in 1932 and received her
degree in occupational therapy from San Jose State College. After college,
she worked at a hospital in Kentucky for several years. In 1963, Fossey
made a six-week trip to Africa. While she was there, she met the British
anthropologist Louis Leakey, who influenced her to do research on mountain gorillas, the rarest and largest of the great apes.

Mountain gorillas live only in the misty rainforests of central Africa on
the slopes of the Virunga range, which consists of eight volcanoes that
border three nations, Rwanda, Uganda, and the Democratic Republic of
the Congo. The highest elevation of the mountain range is about 14,700
feet. Mountain gorillas are social animals who live in close family groups
of around five to forty. A typical family may have an adult male silverback
leader, some adolescent blackback males, and several females with their
infants. A silverback may weigh over 200 kilograms (twice as much as a
female gorilla and three times as much as an average human) and be 1.7
to 1.8 meters tall when standing.

Gorillas are almost entirely herbivorous. They eat berries, roots, shoots,
fruit, leaves, bark, bamboo, wild celery, and occasionally ants. An adult
male can consume up to thirty kilograms of food each day. As well as
having fingerprints, gorillas also have unique noseprints. No mountain gorilla has ever survived in captivity for more than a few years. Much of these
data about mountain gorillas was not available until Fossey's research.

In 1967, after she met with Leakey and determined to work in Africa,

Fossey obtained support from the National Geographic Society and the Wilkie Foundation for a research program on mountain gorillas. At that time, Fossey was also influenced by a popular book written by George Schaller titled *The Year of the Gorilla*. In the same year, she established the Karisoke Research Center in the Virunga Mountains of Rwanda, within the Parc National des Volcans. Her objective was to provide the world's primary focus for conservation of and research on mountain gorillas. In 1967, her friend Alan Root, the wildlife filmmaker who introduced her to gorillas, described her in *Swara* magazine as an occupational therapist with lung problems, a great fear of heights, and no training in animal behavior. Fossey was hardly tailor-made for the job of following gorillas among the steep ravines of a 14,000-foot, rain-shrouded volcano with only a two-day crash course on data collection from Jane Goodall to guide her.

However, over the years, Fossey collected a vast amount of data on mountain gorillas. She recorded everything she saw, and from the beginning she realized that gorillas were doomed unless something was done about the uncontrolled encroachment and poaching that was going on. The mountain gorillas were threatened by poaching, loss of habitat, human disease, and war. Intense observation over thousands of hours enabled Fossey to earn the complete trust of the wild groups she studied and brought forth new knowledge concerning many previously unknown aspects of gorilla behavior. Her work also dismissed many myths about the aggressive behavior of gorillas.

In 1970, her efforts to get the gorillas to accept her presence were finally rewarded when Peanuts, an adult male, touched her hand. This was the first friendly gorilla and human contact ever recorded. Later, several groups accepted her presence in the field. When poachers attacked and killed a young male named Digit to whom she had grown especially attached, she reacted by waging a public campaign against gorilla poaching. *National Geographic* heeded her pleas by placing her photograph on the cover of an issue containing an in-depth article with photos by Bob Campbell. Contributions poured in from around the world, allowing Fossey to establish the Digit Fund (renamed the Dian Fossey Gorilla Fund in 1992) and dedicate the rest of her life to the protection of the gorillas.

Fossey obtained her Ph.D. at Cambridge University and in 1980 accepted a position at Cornell University that enabled her to begin writing *Gorillas in the Mist*, detailing her years of observation and field research. Its publication brought her world fame and helped to focus much-needed attention on the plight of the mountain gorillas, whose numbers had by then dwindled to 250. She returned to Karisoke to continue her tireless campaign to ensure the survival of the mountain gorilla and to stop poaching. Fossey spent twenty-two years studying the ecology and behavior of the mountain gorillas. In time, her study site, Karisoke, became an international center for gorilla research.

In December 1985, Fossey was murdered by an unknown attacker in her cabin at Karisoke. Today, the Dian Fossey Gorilla Fund continues the work started by its founder protecting the mountain gorillas. Political instability and war have caused many setbacks, but as peace slowly returns, the economic potential of conservation of the mountain gorilla has also reemerged. The fund provides training, equipment, and material support for national park rangers, wildlife veterinarians, and local conservation groups, which, along with an effective regional communications network, continue to reinforce the front line of protection of the mountain gorilla. Ranger patrols and law enforcement within the national parks have always been the main deterrent to poaching and the unsustainable use of forest resources.

Presently, there are fewer than 650 mountain gorillas left in the world. However, their population has increased since the 1970s due to the work of Dian Fossey and the Dian Fossey Gorilla Fund International. More work is needed to maintain these rare animals.

BIBLIOGRAPHY

Dian Fossey Gorilla Fund International. *http://www.gorillafund.org*
Fossey, D. (1983). *Gorillas in the Mist*. Boston: Houghton Mifflin.

—*John Mongillo*

DAVID GAINES
(1947–1988)

In the spring 1986 edition of the *Mono Lake Newsletter*, David Gaines wrote, "Let's not change the planet. Let the planet change us." His words are simple and poignant. They are also ironic, as David Gaines did change the planet—he changed it for the better. Inspired by the natural beauty of Mono Lake, challenged by the imminence of its destruction, and compelled to pursue what he knew was right, David Gaines was the spark behind the early, and eventually successful, efforts to save Mono Lake.

David Gaines was born in 1947 to a cultured Jewish family in Los Angeles. Inspired by music, at various times in his life he played the piano, the clarinet, the guitar, the mandolin, and the hammer dulcimer. He listened to everything from the music of Beethoven to the songs of the Grateful Dead. In high school, he belonged to a philosophy club. In college, he drifted into a major in German literature, spent his junior year in Göttingen, Germany, and won a prestigious fellowship to continue his studies at Stanford.

But there was another element in Gaines's life, and after almost a year at Stanford, he realized that his ecological interests were stronger than his literary ones. He transferred to the University of California at Davis and to a program that would allow him to earn a master's degree in ecology.

Photo by Jim Stroup. © 1998 Mono Lake
Committee.

Gaines's parents owned a condominium in Mammoth Lakes, and after
receiving his master's degree in ecology, he stayed there to help do an
inventory of Mono County for the California Natural Areas Coordinating
Council. He was instantly captivated by Mono Lake and alarmed by the
changes he saw taking place there.

The beginnings of the Mono Lake Committee relied heavily on Gaines's
amazing ability to communicate the Mono Lake message—that the lake
was in serious trouble, and that it was up to each and every one of us to
stop its demise. He wanted to make sure that everyone in California had
heard about Mono Lake—if possible, straight from him. For eighteen
months, from late 1978 to early 1980, he spent more than half his days
on the road with a Mono Lake slide show. He hit schools and colleges. He
hit every Sierra Club and Audubon chapter in the state, and every Lions
Club that would have him. He paid special attention to Sacramento, where
the legislators were, and to his hometown of Los Angeles, where the water
went. No Californian who even glanced at the environmental press could
have failed to get the message, by 1980, that Mono Lake was in trouble,
and that someone was offering to save it.

A decade after founding the Mono Lake Committee, David Gaines was
killed in an automobile accident during a winter storm. Although many
organizations and individuals contributed significantly to achieving protection for Mono Lake, without Gaines's passion, focus, and original vision,
it could not have been done. Today, Sally Gaines cochairs the Mono Lake
Committee's Board of Directors, and the family is still very much involved
with the committee's continued efforts.

Today, the Information Center and Bookstore that David began still thrives in the small town of Lee Vining, and the committee is still working hard. The 15,000-person membership supports continued protection of, restoration of, and education about Mono Lake. Mono's waters are rising, and the city of Los Angeles is using the same amount of water today as it did a decade ago, despite population growth. The story of Mono Lake is known by children, community-based organizations, environmentalists, and politicians all over the state of California and throughout the West—and the message continues to spread. The message is not only that co-operative solutions can provide for both human needs and the needs of the environment, but also that individuals do have the power to change things for the better.

—*Mono Lake Committee*

LOIS GIBBS
(1951–)

The least we can expect is that corporations, like private citizens, act responsibly to *prevent* injury. The double standard that puts corporations above responsibility for the harm they cause must be eliminated. Corporations created by laws to serve us must at least meet the standards that we the people do.

—Lois Gibbs

Lois Gibbs is an environmental activist and founder and executive director of the Center for Health, Environment, and Justice (CHEJ), a clearinghouse for community groups seeking information and assistance in fighting existing or proposed environmental threats. She is well known for her efforts in exposing a hazardous toxic-waste dump in Love Canal, New York, in the late 1970s. About 20,000 tons of buried chemicals polluted the ground and the water in a residential area of 800 families. Her grassroots efforts influenced the national public awareness of toxic wastes in their environment, and her actions led to the formation of the Superfund program.

Lois Marie Gibbs was born on June 25, 1951, in Buffalo, New York. She grew up in Grand Island, New York, a city that is on an island of the same name in the middle of the Niagara River between Buffalo and Niagara Falls. Her family home was located at 1547 Love Road, so the word "Love" is clearly a fixture in her life.

Gibbs was the third-oldest of six children in a large working-class family with little spare money. Her father was a bricklayer, and her mother stayed at home to care for the family. Gibbs was a shy "middle child" with little interest in outside activities and was not involved in clubs, sports, or other

Photo courtesy of Lois Marie Gibbs.

things. But she liked crafts and spent a great deal of her younger years sewing and making things.

She attended elementary school and Grand Island Junior-Senior High School. She did not go to college or attend other educational institutions after high school. She spent her summers at home, but she was near two very popular beaches on Grand Island. Young people would come from miles away to swim and sunbathe. Gibbs describes that most of her time was spent at Beaver Island Beach hanging out with her friends and flirting with the boys.

Her first job was babysitting in the neighborhood. She recalled that one time she was babysitting for a family with six children. The parents left the children in her care overnight when Gibbs was only twelve years old. As Gibbs remembers, "Not something one can do now a days!" She worked for several years as a cashier at Tops, the only supermarket on the island, advancing to head cashier—at that time a big deal. After high school, she worked at JC Penney in Buffalo. She had to live with her aunt, however, because there was no public transportation off the island to get her to work.

Until the problems at Love Canal, Gibbs believed that she would have followed her mother's footsteps and become a full-time homemaker. But when she discovered the dump that was poisoning her children, she got involved in social justice issues. Gibbs believes that her major accomplishment was to lead about 900 families at Love Canal through two years of struggle to achieve relocation of all families who wished to leave. The work of her organization resulted in the passage of the federal Superfund law. Superfund is the common name for the Comprehensive Environmental Response, Compensation, and Liability Act (CERCLA) of 1980, a federal law whose mission is to clean up the worst hazardous and toxic-waste sites on land and water that constitute threats to human health and the environment.

When Gibbs moved to Love Canal, Niagara Falls, New York, she did not know that the chemicals were buried there. The area was a chemical dump site for the Hooker Electrical Company founded in 1905, which used the area to dump about 21,800 tons of chemicals between 1942 and 1952. In 1954, the school committee bought the site to erect a school building, and houses were also built in the area. During the winter of 1975 and 1976, disposed drums began surfacing to the top of the site, forming chemicals in outdoor pools and in basements. Meanwhile, Gibbs noticed changes in her family's health. One of her children got sick. The neighborhood also had a high rate of birth defects, miscarriages, and cancer. Gibbs began a door-to-door petition to close the school. According to Gibbs, the New York Health Department was not very helpful, nor were federal, state, and local governments. During this time, Gibbs also contacted Beverly Piagen, a cancer researcher, to look at the health problems of her community. In 1978, she organized the residents and founded the Love Canal Homeowners Association. The association demanded evacuation, restitution, cleanup of the site, and a determination of the extent of chemical migration. It conducted petitions and pickets and appeared on the media. The association faced fierce opposition from Occidental Petroleum, a chemical manufacturer that believed that the toxic materials were not the cause of the neighbors' high levels of disease. Finally, in 1978, the New York Health Department ordered temporary evacuation of pregnant women and children under the age of two from the homes immediately next to the canal. Later that year, Governor Hugh Carey announced that all families, about 240, who lived in houses next to the canal site, would be evacuated. In 1980, President Jimmy Carter declared the area a health emergency and announced the evacuation of the rest of the families from the hazardous site.

Through the Love Canal success, Gibbs gained considerable skills and experience, as well as widespread notoriety. She recalls that after Love Canal was fought and won, people from across the country who were experiencing similar problems in their communities called her volunteer offices to ask her how they too could fight the pollution in their backyards. For this reason, Gibbs would have to say that it was the parents faced with an environmental threat to their families who caused her to go beyond Love Canal into working nationwide.

Gibbs is described as motivational, fair to work with, creative, and very strategic in her work. Over the past twenty years, her organization CHEJ has been a major player in the formation of a new social justice network commonly referred to as the environmental justice network. This new movement crosses all lines of class and color, as well as urban and rural, gender, and geographic boundaries. It is one of the fastest-growing networks in this country.

Gibbs believes that the poisoning of American families is the most obvious sign that we have lost most of our ability to be a democracy. She believes that people are now being controlled and even their destinies are being held by large corporations. The same corporations that were at the founding of this great country are supposed to serve the people. Now the people serve the corporations. Gibbs is not against corporations, because they are needed for a strong economic base, but she is opposed to the tremendous power of the corporations over the people and their environment. She states that one needs only to look at the increases of disease in the United States to see the human health problems. Children's cancer is increasing, as are infertility and asthma. The natural environment and its creatures are suffering the same. Gibbs states, "There needs to be a way or ways to balance our country and give back to the people their governance. If we do not then I believe the future is bleak at best and could cause serious political uprising with both a loss of human life and property damage at worse. We've gone from King George to King DOW or EXXON or who ever you want to name. It's scary."

Gibbs's advice to young people is to get involved around what you care about, but get involved. The scary future she speaks of is yours, not hers. She explains that you can stop the damage before it is too late, but only if you act. You need to look back at other social justice issues in this country and the key role students and young people played. Gibbs states that it was the students sitting at the lunch counter who made a difference in the civil rights struggle. Students on campuses across the country made the difference in the antiwar and women's rights movements. Gibbs believes that we need young adults in this movement in the same way.

In 1981, Gibbs's activism led her to found the Citizens' Clearinghouse for Hazardous Waste, now called the Center for Health, Environment, and Justice (CHEJ). The organization provides environmental and organizing information to support grassroots activism. The organization has three main campaigns going at this time. The Children and Health Project works with groups concerned about children to eliminate their exposures to environmental chemicals such as pesticides in schools, parks, and playgrounds. Health Care without Harm is working within communities to stop the burning of medical wastes in incinerators, change the waste-management practices of health-care institutions, and change their purchasing to nontoxic items. The third campaign is Stop Dioxin Exposure, which works to pass policies at the state level to eliminate the sources of dioxin in the environment that ends up in our food, water, and air. CHEJ also works on a day-to-day basis with groups across the country that need help on their local environmental problems. It is presently working with a network of 8,000 groups.

—*Lois Gibbs*

CHELLIS GLENDINNING
(1947–   )

Chellis Glendinning has been a political activist most of her life. She is also a psychologist, and her focus has consistently been the intertwining of psychological well-being, political power, technological development, and the natural world. She is the author of four books and has contributed to many magazines, newspapers, and journals, including the *San Francisco Chronicle*, the *San Francisco Bay Guardian, Utne Reader, Mother Jones, Orion, Not Man Apart*, and *Earth First! Journal*. Her work has also been included in seventeen anthologies.

Born in 1947 in Cleveland, Ohio, the youngest daughter of an Irish-Scottish-Welsh doctor and a Dutch-French civil rights activist, Glendinning attended Smith College, earned her bachelor's degree in 1969 from the University of California at Berkeley, and was inducted into Phi Beta Kappa. She received her doctorate in psychology from Columbia Pacific University in 1984, is licensed in the state of New Mexico, and has a small psychotherapy practice.

Glendinning was active with her mother in the civil rights movement and continued her passion for justice at Berkeley after her blood brother, a Lakota musician from the Standing Rock Reservation in South Dakota, was killed at Da Nang in South Vietnam. She protested the war and was arrested, along with some 365 others, in the famed "mass arrest" of Peoples' Park protestors. The battle over a communal plot of land celebrated and gardened by the community versus the university's centrist development was seminal, foreshadowing Glendinning's later dedication to the interweaving of community, sustainability, land, and human well-being.

At that time, she was also influenced by the work of Wilhelm Reich, who linked psychological rigidity and physical illness with the demands of mass social forms; Lewis Mumford, who recognized the metaphor of mass society as the megamachine and championed sustainable regionalism; her mother, Hooker Daoust Glendinning, who, increasingly concerned for the

Photo © Lindsay Holt II.

state of the earth, was becoming active against corporate development and nuclear power; and the women's movement, which revealed a prepatriarchal era when her European ancestors lived in sustainable harmony with the natural world and women were powerful leaders, thinkers, and healers. Glendinning became a feminist and began her training as a psychotherapist. She opened a private practice in 1974, taught at Esalen in San Francisco, produced the first major conference on women's health, politics, and spirituality at the University of California at Berkeley, and became a writer for *Chrysalis: a magazine of women's culture* in Los Angeles.

In 1979, Glendinning also began to write for the *San Francisco Bay Guardian*, specializing in women's, environmental, and nuclear issues. That same year, after Three Mile Island, she joined with other mental health professionals to develop psychological approaches to breaking through the "psychic numbing" and denial widespread in relation to the nuclear arms race. Along with Carol Wolman, Fran Peavey, Joanna Macy, and others, Glendinning synthesized the resulting "Despair and Empowerment Work" and helped guide thousands to environmental and peace activism in the 1980s. She was a cofounder of the international network Interhelp and Psychotherapists for Social Responsibility and the founder of Waking Up in the Nuclear Age, an institute that offered trainings throughout the United States. Glendinning's first book, *Waking Up in the Nuclear Age* (1987), is a description of the process.

In 1990, Glendinning's second book came out, an outgrowth of this earlier work. *When Technology Wounds: The Human Consequences of Progress* is a psychological study of technology survivors—asbestos workers, atomic veterans, downwinders, electronics-plant workers, Dalkon Shield users, families harmed by toxic waste—in the first overview of the psychological ecology of technological society. The book was nominated for a Pulitzer Prize in general nonfiction in 1991. Living at that time in Tesuque, New Mexico, Glendinning was simultaneously active in fighting the opening of the Waste Isolation Pilot Project in Carlsbad, New Mexico, the nation's first underground repository for low-level radioactive/hazardous chemical waste.

In 1991, she was invited to join the Board of Listeners at the World Uranium Hearing in Salzburg, Austria, an international testimonial for indigenous peoples on the cultural, ecological, and health effects of nuclear development on their homeland. Here she linked her connection with native land-based community—lost to grief at the time of her blood brother's death—with her passion for the environment. From 1991 to 1994, Glendinning worked with uranium miners of the Navajo Nation and Laguna Pueblo developing intratribal strategies, putting on educational forums, and writing articles on the environmental and health problems incurred from nuclear development on native lands. She was guest editor for the environmental justice journal *Race, Poverty, and the Environment* issue "Burning Fires: Nuclear Technology and Communities of Color," put out by Urban Habitat in San Francisco.

Glendinning's third book came out in 1994. *My Name Is Chellis and I'm in Recovery from Western Civilization* is a primer for what was becoming known in the wake of the publication of Theodore Roszak's book *The Voice of the Earth* as ecopsychology, a field of study and activism revealing the connections between the well-being of the planet, the well-being of the human community, and the well-being of the human psyche. In this book, she suggests that European-based societies suffer from posttraumatic stress disorder, a disease resulting from the separation of human activity from the natural world that first occurred in the Neolithic period and led to the expansionism of Western imperialism and the Industrial Revolution and eventually to the madness of today's technologically based global civilization.

At this time, Glendinning moved to the northern New Mexico village of Chimayó, home to a land-based Chicano people, part Mexican, part Tewa, part Navajo, part Hispanic. She became active as an advocate for the preservation of the culture—under assault from the external forces of urban development, tourism, and corporate logging and mining—and the reconstruction of sustainable communities living on communal lands. Witness now to a culture in acute distress from past and ongoing colonization, she wrote *Off the Map (An Expedition Deep into Imperialism, the Global Economy, and Other Earthly Whereabouts)*. Published in 1999, this is Glendinning's first effort at creative nonfiction and an addition to the literature of imperialism. It juxtaposes the forces of both classical European empire and today's rendition of empire, economic globalization, against the more ecological and sane ways of decentralized, land-based communities. The book also links child abuse with abuse of the planet.

Through the years, Glendinning has lectured at such institutions as Harvard University, Stanford University, the University of California at Berkeley, the American Psychological Association, the International Society of Political Psychology, the Association for Humanistic Psychology, the Environmental and Outdoor Education Council of Canada, Langley Porter Psychiatric Institute, the E. F. Schumacher Society, Esalen, Naropa Institute, Prescott College, and Carroll College. In 1983, she won the *San Francisco Examiner*'s Billy Award for her antinuclear activism. In 1997, she won the Zero Injustice Award given by the Rio Arriba County Commissioners in recognition of her "tireless efforts to show that environmental concerns need not be synonymous with destruction of indigenous cultures." In 2000, her book *Off the Map* was awarded first place for nonfiction by the National Federation of Press Women, and she is recognized by the International Biographical Centre as one of the 2,000 outstanding writers of the twentieth century.

Glendinning is a founding member of the neo-Luddite Jacques Ellul Society in Washington, D.C. She has served on the advisory boards of Earth Island Institute in San Francisco, EarthWays Foundation in Malibu, the Peace and Conflict Studies Department at the University of California at Berkeley, Concerned Citizens for Nuclear Safety in Santa Fe, New Mexico, and the Loka Institute in Amherst, Massachusetts. She was also a director at Fritjof Capra's ecological think tank, the Elmwood Institute, in Berkeley, California.

Glendinning lives in Chimayó, New Mexico, and drives a 1977 Honda Civic with Subcomandante Marcos airbrushed on the hood.

BIBLIOGRAPHY

Glendinning, C. (1987). *Waking Up in the Nuclear Age*. New York: Morrow, William and Company.

Glendinning, C. (1990, March/April). Notes toward a new-Luddite manifesto. *Utne Reader*, pp. 50–53.

Glendinning, C. (1990). *When Technology Wounds: The Human Consequences of Progress*. New York: William Morrow and Company.

Glendinning, C. (1993). *Waking Up in an Endangered World* [Audiocassette]. Sound Photosynthesis.

Glendinning, C. (1994). *My Name Is Chellis and I'm in Recovery from Western Civilization*. Boston: Shambhala Publications.

Glendinning, C. (1995). Technology, trauma, and the wild. In T. Roszak, et al. (Eds.), *Ecopsychology*. San Francisco: Sierra Club Books.

Glendinning, C. (1999). *Off the Map (An Expedition Deep into Imperialism, the Global Economy, and Other Earthly Whereabouts)*. Boston: Shambhala Publications.

Mills, S. (Ed.). (1997). *Turning Away from Technology*. San Francisco: Sierra Club Books.

Sewall, L. (2000, Spring). Off the map: A tour of Chimayó with Chellis Glendinning. *Orion* 19(2), pp. 40–49.

—Chellis Glendinning

CHARLES R. GOLDMAN
(1930–   )

Charles R. Goldman is a professor of limnology in the Department of Environmental Science and Policy at the University of California, Davis (UCD), and has been at the university since 1958. He developed the first courses in limnology and oceanography at UCD, served as chair of the Division of Environmental Studies from 1988 to 1992, and was founding director of the Institute of Ecology, serving from 1966 to 1969 and again in 1990–1992.

Prior to his tenure at UCD, he earned bachelor's and master's degrees from the University of Illinois and a Ph.D. in limnology and fisheries from the University of Michigan. He has supervised eighty-six graduate students and thirty postdoctorals during his forty years at UCD.

Goldman's many prestigious awards include an NSF Senior Postdoctoral Fellowship in 1964 for limnological research in the Arctic (Lapland), a Guggenheim Fellowship in northern Italy in 1965, the Goldman Glacier in Antarctica, named in 1967, and the Antarctic Service Medal by Congress in 1968. He served as president of the American Society of Limnology and Oceanography in 1967–1968 and was elected a fellow by the California Academy of Sciences in 1969. In 1973–1974, he was elected vice president of the Ecological Society of America. He accepted a Fulbright Distinguished Professorship to Yugoslavia in 1985 and was awarded the Vollenweider Lectureship in Canada in 1989, the Chevron Conservation Award and Culver Man-of-the Year in 1991, the Earle A. Chiles Award in 1992, and the UC Davis Distinguished Public Service and Research Lecturer awards in 1993. He was elected vice president of the International Society of Limnology (SIL) for 1992–1998 and presented the prestigious Baldi Lecture at the triennial SIL Congress in Ireland in August 1998.

Goldman has published four books and over 400 scientific articles and has produced four documentary films that are in worldwide distribution. He has served on many national and international committees and is fre-

Photo courtesy of Charles Goldman.

quently sought for consultation and research missions to foreign countries on major environmental problems.

In 1990, Goldman was a member of a UNESCO team to qualify Lake Baikal as an International Heritage Lake and served as senior scientist for the National Geographic Baikal Project. His single most important and sustained contribution is the forty years of research on Lake Tahoe. Goldman is director of the Tahoe Research Group and has pursued long-term ecological research simultaneously at Lake Tahoe and Castle Lake, California, since 1958. He successfully combined effective research and social action with his pioneering studies of lake eutrophication. These have been directly applied to engineering solutions, social needs, and legal decisions. This work has recently included the development of artificial wetlands and research on alternatives to conventional road salt for deicing highways. This relationship of basic science to political change has been of particular importance to the Lake Tahoe basin.

During the summer of 1997, Goldman hosted President Clinton and Vice President Gore aboard the UCD research vessel *John Le Conte* during the Lake Tahoe Presidential Forum. Similar studies have extended Goldman's research and social action efforts to analysis of lakes like Baikal in Russia and hydroelectric impoundments throughout the world. Thus, while aggressively pursuing basic research on lake dynamics, he has also been able to translate the findings directly to state, national, and international policy decisions, contributing decisively to the conservation and judicious use of aquatic resources from the Antarctic to the lakes and wetlands of South and Central America, New Guinea, Africa, Asia, Europe, and the United States.

Goldman's career work has now been honored with his most prestigious award yet: he received the 1998 Albert Einstein World Award of Science in a formal ceremony in New Zealand. The Einstein Award, bestowed annually on a single individual by a council of eminent scientists that includes twenty-five Nobel laureates, recognizes those who have accomplished scientific and technological achievements that have advanced scientific understanding and benefited humanity.

—Charles R. Goldman

GEORGE BIRD GRINNELL
(1849–1938)

George Bird Grinnell was a naturalist, conservationist, ethnographer, and newspaper editor. He was born in Brooklyn, New York, on September 29, 1849. Grinnell was the oldest child in an upper-class family, the son of New York industrialist George Blake Grinnell and Helen Lansing Grinnell. In 1857, while his family lived in upper Manhattan, Grinnell became a student of John James Audubon's widow, Lucy, and a friend of her grandchildren. He graduated from Churchill Military School in Ossining, New York, in 1866 and earned a B.A. in 1870 from Yale University.

The young Grinnell spent several months as a volunteer with Yale professor Othniel C. Marsh in the western United States in 1870, which marked his beginning as a leading interpreter of the West. On the 1870 Marsh expedition, Grinnell met the famous western figure William F. "Buffalo Bill" Cody and the Pawnee scouts Frank and Luther North. In 1872, he accompanied the Pawnee on one of their last tribal buffalo hunts. Grinnell's first publication, "Elk Hunting in Nebraska," in the October 1873 issue of *Forest and Stream* was based on an expedition with Luther North. During the 1870s, Grinnell participated in government expeditions, serving as a naturalist with George Custer's Black Hills Expedition and Colonel William Ludlow's reconnaissance of Yellowstone National Park. He worked also as an assistant in osteology at Yale's Peabody Museum, a cattle rancher, and a big-game hunter with Cody and North in western Nebraska.

In 1880, after receiving his Ph.D. in paleontology from Yale, Grinnell became editor-in-chief of *Forest and Stream*. Beginning in 1885, he made hunting trips to the St. Mary's Lake region of northwestern Montana, where he discovered the glacier named for him. This area became Glacier National Park in 1910 as a result of his early efforts. In 1899, Grinnell participated in the Harriman Alaskan Expedition.

Posthumously called "the father of American conservation," Grinnell had a great impact on the early national legislation relating to wildlife, parks, and forests. In particular, he influenced the future president Theodore Roosevelt's conservation philosophy. He was a founder of the first Audubon Society in 1886 and later a director of the National Audubon Society. Grinnell was also a founder and trustee of the New York Zoological Society. Presidents Grover Cleveland and Theodore Roosevelt appointed Grinnell as an emissary to treaties and other negotiations with the Blackfoot, Fort Belknap, and Standing Rock Sioux Indians.

In the field of wilderness conservation, Grinnell held several notable positions. He was a member of the first advisory board on the federal Migratory Bird Law and a fellow of the American Ornithologists' Union. Also, he was a trustee of the American Museum of Natural History. Beginning in 1911, he helped to found and served as director of the American Game Association. He was chairman of the Council on National Parks, Forests, and Wildlife.

As the anonymous editor of *Field and Stream*, he used the power of this medium to protect Yellowstone National Park, end commercial hunting, enact enforceable game laws, and urge the federal government to adopt scientific European forestry methods. The Yellowstone Park Protection Act (1894), a landmark in national park legislation, and the Migratory Bird Treaty with Great Britain (1916) were results of his efforts. Grinnell was known as a modest, self-effacing individual who preferred to work behind the scenes and sometimes did not receive credit for his initiatives.

Grinnell's first book, *Pawnee Hero Stories and Folk Tales*, appeared in 1889 and was followed by nine other ethnographic works. Among his best-known books is the two-volume *Cheyenne Indians* (1923). Anthropologists Margaret Mead and Ruth Bunzel contended that no work on any other tribe "comes closer to their everyday life than Grinnell's classic monograph on the Cheyenne" (Sterling et al.). He also edited a series of books on hunting, natural history, and conservation. His *American Duck Shooting* (1901) and *American Game-Bird Shooting* (1910) became classics among sportsmen and naturalists. Grinnell also wrote seven books for young readers in the so-called Jack Series between 1899 and 1913, which were based on true experiences.

In 1925, Grinnell became president of the National Parks Association and received the Theodore Roosevelt Gold Medal of Honor for distinguished service in promoting outdoor life and conservation causes. He died in New York on April 11, 1938.

BIBLIOGRAPHY

Diettert, G. A. (1992). *Grinnell's Glacier: George Bird Grinnell and Glacier National Park*. Mountain Press Publishing Company.

Grinnell, G. (1910). *American Game-Bird Shooting*. New York: Forest and Stream.

Grinnell, G. (1923). *Cheyenne Indians*. New Haven, CT: Yale University Press.

Grinnell, G. (1990). [1889]. *Pawnee Hero Stories and Folk Tales*. Lincoln: University of Nebraska Press.

Grinnell, G. (1991). [1901]. *American Duck Shooting*. Mechanicsburg, PA: Stackpole Books.

Sterling, K. B., Harmond, R. P., Cevasco, G. A., & Hammond, L. F. (Eds.). (1997). *Biographical Dictionary of American and Canadian Naturalists and Environmentalists*. Westport, CT: Greenwood Press.

—*John Mongillo*

JUANA BEATRÍZ GUTIÉRREZ
(1932–)

Juana Beatríz Gutiérrez is the cofounder and president of Madres del Este de Los Angeles Santa Isabel (Mothers of East Los Angeles Santa Isabel). Juana and her husband Ricardo have lived in Los Angeles for forty-three years. They have nine children, twenty-one grandchildren, and one great-grandchild. Juana's work as an activist has made her the subject of countless media stories and a source for numerous scholarly works. Her work in East Los Angeles has gained recognition by media agencies and grassroots organizations in London, the former Soviet Union, Mexico, Australia, and South Africa. Nationally, Gutiérrez's work has been featured in *People, Time, Rolling Stone, L.A. Style*, and various local and regional publications. She is the recipient of numerous awards, including the Mujer Award from the National Hispana Leadership Institute (1995), the National Association of Bilingual Education Citizen of the Year Award (1994), the Women of Courage Award from the Los Angeles City Commission on the Status of Women (1994), and Mexican Mother of the Year, Los Angeles (1993). She has been honored for her community work by citations and commendations from President Bill Clinton (1995) and Pope John Paul II (1992) as well as numerous city, state, and federal officials.

Gutiérrez was born on May 6, 1932, in Lodemena, Municipio de Sombrerete, Zacatecas, Mexico, the daughter of Jesus Escamilla and Soledad Alamos who were *campesinos*. Her family, which included four sisters and five brothers, moved from rural Sombrerete to the pueblo of Cantuna, Zacatecas, and finally to Ciudad Juarez, Chihuahua, across the border from El Paso, Texas.

As the third oldest in the family and the oldest among the women, Juana

Photo courtesy of Juana Gutiérrez.

helped her mother with the household work and did not go to school every day. She started school in Cantuna and entered but did not finish the sixth grade in Ciudad Juarez. At school, Gutiérrez considered herself to be "rebellious." She did not allow herself to be taken advantage of, defended herself as best she could, and even defended her older brothers. The school administrators would always send notes home to her parents, and her parents' punishment was always harsher than the teacher's.

Juana and Ricardo were married on June 9, 1954, in Las Cruces, New Mexico. In May of 1956 they moved to Los Angeles, once they were assured that Ricardo, who had enlisted in the Marine Corps, would remain in the United States rather than being sent overseas.

Gutiérrez's experience in Los Angeles included the arduous task of child rearing. Noting that caring for her children required community involvement, she became involved in community affairs. At first, she invested much time in her children's education. Gutiérrez believed that since she had not enjoyed a formal education, she wanted her children to have it. This sentiment was evident in a speech she gave to the Esperanza Foundation in Santa Barbara, California, in which she stated, *"Para nosotros los pobres la unica herencia que les podemos dejar a nuestros hijos/as es su educación*

[For those of us who are poor, the only inheritance we can leave for our children is their education]."

Her belief that education was crucial to success in the United States encouraged Gutiérrez to become an active member in the Parent Teacher Association (PTA) and later in sports boosters' clubs. She became vice president of her PTA and organized the mothers to do kitchen work for the annual festivals, dances, and sports programs. Much of her involvement included raising funds for educational supplies and sports equipment as well as awards banquets.

All of Gutiérrez's children attended Santa Isabel Elementary School in the Boyle Heights section of Los Angeles. The women went to Sacred Heart of Jesus High School in Lincoln Heights, and the men went to Cathedral High School in Los Angeles. They all went to different colleges and universities when they graduated from high school. Professionally, their careers include an attorney, a director of public relations based in Washington, DC, an executive director for a community-based organization, an environmental and safety supervisor, a political consultant, a graphics designer, a university professor, and a high school teacher.

Before long, Gutiérrez noticed various issues and problems that faced her community. As her children grew older, their interaction with other youth throughout the neighborhood and city parks brought to her attention the existence of drug dealers and others who threatened the well being of her children and their friends. When she moved to her current home in Boyle Heights in the early 1970s, Juana and her husband initially confronted the drug dealers themselves. Realizing that they needed an organized effort to deal with these issues, they became active and helped to found community-based organizations in the early 1970s. Ricardo, for instance, was one of the co-founders of United Neighborhood Organization (UNO), and Juana helped to start the Neighborhood Watch Program on her block and later in other portions of her community.

Before long, Gutiérrez became a leading force in an emerging network of women who were advocating community improvement and positive programs for youth. While threats against Gutiérrez and her family caused consternation and concern, they did not affect her resolve. In fact, they made her more resilient as she became convinced that she would not allow drug dealers to take control of the parks and the community.

However, due to some health problems in the late 1970s and early 1980s Gutiérrez's involvement in community issues was affected. Diagnosed with diabetes, she was placed on a rigorous diet for about six months. The diet, the planning of her twenty-fifth wedding anniversary, the untimely death of a brother-in-law, and several other events caused her, in her words to "look like *la muerte* (death)". Feeling herself physically and emotionally drained, she told herself, "If I'm going to die, I am not going to die this way." She stopped the diet and stopped taking her insulin. In place of the insulin, she followed her mother's advice to cure herself with *liquados de savila con nopal* (aloe vera and cactus smoothies). In the twenty years since then, Gutiérrez has revitalized her health and has recommitted herself to bringing about positive change for her community.

The network of women that Gutiérrez helped to form in Los Angeles became a driving force for some of the major victories that the Latino/a community has enjoyed. In May 1984, Gutiérrez and several other community leaders were made aware of the attempt of the California Department of Corrections (DOC) to place another prison in East Los Angeles. By that time, Juana and Ricardo had already introduced the Neighborhood Watch Program in other areas and were working with various block captains. A meeting was called and held in the Gutiérrez's living room in which a representative of then Assemblywoman Gloria Molina briefed community leaders on the DOC's intent. Following the meeting, those in attendance set out to inform the community after Sunday mass at Santa Isabel Church. This effort included a petition drive against the prison. Three mothers collected signatures at Santa Isabel, and Ricardo went to Saint Mary's Parish. Combined they collected 900 signatures at both churches.

The community was informed through different avenues such as school meetings (PTAs) and senior-citizen meetings. Four months after the campaign against the prison began, Father John Moretta from Resurrection Church asked Father Luis Garbo of Santa Isabel Parish about what was to be done against the prison. Father Garbo directed him to Juana, who had already been active in the antiprison struggle. Father Moretta, who was involved with the Chambers of Commerce, visited Juana to compare notes. About this time, the Coalition against the Prison emerged from an assortment of organizations and business groups.

Mothers from Santa Isabel, Resurrection, and Dolores Mission parishes attended the first meeting of the Coalition against the Prison, where Father Moretta asked if they would like to be called the Mothers of East Los Angeles. Thus, Father Moretta coined the name of the organization. The Mothers of East Los Angeles became the rallying point of the Coalition against the Prison. The struggle, which included marches and lobbying efforts at the state capitol, lasted eight years before the state abandoned its plans.

In 1987 Gutiérrez's community was confronted with an attempt to build a toxic waste incinerator in the neighboring city of Vernon. Vernon is an industrial city adjacent to Boyle Heights, where large numbers of the work force for companies in Vernon live. Assemblywoman Lucille Roybal-Allard, who was later elected to Congress, brought this attempt to the attention of community leaders. Strategies similar to those used in the antiprison struggle were employed against the proposed Vernon incinerator.

During the incinerator issue, Gutiérrez and her group of mothers separated themselves from the group at Resurrection after organizational disputes emerged regarding leadership and gender-equity issues as well as the intent of the loosely structured organization. Father Moretta had named Gutiérrez as the spokesperson for the group to address only the Spanish media. Moretta and another male philanthropist and developer from the Montebello Chamber of Commerce prepared press releases and expected that every time Gutiérrez spoke in public, she would say only what had been scripted by them beforehand. Regarding this experience, Gutiérrez recalls, "Since I did not comply, I had problems with them. When I speak, I don't write my speeches ahead of time. Much less am I going to accept that somebody else write them for me. Since I did not permit them to treat me this way, they nicknamed me 'La Chingona.' I actually liked that nickname, though it wasn't meant as a compliment." After that encounter, Gutiérrez was asked not to bring any more women from Santa Isabel and Dolores Mission parishes to the meetings. Gutiérrez and several other women formed Madres del Este de Los Angeles–Santa Isabel (MELASI) whose intent was not solely to fight against the prison and incinerator but to take a proactive approach to community improvement.

MELASI coalesced with national environmental groups to oppose the proposed Vernon incinerator and likewise lent its support to similar struggles throughout the state and the Southwest. MELASI began to address issues of quality of life and became a proactive organization committed to working within the environmental movement, on issues of education, and for community economic self-reliance. This proactive approach led MELASI to network with activists from several parts of the country on issues including the struggles against the proposed incinerator in Kettleman City, the toxic dumps in Casmalia and Button Willow, and the oil pipeline in East Los Angeles as well as those issues pertinent to the United Farmworkers of America (UFW, AFL-CIO).

MELASI initiated a water conservation project with Corporate Technology Systems International (CTSI) and the Mono Lake Committee. In an effort to preserve natural habitats from which southern California draws its water resources, this project focused on reducing the level of water consumption by southern California residents. Economically, consumers' water bills were reduced by the ultra-low-flush toilets that were distributed free of charge. In addition, twenty-eight community people were employed full-time with benefits through this program. The revenue gained from the water-conservation program provided funds for a scholarship program which since its inception has awarded over $350,000. Also, a lead-abatement program, an antigraffiti program, and an immunization program were started to address those concerns.

As part of the water-conservation program MELASI takes groups of young East Los Angeles residents to Mono Lake, where the Mono Lake Committee provides a hands-on environmental education program. This program includes overnight camping and educational field trips during the day. In addition, MELASI converted a former vacant lot underneath the infamous East Los Angeles Interchange into the Eastside Community Garden and Education Center. El Jardín (The Garden) provides local school-children and teachers with a learning facility. Several events are held throughout the year at the Eastside Garden to promote culture and environmentalism as well.

Juana Beatríz Gutiérrez remains committed to social, cultural, economic, political, and gender justice. She maintains, "As a mother and resident of East L.A., I will continue fighting with perseverance, so we will be respected. And I will do this with much affection for my community. I say 'my community' because I am part of it; I love my raza [people] as my family, and if God allows, I will continue fighting against all the governors who try to take advantage of us." Noting that her activism began on behalf of her children when they were young, she asserts, "I once said that I was going to get involved for my children and neighbors, but now it is also for my grandchildren and because of all the injustices against our raza, especially those brought about by the government and corporations."

BIBLIOGRAPHY

Acuña, R. F. (1996). *Anything But Mexican: Chicanos in Contemporary Los Angeles*. London: Verso Press.

Acuña, R. F. (2000). *Occupied America: A History of Chicanos*. 4th ed. New York: Longman Press.

Bullard, R. D. (Ed.). (1993). *Confronting Environmental Racism: Voices from the Grass Roots*. Boston: South End Press.

Gutiérrez, G. (1994). Mothers of East Los Angeles strike back. In Bullard, R. D. (Ed.), *Unequal Protection: Environmental Justice and Communities of Color* (pp. 220–233). San Francisco: Sierra Club Books.

Gutiérrez, J. B. Interview. June 14, 1999.

Lerner, S. (1997). *Eco-Pioneers: Practical Visionaries Solving Today's Environmental Problems*. Cambridge: MIT Press, pp. 277–286.

Pardo, M. S. (1998). *Mexican American Women Activists: Identity and Resistance in Two Los Angeles Communities*. Philadelphia: Temple University Press.

Pardo, M. S. (1991, Spring/Fall). Creating community: Mexican American women in East Los Angeles. *Aztlan: A Journal of Chicano Studies 20* (1, 2), pp. 39–72.

Pulido, L. (1996). *Environmental and Economic Justice: Two Chicano Struggles in the Southwest*. Tucson: University of Arizona Press.

Shorris, E. (1992). *Latinos: A Biography of a People*. New York: Norton.

—Gabriel Gutiérrez

DAVID HAENKE
(1945-   )

To do all things ecologically, everywhere . . .

—David Haenke

David Haenke is the coordinator of the Bioregional Project of Ozarks Resource Center, Inc., now known as the Watershed Economics Project, in Missouri. The Ozarks Resource Center, a nonprofit organization located in the south central Missouri Ozarks, was incorporated in 1978. Haenke is also the manager of Alford Forest, Ozark County, Missouri, where he oversees a 4,500-acre oak, pine, and hickory forest using ecoforestry methods that provide value-added forest products.

Haenke is a bioregionalist. He defines the term "bioregion" as an area that is simply defined by nature in the natural world. Its boundaries are not political but ecological boundaries. The bioregions are also defined by cultures that ecological conditions within those boundaries have helped to shape. People are changed by the nature of regions they live in, although this is less so now than in the past. The native peoples and their cultures were shaped by the plants and animals and the rivers and climate of the particular bioregion. Haenke lives in the Ozarks, which form a very well defined bioregion. It is an ancient uplift and is very clearly defined by rivers that surround it and by all kinds of flora and fauna that make it distinct from the areas around it.

Haenke was born in 1945 in Mount Pleasant, Michigan, and he grew up around Saginaw, Michigan. He believes that there were a couple of defining points in his childhood that put him on the path that he is on now. One occurred when he was seven or eight and went for a walk in the woods near his house with friends his age. He had an experience he can only describe as being enchanted by the forest during the fall. To him, the scene was an ecological and religious experience that he has never forgotten. Another experience occurred when he was around nine or ten. It became apparent to him that a local river that was close to his hometown was polluted by a local chemical company, the largest single chemical plant in the world at that time. The river never froze even though it got very cold in Michigan because of the pollutants in it. The river would just steam throughout the winter. He became aware that it was polluted, and he was outraged.

Haenke went to the grade schools and high school in his hometown. After graduating from high school, he went to Michigan State University and graduated in 1967 with a degree in English and education. He then taught school for a year and did a variety of other jobs that included working in an automobile factory and in a grain elevator. In 1971, he went down to the Ozarks in southern Missouri to start an ecological community. He describes the move as an attempt to put into practice the things that he had been thinking about in terms of ecology. He wanted to get involved in organic agriculture, solar energy, and alternative technologies to put the theory into practice so he could live a more authentic life in the countryside as opposed to living in suburban areas or cities. He started co-ops and used organic agriculture to grow food. In 1978, he participated in the incorporation of an organization called New Life Farm, a technology research center that did research on solar energy, hydraulic dam water pumps, and ecological agriculture.

While he was farmsteading, he lived and worked with his family on forty wooded acres in the Ozarks, doing an organic garden, tending farm animals, writing how-to articles for magazines, and working at sawmill and railroad labor for supplemental income. However, he continued to network with other communities throughout the Ozarks in the areas of solar energy, appropriate technology, ecological agriculture, and cooperative economics. Because of his interest in organically grown food and co-ops, he joined and then became a board member (1978–1983) of the New Destiny Cooperative Federation, a south central U.S. food-warehousing and distribution company.

In 1984, Haenke was the coordinator of the first North American Bioregional Congress (NABCI), held in May 1984. NABCI was an ecologically based, interdisciplinary convening of 200 people from thirty-two U.S. states, six nations, five native tribes, and 130 organizations. Along with the sixty-five informational and cultural workshops there, fourteen working committees drafted statements and resolutions on bioregional policy issues for presentation, review, or ratification by the congress's plenary session.

In 1985, Haenke became the director of the Ecological Society Project (ESP) of the Tides Foundation founded in San Francisco, California, until the project ended in 1999. ESP was engaged in broad-range, whole-systems ecological work serving a larger constituency than the Bioregional Project and bioregionalism. In determining the nature of problems and solutions, ESP used an ecological analysis of the structure and function of virtually all human systems and modalities. These include economics, business and investment, technology, resource and energy use, community planning and design, agriculture and food, land tenure, government, health, education, and more. ESP was involved in a number of initiatives, such as the North American Conference on Christianity and Ecology (NACCE), of which Haenke was the founder. NACCE's 1987 founding event drew over 600 people and was the first major convening of the Christian ecology movement. In an annual report in 1992, Haenke summed up the philosophy of the Ecological Society Project:

The Ecological Society does not just include human beings. In any given life place, the definition of society and community is expanded to fully include ecosystems, watersheds, bioregions, animals, plants, trees and forests, rivers, birds, lakes, all living things and ecological entities, including human beings. Further, all living things and ecological entities in this larger society deserve rights equal to those of humans to life, necessary habitat, and the integrity of their place in the larger community or society. Such a consideration goes beyond ethics. It is the prime factor in the continuance of our own species, as well as that of the countless others that we depend on.

BIBLIOGRAPHY

Haenke, David. (1991, Fall). Spirit of the waters. *Synapse* (Traverse City, MI), *17*, 11.

Haenke, David. (1992, Winter). A foundation for ecological economics. *Firmament* (San Francisco, CA), 2 (2), 8–10.

Haenke, David. (1993a, March). In the beginning: A history of the Ozark Area Community Congress. *Talking Oak Leaves*.

Haenke, David. (1993b, July). The ecological economics of Missouri's public forests. In Alan R. P. Journet & Henry G. Spratt, Jr. (Eds.), *Towards a Vision for Missouri's Public Forests: Conference Proceedings* (pp. 119–128). Cape Girardeau: Southeast Missouri State University Press.

Haenke, David. (1993c, Fall). Ecological economics as an essential component of the Wildlands Project. *Wild Earth*.

Haenke, David. (1993d, Fall). What is deep about deep ecology. *Institute for Deep Ecology Education Newsletter*, p. 4.

—*John Mongillo*

GARRETT HARDIN
(1915–   )

In terms of the practical problems that we must face in the next few generations with the foreseeable technology, it is clear that we will greatly increase human misery if we do not, during the immediate future, assume that the world available to the terrestrial human population is finite.

—Garrett Hardin

Garrett Hardin is an ecologist and a microbiologist who established the field of human ecology, with the viewpoint that "ecology is the overall science of which economics is a minor specialty." Hardin is best known for his 1968 article "The Tragedy of the Commons," which argued for strict controls on reproductive choice to solve the global population problem. "Throughout most of history there's been no need for concern about population control," Hardin says. "Nature would come along with epidemic diseases and take care of the matter for us. Disease has been the primary population controller in the past. Because widespread disease and famine no longer exist, we have to find other means to stop population" (Carnell).

Hardin was born in Dallas, Texas, and was educated at the University of Chicago and Stanford University, where he received his doctorate in 1941. He joined the faculty of the University of California at Santa Barbara in 1946. Originally a plant biologist, Hardin became increasingly interested in genetics, evolution, and the problems of pollution and population growth. As a professor of human ecology at the University of California at Santa Barbara for more than thirty years, Hardin publicly debated the issues of abortion, population control, foreign aid, nuclear power, and immigration. In the 1960s, he campaigned for legalized abortion in the United States and was a founder of Planned Parenthood.

Hardin's "commons" refers to common resources that are owned by everyone. The "tragedy," according to Hardin's theory, occurs as the result of everyone's having the "fatal freedom" to exploit the commons. Using as his model a scenario involving commonly owned pasturelands, Hardin was the first to hypothesize clearly and convincingly the tragedy of the commons. His analysis showed how rational human self-interest within a system of common ownership or usage results ironically, but predictably, in a loss to everybody within it. Unmanaged, he asserted, the common pastureland must inevitably be ruined by overgrazing, and competitive individualism will be helpless to prevent this social disaster.

Although Hardin's article described exploitation in an unregulated public pasture, the pasture served as a metaphor for our entire society: our communities are the commons, our schools are the commons, as are our roads, air, and water; in some senses, even we ourselves are the commons. Most important, Hardin illustrated the critical flaw inherent in the concept of freedom in the commons: all participants must agree to conserve the commons, but anyone can force the destruction of the commons. Thus, as long as humans are free to exploit the commons, we are locked into a paradoxical struggle against ourselves—a terrible struggle that must end in universal ruin.

The tragedy of the commons comes in many different versions, Hardin believes. He points to numerous instances in which the environment is degraded because nobody owns the land or air or water, and so nobody has much incentive to prevent pollution. "The National Parks present another instance of the working out of the tragedy of the commons," Hardin says. "At present, they are open to all, without limit. The parks themselves are limited in extent—there is only one Yosemite Valley—whereas population seems to grow without limit. The values that visitors seek in the parks are steadily eroded. Plainly, we must soon cease to treat the parks as commons or they will be of no value to anyone" (Hardin 1968). Taking the argument a step further, he notes that resource subsidies are a disguised form of "commonization": the greater the subsidy, the less the nominal owner of an asset really has at stake.

Hardin's article has become a classic in environmental literature, and since its first publication, it has been reprinted in over one hundred anthologies. It became required reading for a generation of students and teachers seeking to merge multiple disciplines in order to come up with better ways to live in balance with the environment. It has been widely accepted as a fundamental contribution to the fields of ecology, population theory, economics, and political science. Hardin's application of his theory to the problems of population control, foreign aid, and immigration was worked out in his 1974 article "Living on a Lifeboat" and was more recently elucidated in 1994's "The Tragedy of the Unmanaged Commons: Trends in Ecology and Evolution."

The overall message of Hardin's work on population control is that reproductive decisions cannot be left in the hands of individuals, but must rest with society. He notes that science has long understood that our environment has limits—a carrying capacity—and yet economists consistently turn a blind eye to one feature we share with all of our planet's inhabitants: the potential for irreversible environmental damage through overcrowding. "In terms of the practical problems that we must face in the next few generations with the foreseeable technology, it is clear that we will greatly increase human misery if we do not, during the immediate future, assume that the world available to the terrestrial human population is finite" (Hardin 1968). In 1996, he said, "The only answer is that family size cannot be left to individual decisions. You don't have to be brutal about it. You can use incentives. But control of population will have to take the form of mutual coercion, mutually agreed upon" (Meile).

The common thread running through all of Hardin's work is an interest in bioethics, which, in his view, is more than merely ethics as applied to biological problems. Rather, it is ethics, which he sometimes calls "tough-love ethics," built onto a biological foundation. Essential elements of such ethics are relative quantities, feedback processes, and the changes that develop over time as the unforeseen consequences of actions that have been taken. In his 1985 book *Filters against Folly*, he argues that ethical theory, to be useful, must employ three intellectual filters: literacy, the correct use of words; "numeracy," the appreciation of quantities; and "ecolacy," the study of relationships over time.

Today, Hardin's commons theory has been coming under increasing criticism, particularly from advocates of cooperative management. *Cultural Survival*, a quarterly journal that focuses on the rights of indigenous peoples and ethnic minorities, devoted its spring 1996 issue to attempting to disprove Hardin's logic, arguing that "effective use of common resources by local communities is in our day often the most efficient way of ensuring that modern, industrialized economies promote growth with equity and minimal environmental degradation" (Maybury-Lewis). Hardin believes that sustainable growth is an oxymoron. "Don't go for growth, go for a steady state, a term that comes from biological study. There's always an increase, but there's also a decrease around a mean value" (Meile).

Among other awards, in 1986, Hardin received the American Institute of Biological Sciences Distinguished Service Award for his contributions in the field of ecology and his longtime efforts to apply scientific methods to the ethical and political dilemmas posed by population growth and resource depletion. He was president of the Environmental Fund from 1980 to 1981. Hardin's other books include *Living within Limits: Ecology, Economics, and Population Taboos; The Immigration Dilemma*; and *Population, Evolution, and Birth Control*. Since Hardin's 1978 retirement from the University of California, he has been professor emeritus of human ecology there and has devoted himself to writing and speaking. "Lecturing is a difficult art," he notes. "You have to get people willing to be lectured to. My basic position is, never globalize a problem if it can possibly be dealt with on a local basis ... and reproduction is a very localized problem" (Meile).

BIBLIOGRAPHY

Biography Online. Hardin, Garrett (James). *http://www.biography.com/cgi-main/biomain.cgi*

Carnell, B. Overpopulation.com. Garret Hardin. *http://www.overpopulation.com/garrett_hardin.html*

Commoner, B. (1975, August/September). How poverty breeds overpopulation (and not the other way around). *Ramparts*, p. 21.

Crowe, B. 1969. The tragedy of the commons revisited. *Science, 166*, pp. 1103–1107.

Fletcher, J. (1991, Spring). Chronic famine and the immorality of food aid: A bow to Garrett Hardin. *Population and Environment, 12*. Available: *http://dieoff.org/page91.htm*

Hardin, G. (1968). The tragedy of the commons. *Science, 162*, 1243.

Hardin, G. (Ed.). (1969). *Population, Evolution, and Birth Control: A Collage of Controversial Ideas*. New York: W. H. Freeman and Co.

Hardin, G. (1974, October). Living on a lifeboat. *Bioscience*, p. 561.

Hardin, G. (1986a). Cultural carrying capacity. Stalking the Wild Taboo online. *http://www.lrainc.com/swtaboo/stalkers/gh_ccarr.html*

Hardin, G. (1986b). [1985]. *Filters Against Folly: How to Survive Despite Economists, Ecologists, and the Merely Eloquent*. New York: Viking Penguin.

Hardin, G. (1994). The tragedy of the unmanaged commons. *Trends in Ecology and Evolution, 9*, pp. 199–203.

Hardin, G. (1995a). *The Immigration Dilemma: Avoiding the Tragedy of the Commons*. Washington, DC: Federation for American Immigration Reform.

Hardin, G. (1995b). *Living Within Limits: Ecology, Economics, and Population Taboos.* New York: Oxford University Press.

Hardin, G. (1998). Extensions of the tragedy of the commons. *Science, 280,* pp. 682–683.

Henderson, D. R. (1985, November 25). An ecologist has some ideas, mostly sensible, about how to see through nonsense on policy issues. *Fortune,* p. 203.

Maybury-Lewis, D. (1996, Spring). Voices from the commons—evolving relations of property and management. *Cultural Survival Quarterly 20* (1), pp. 24–71.

Meile, F. (1996). Evolutionary ethics [interview with Garrett Hardin on ecology, economy, and ethics]. *Skeptic, 4* (2), 42.

Oxford University Press–USA online. (1999, December). *The Ostrich factor: Our population myopia,* Garrett Hardin. *http://www.anb.org/docs/0195122747. html*

University of California, Santa Barbara, online. Garrett Hardin in brief. *http:// www.ucsb.edu/Veritas/SPEAKERS/Hardin.html*

—Bibi Booth

PAUL HAWKEN
(1946–)

Paul Hawken is an environmentalist, educator, lecturer, entrepreneur, journalist, and best-selling author. He is known around the world as one of the leading architects and proponents of corporate reform with respect to ecological practices. Several companies have transformed their internal corporate cultures and business philosophies toward environmental restoration as a result of his work.

Hawken has published six books in many languages including *Natural Capitalism: Creating the Next Industrial Revolution* written with Amory and L. Hunter Lovins, and national best-sellers *Ecology of Commerce* and *Growing a Business.* He has also written dozens of feature articles on subjects ranging from entrepreneurism and business management to sustainability, ecological tax reform, and corporate globalization. For more than thirty-five years, Hawken has championed socially and environmentally-responsible organizations and companies.

Since the early 1990s, Hawken has been involved with The Natural Step, an organization aimed at creating shared frameworks for understanding sustainable development. TNS was founded in 1989 in Sweden by scientist and oncologist Dr. Karl-Henrik Robèrt, and Hawken founded and chaired the United States chapter from 1994 to 1998, and has served as cochairman of The Natural Step International between 1996 and 1998. The purpose of The Natural Step is to teach and support environmental systems–thinking in corporations, cities, government, unions, and academic institutions through a dialogue process rooted in basic science. These teachings are sequenced scientific principles that provide a comprehensive basis for understanding the requirements for life on earth and, in particular, how individuals, organizations, and companies can act so that these requisites are maintained and enhanced. The curriculum was developed by a wide spectrum of scientists of varying disciplines who worked together to create a consensus description of the living systems that drive our economy and culture. Today, The Natural Step has organizations in nine countries: England, Canada, New Zealand, Australia, The Netherlands, Sweden, the United States, South Africa, and Germany.

Hawken was also the founder and CEO of Smith & Hawken from 1979 to 1991, a catalog and retail company specializing in garden and horticultural products. It began as a non-profit offshoot of Ecology Action, offering hand tools used specifically in French intensive/biodynamic gardening to the public, and grew to change the "landscape" of gardening in America by introducing European tools, techniques, plant varieties, and literature. Smith & Hawken has been cited as being one of the most environmentally-innovative companies in the United States.

During the years 1966 to 1973, Hawken founded and served as president of Erewhon Trading Company in Boston Massachusetts, the first U.S. natural-foods company to rely solely on organic, sustainable, agricultural foods and methods. Erewhon was the first U.S. company to produce organically-grown rice, and the first to manufacture organically-grown seeds for oils, pasta, nut butters, cereals, and dozens of other products.

In 1965, Hawken worked with Martin Luther King Jr.'s staff in Selma, Alabama, prior to the historic March on Montgomery. As press coordinator, Hawken registered press, issued credentials (he describes it as a battle zone in which people needed to be identified), gave dozens of updates and interviews on national radio, and acted as marshal for the final march. That same year, Hawken worked in New Orleans as a staff photographer for the Congress of Racial Equality, focusing on campaigns in Bogalusa, Louisiana, the panhandle of Florida, and Meridian, Mississippi, after three civil rights workers were tortured and killed.

Hawken has spoken, conducted research, and traveled extensively throughout the world, undertaking journeys into insurgent-held territories of Burma to research tropical teak deforestation, as well as a 1999 humanitarian/photojournalistic trek to war-torn Kosovo and Macedonia. He has served on dozens of boards of socially and environmentally-responsible organizations, received honorary doctorate degrees from Oregon State University and Northland College, named one of UtneReader's "100 Visionaries Who Could Change Our Lives" and received numerous awards over the past 15 years for excellence in business, writing, and activism. Key books and articles written by and about Paul Hawken follow.

BIBLIOGRAPHY

Bleyer, P., Cavanagh, J., Hawken, P., Hayes, R., Henderson, H., Kelly, M., Korten, D., Korten, F., Lee, T., & Wallach, L. (2000, March/April). Globalization after Seattle: A roundtable. *Tikkun.*

Hawken, P. (1983). *The Next Economy.* New York: Henry Holt and Company.

Hawken, P. (1987). *Growing a Business.* New York: Simon and Schuster.

Hawken, P. (1990, November/December). The junk (mail) stops here. *Utne Reader*

Hawken, P. (1992, April). The ecology of commerce. *Inc.*

Hawken, P. (1993, September/October). Let them eat rainforest crunch. *Utne Reader.*

Hawken, P. (1994). *Ecology of Commerce. A Declaration of Sustainability.* New York: Harper Business.

Hawken, P. (1997, March/April). Natural capitalism. *Mother Jones.*

Hawken, P. (2000, Spring). N30: On the streets of Seattle. *Amicus Journal.*

Hawken, P., Bernow, S., Costanza, R., Daly, H., DeGennaro, R., Erlandson, D., Ferris, D., Hoerner, J. A., Lancelot, J., Marx, T., Norland, D., Peters, I., Roodman, D., Schneider, C., Scyamsundar, P., & Woodwell, J. (1998, March). Ecological tax reform. *BioScience.*

Hawken, P., Costanza, R., Daly, H., Folke, C., Holling, C. S., McMichael, A. J., Pimentel, D., & Rapport, D. (2000, February). Environmental portfolio. *BioScience.*

Hawken, P., Lovins, A., & Lovins, L. H. (1999, May/June). A road map for natural capitalism. *Harvard Business Review.*

Hawken, P., Lovins, A., & Lovins, L. H. (1999). *Natural Capitalism: Creating the Next Industrial Revolution.* New York: Little Brown and Company.

Hawken, P., McKibben, B., Ehrlich, A., & Ehrlich, P. (2000, January/February). Getting it right. *Sierra.*

Hawken, P., Schwartz, P., & Olgilvy, J. (1980). *Seven Tomorrows. Toward a Voluntary History.* New York: Bantam Books.

Van Gelder, S. (1999, Summer). Corporate futures: Interview with Paul Hawken and David Korten. *YES! A Journal of Positive Futures.*

—*John Mongillo*

DENIS HAYES
(1944–   )

Denis Hayes has been an environmental lawyer, a professor of engineering, and the coordinator of Earth Day and is now the president of the Bullitt Foundation, an environmental organization. Hayes was born in 1944 and grew up in a paper-mill community nestled in the spectacular beauty of the Pacific Northwest. During high school, he worked one summer for the U.S. Fish and Wildlife Service and spent another summer at one of the first "ecology" seminars in the nation, sponsored by the National Science Foundation.

Photo courtesy of Tara Middlewood.

After Hayes's sophomore year in college, he took off for three years of hitchhiking around the world. On his travel route, he went aboard the Trans-Siberian Railway and across what was then the Soviet Union. Hayes traveled all over Africa, the Middle East, Southwest Asia, and even through war zones in Southeast Asia. He believes that he scared his parents to death. By the time he returned to school, he had decided to spend his life trying to solve some of the huge problems he had seen—problems of people literally destroying the very earth that gave them sustenance.

By the time Hayes graduated from Stanford University in 1969, he had experienced most of the passions of the 1960s. As president of the student body, he had sat in and marched and inhaled. He had watched dusty-footed flower children slip long-stemmed daisies down the rifle barrels of blushing National Guardsmen who were barely into puberty. He had witnessed out-of-control cops truncheoning unarmed, nonviolent teenagers seeking to affirm their civil rights.

As Hayes describes it, he and his friends displayed a lot of generational arrogance, nurtured by politicians and professors and others who pandered to their egos. They were the first generation raised on television, the first generation with strontium 90 in its bones. They believed that their unprecedented upbringing made them fundamentally different from earlier generations, and they proclaimed, "Never trust anyone over 30." (Hayes personally abandoned this particular view some decades ago.) Hayes noted, "We had a smug confidence that we could—and would—pass on to our kids a far better world than the one we were inheriting. Perhaps we were the last genuine optimists. Nothing was neat and tidy in the 1960s. Movement after youthful movement strode across the political stage. But occasionally we scored meaningful victories. A mobilized generation eventually brought the Vietnam War to an end. We pressured a Democratic White House to begin enforcing civil rights over states' rights, knowing it would cost the Democrats their political hegemony in the South. The lesson we took from all this was that protest worked."

Hayes believed that Earth Day was a protest, too, but a protest with a difference. Earlier movements had evolved in a direction that had begun tearing society apart. Branches of antiwar activists resembled the Red Brigades and the Bader Meinhoff Gang, blowing up buildings and attacking foes with baseball bats. Branches of the Black Power movement armed themselves, dismissed old allies, and advocated a goal of total racial segregation.

Earth Day, without compromising any core values, consciously set out in 1970 to pull people together. Not everyone saw it that way. Maurice Stans, then secretary of commerce, denounced Earth Day as the latest wave of wild-eyed radicalism by kids who had lost touch with reality. The Daughters of the American Revolution found it sinister that Earth Day happened to fall on Lenin's birthday. But for most Americans, the Earth Day message smacked of simple common sense. Earth Day staff members put a huge amount of energy into enlisting schools, colleges, churches, garden clubs, civic organizations, businesses, and most important labor organizations. Hayes comments that in today's era of "jobs versus owls" bumper strips, few remember that organized labor was the largest source of funding for the first Earth Day. Enlightened labor leaders understood that factories were the most polluted places in America. (The United Auto Workers also did as much to support the passage of the Clean Air Act of 1970 as any traditional conservation organization. Walter Reuther, the legendary president of the UAW, became convinced long before the automobile industry that cities would not long tolerate levels of pollution in which merely breathing was the equivalent of smoking two packs of cigarettes a day. "My guys can build clean cars," he once told Hayes. "Hell, we can build subways. My job is to make sure we are building things that have a future. And unless we clean up our cars, they don't have a future." The post–Earth Day retreat at which organizers from around the country decided that the Clean Air Act would be their next major objective was hosted by the UAW at its new conference center in Black Lake, Michigan.)

The message of that first Earth Day was broad enough to be broadly inclusive. Nevertheless, it was genuinely revolutionary. America was growing wealthier, but the quality of life was getting worse. There was too little connection between the way people measured progress and things people really cared about. There was a growing gap between what statisticians counted and what really counts. Earth Day organizers in different cities protested against air pollution, oil spills, vanishing wilderness, eight-lane freeways cutting through their neighborhoods, DDT, lead poisoning in America's ghettos, and dozens of other specific blights. But Earth Day was not just a collection of piecemeal protests. Earth Day was in effect a huge town meeting, asserting that there is more to the American Dream than ever-increasing consumption.

Until 1970, economists had completely dominated most domestic policy debates with theories about the impact of various policies on economic growth, capital formation, jobs, productivity, and the like. Arguably, Earth Day's most important achievement was to forcefully introduce biology into public policy. Earth Day stood for the proposition that the American Dream could not be sustained unless it reflected ecological principles. For example, Kurt Vonnegut spoke for the most of the attendees that first Earth Day when he said of President Nixon: "I am sorry he's a lawyer; I wish to God that he was a biologist. He said the other night that America has never lost a war, and he wasn't going to be the first American President to lose one. He may be the first American President to lose an entire planet." Aided by stunning National Aeronautics and Space Administration (NASA) photographs of the earth from space, Earth Day's organizers argued that humans needed to pay more attention to the web of life and to our role in it. Human well-being is linked more closely than most people realize to the great marine and terrestrial ecosystems.

Hayes believed that that first Earth Day sounded a lot like what is now termed "industrial ecology." Carbon emissions must not exceed nature's capacity to withdraw carbon dioxide from the atmosphere and fix the carbon. Logging cannot exceed the rate of tree growth and must be sensitive to the needs of complex forest ecosystems. Fish catches cannot exceed sustainable yields, and spawning habitats must be protected. Soil erosion must not be faster than soil formation. Water must not be pumped out of aquifers faster than they recharge. Hayes's standard 1970 stump speech emphasized the linkage of humans with the rest of nature:

The environment is not something "out there." We eat the environment. We drink the environment. We breathe the environment. We are part of the environment, and it is a part of us. We cannot destroy the environment without destroying ourselves.

We have been breaking a lot of little laws for a long time, and now the big laws, Nature's Laws, are starting to catch up with us. We *can't* break Nature's Laws. We can only prove them.

On that first Earth Day, a huge cross section of America came together around a common set of values, prompted perhaps by the most basic biological drive—the drive for survival. Twenty million people took part in that first Earth Day, and it created the impetus for the creation of the Environmental Protection Agency (EPA) and the passage of the Clean Air Act, the Clean Water Act, the Endangered Species Act, and most of the rest of the political framework of modern environmental policy.

In the years since then, Hayes has accomplished many different things. He has been an engineering professor at Stanford University, a lawyer in Silicon Valley, and the director of the federal Solar Energy Research Institute for the Carter administration. He has been a visiting scholar at the Smithsonian Institution and the Worldwatch Institute. Currently he is president of the Bullitt Foundation, a $100-million environmental foundation in Seattle, and chairman of the nation's largest energy-oriented philanthropy, the Energy Foundation, in San Francisco. He believes that he has fought dozens of important battles to make the world a better place and has won a fair number of them.

The values that have guided Hayes's entire adult life took root during his childhood spent in the great outdoors. They were tempered during three years of travel through the Earth's most destitute countries. They were forged into a solid framework amid the tumultuous campaign to use Earth Day to create a powerful new constituency. By and large, they have served Denis Hayes very well.

BIBLIOGRAPHY

ABC News Transcript of an interview with Denis Hayes. *http://www.abcnews.go. com/sections/science/DailyNews/chat_earthday990422.html*

Bullitt Foundation. *http://www.bullitt.org*

Hayes, D. Fight global warming. Go solar. Available: *http://www.energyfuturesnews. org/html/page17.html*

Hayes, D. How to create a solar economy in four years. Available: *http://www. wenet.net/users/verdant/solarpossible.htm*

Hayes, D. Reindeer, cars & Malthus: Population, consumption and carrying capacity. Available: *http://www.spiritone.com/~orsierra/rogue/popco/sustain/ reindeer.htm*

Hayes, D. (1977). *Rays of Hope*. New York: Norton.

Hayes, D. (1998, July 30). Energy for the new millennium; The solar imperative. Basel, Switzerland: Sun 21. *http://www.sun21.ch/98/Denis%20Hayes.htm*

Hayes, D. (1998, October 7). Earth Day 2000: New energy for a new era. *http:// www.ccearthday.org/ed2000newera.htm*

Hayes, D. (1999). Baked Alaska. Bullitt Foundation 1999 President's Report. *http:// www.bullitt.org/bottom_pres.htm*

Hayes, D. (1999, June 15). Earth Day 2000 dawn of the solar age. American Solar Energy Society conference, Portland, ME. Available: *http://www.ases.org/ conference/solar99/hayes.htm*

Hayes, D. (2000). *The Official Earth Day Guide to Planet Repair*. Washington, DC: Island Press. Available: *http://www.islandpress.org/earthday/kit/hayes.html*

Hiway9.Com. (1995). Denis Hayes, U.S. delegate to the United Nations Conference on Population and Economic Development. *http://www.hwy9.com/psp/dhayes.htm*

Outside Magazine. Online chat. *http://www.outsidemag.com/disc/guest/hayes/jonah.html*

Pilau's Environmental Heroes Collection stamps. *http://www.sites.netscape.net/planetpatriot/stamps/palau_env_heroes_4throw.html*

Time Magazine's Heroes for the Planet. *http://www.time.com/time/reports/environment/heroes/heroesgallery/0,2967,hayes,00.html*

Washington Post National Closeup. Earth Day, focus on sustainable energy. *http://www.washingtonpost.com/wp-srv/zforum/99/nation/earthday042199.htm*

—*Denis Hayes*

RANDY HAYES
(1950–   )

Randy Hayes is the founder and president of the Rainforest Action Network (RAN) in San Francisco, California. His ecological training ground was growing up in the forests of West Virginia and the swamps of Florida, and his activist training ground was documentary filmmaking in college. Hayes was born in 1950 in East Liverpool, Ohio, where he grew up and attended local schools. Hayes went to Bowling Green State University in Ohio and in college had several jobs as a truck driver, bouncer, furniture mover, and carpenter.

Hayes graduated from college in 1973 with a major in psychology. He then went to San Francisco State College and graduated with a master's degree in environmental planning at San Francisco State University. Hayes's master's thesis was the award-winning film he did with two partners titled *The Four Corners, a National Sacrifice Area*. The film documents the tragic effects of uranium and coal mining on Hopi and Navajo Indian lands in the American Southwest. This film won the 1983 Student Academy Award for the best documentary. Hayes also won the Focus Film Festival and at least five other film awards. He traveled throughout the Southwest with Indian leaders showing the film. He was asked to show the film in the U.S. House of Representatives and at the United Nations.

Hayes also coproduced the powerful short film *The Cracking of Glen Canyon with Edward Abbey and Earth First!* This film has had more than a thousand grassroots showings across the United States and in many other countries. Other film or video projects he has participated in include *Laughs, Nuclear Shadow: What Can the Children Tell Us?* and *The Cry of the Condor.*

Photo courtesy of Randy Hayes. © Rainforest Action Network.

From working with American Indians, Hayes learned about the plight of
native people in tropical rainforests. Now, as director of Rainforest Action,
he is a leader in the effort to halt the destruction of tropical rainforests and
to fight for the rights of indigenous people. In 1985, he gathered the who's
who of rainforest activists from around the world to create a more effective
global movement. In 1986, he spearheaded the first environmental demonstration against the largest financial institution on the planet, the World
Bank. In 1987, he led the first nonviolent civil disobedience against that
institution for failing to stop its destructive loans. Since that time, he has
been arrested three times in front of the bank doing nonviolent protests.
He has also joined native Hawaiians and has been jailed for protesting to
protect American rainforests. He has raised more than half a million dollars
to give away to Third World indigenous peoples and Japanese grassroots
groups to help them in their fight for survival.

Hayes has traveled, lectured, written, and demonstrated extensively on
tropical rainforests and activism. His chapters on activism and rainforests
appear in the books *Call to Action: Handbook for Ecology, Peace, and
Justice* and *Lessons of the Rainforest*, both published in 1990. He has spoken on National Public Radio nationwide and the CBS morning show and

at the Cleveland City Club, the Commonwealth Club, the Ford Hall Forum, the Harvard campus, and Stanford Business School. He works with organizers and regional networks in Africa, Australia, Latin America, Southeast Asia, Japan, North America, and Europe in building a world rainforest movement. He brings a positive approach, showing how a worldwide network of active people can protect much of the life-giving rainforest.

Hayes's message points out that second to nuclear warfare, rainforest problems are the most important issue today. He explains how Japanese, European, and U.S. overconsumption of rainforest imports, ill-spent foreign aid, and irresponsible multinational corporations are destroying rainforests at an alarming rate. More important, Hayes stresses what we can do to stop this madness. He urges communities and colleges to set up local rainforest action groups. RAN then provides ongoing support by supplying updated information and suggestions for action. Hayes is also leading a national campaign to confront the overconsumption of wood use, calling for a 75 percent reduction of wood and wood-based paper in the United States within a decade.

In a little over five years, Hayes has built one of the better-known environmental organizations. With more than 40,000 members and 150 affiliates, he can make governments and multinational corporations take notice of their actions or lack of actions. In conference after conference and local presentation after presentation, Hayes explains how even the United States has tropical rainforests and how RAN's campaign succeeded in protecting them. The slides portray the importance of rainforests to our everyday life, the beauty of plants and animals, the elements of destruction, and efforts to conserve forests.

Hayes believes that this is one of the most winnable issues around, but a massive grassroots movement is needed. This issue will sweep the college campuses in the manner of Vietnam and antiapartheid. "We must take it to the streets to make changes fast enough to halt this epidemic extinction." Hayes's presentation is more than informative; it is a call to action. That is what his life stands for.

The 1992 Columbus quincentennial celebrations spurred many to focus on the last 500 years. Hayes thinks that it is time to plan specifically for the next 500 years. Currently he is working on a 500-year plan for the future. *Smart* magazine reported that "of all the celebrities and activists in the fight, it is Hayes who is making the biggest difference."

BIBLIOGRAPHY

Erickson, B. (Ed.). (1990). *Call to Action: Handbook for Ecology, Peace, and Justice*. San Francisco: Sierra Club Books.

Head, S., & Heinzman, R. (Eds.). (1990). *Lessons of the Rainforest*. San Francisco: Sierra Club Books.

Rainforest Action Network Web site. *http://www.ran.org*

—*Randy Hayes*

HAZEL HENDERSON
(1933–   )

Hazel Henderson is an independent futurist, an environmentalist, the author of six books, a worldwide syndicated columnist, and a consultant. She serves on many boards including the Worldwatch Institute and in 1996 shared the Global Citizen Award with Nobelist A. Perez Esquivel of Argentina.

Hazel Henderson grew up in a small town of 3,000 people on the west coast of England, near Bristol. The community was based on farmland and fishing. Very early in her life, she realized that everything that people consume comes from nature. Her mother grew all of her family's fruits and vegetables without pesticides in their garden. Henderson remembered as a child being given pans and being told to go down to the garden and pull so many carrots and dig up so many potatoes. Also, her family got its milk from the farm next door and had its own chickens. They went almost every day to get fresh fish from the local fishers who were selling fish off the pier. She said that it was impossible to grow up in an environment like this and not be deeply respectful of nature. She loved riding with her pony to the top of the hills nearby where there had been a Roman fort and one could dig in the ground with bare fingers and find coins that were 2,000 years old. For Henderson, it was a very rich and stimulating environment.

Henderson did not take any further formal education after receiving her baccalaureate, which in Britain was equivalent to two years of college. She had no desire to go on to university. As she states, she could not wait to get out there in the big university of life. She began her education with all kinds of entry-level jobs, everything from being a switchboard operator to being a salesperson in a ladies' dress shop and a hotel desk clerk.

In the early 1960s, Henderson moved to New York, where she worked for American Airlines. She married and settled down in New York City. While she was raising a family there, she became very aware of New York City's air pollution after having grown up in a reasonably clean air environment. When her first child was born, she became extremely worried about the air quality, so she organized the women in the local play park in New York City and helped start a group in 1964 called Citizens for Clean Air. She believes that it was the first environmental group east of the Mississippi. Another group called Stamp Out Smog in Los Angeles was also formed by women worried about their children's environmental health. Henderson helped build Citizens for Clean Air from a very small group to

Photo by Olan Mills. Courtesy of Hazel Henderson.

a membership of 40,000. She got the mass media to run ads for the group about cleaning up New York City's air. She found herself thrown onto radio and television programs to debate representatives from the polluting companies—power companies and automobile companies. By that time, she was doing much research about pollution and particularly about what caused the air pollution in New York City. She was constantly having to deal with economists who worked for companies who would say, "Henderson's a very nice lady, but she doesn't understand economics."

First of all, she tried to find university economics courses and figured that she would go back to school and get her degree in economics. She applied to a number of different universities wanting to know if they had courses that were not just straight economics but examined how economics related to the environment and to society. There were no courses like that, so she began to study economics on her own. The more she read about economics, the more she realized that the discipline needed an overhaul in order to embrace the environment and to deal with what she later came to call the nonmoney economy, or the "love economy," to deal with all the volunteer work, all the productive work that economists did not count. The more she read, the more outraged she became. In the meantime, because of her organizing and civic activity for Citizens for Clean Air, she was awarded the Citizen of the Year award by the New York County Medical Society in 1967. She also found time to write an article of about 10,000 words that braided together her concerns about the social responsibility of corporations with their economic model of free trade that included free air and water as well as the free work of women taking care of children and elders and serving on school boards. She sent the article to the *Harvard Business Review* with a very belligerent letter to the only woman who was on the masthead. To her amazement, the article was accepted and was published in 1968. She then wrote two more articles for the *Harvard Business Review*, one on how to perfect the tools of democracy using the new communications tools people had and the other on economists versus ecologists. That article came out in 1973. By that time, the whole issue of whether to go back to school or not had been finessed. There was no point. Henderson was able to publish her thoughts quite easily.

Henderson has been associated with several mentors in her life. She believes that her chief mentor was her mother. Henderson said that her mother really "walked her talk." Her mother served on numerous volunteer activities. She always had time for the rest of the family. Her mother demonstrated how to be a good person and the kind of person Henderson wanted to be. As she was teaching herself economics through reading books, most of the economists whose work was useful to her, like Kenneth Boulding and others, were very willing to guide her over the phone. She would send her articles or drafts of her articles and say, "Am I quoting you correctly?" Henderson thought that Nicholas Georgescu-Roegen, a professor at Vanderbilt University, was really the new Keynes, even though nobody would accept him. In Henderson's opinion, Georgescu-Roegen had written what is probably the most influential book for economists today, *The Entropy Law and the Economic Process*. She became very friendly with his most famous student, Herman E. Daly, who has written books about sustainable economics and also wrote a book called *For the Common Good* with John Cobb. Henderson was also mentored by Barbara Ward, in Britain, who wrote *Spaceship Earth* in 1964, and E. F. Schumacher, who was an economist with the British Coal Board and wrote *Small Is Beautiful* in 1973.

Henderson concluded 25 years ago that the present global economy was extremely unsustainable because it uses up natural resources and converts them into undigestible wastes. She noted that nature can deal with a certain amount of human waste, but nothing like the amount we are throwing into the environment. The other problem is that the current economy widens the gap between the rich and the poor. Since publishing *Creating Alternative Futures* in 1978 and *The Politics of the Solar Age* in 1981, many other economists and other writers have echoed her analyses.

Henderson's *Beyond Globalization* is part of a follow-up to the Jubilee

2000 campaign. The goal of the group was to cancel the debt of the 40 most indebted countries in the world. These debts were quite often engendered by dictators or undemocratically elected governments working in crony situations with multinational corporations. They are what we now refer to as odious debt, and there is no justification in asking the people of these countries to pay these debts. The most realistic thing for the World Bank and the International Monetary Fund and all the lenders to do is what the private banks have already done, to write the debts off. A private bank that has made a dud loan does not go on lending money to the debtor to pay interest on that loan, which would be usury. It writes the debt off and takes a hit. This is what the World Bank and the IMF need to do.

In her book *Beyond Globalization: Shaping a Sustainable Global Economy*, which is geared for high-school and junior-college students, Henderson outlines an agenda for a sensible economic system within environmental limits. She reports that economic policy changes must center around the global human population and how it interacts with the biosphere and other species. She states that the only way we are going to reduce the human population is to give women full empowerment over their lives economically, socially, and educationally. The moment women get that kind of power over their lives, they reduce their fertility and their number of children to replacement level or even below.

Henderson also believes that the work of the United Nations is very important. She says that unfortunately, the United Nations gets caught in the cross fire between isolationists and national governments who do not want the United Nations to be fully empowered because the worldwide organization has conducted extensive research on sustainability. All of the United Nations treaties and conventions on climate change, biodiversity, forest and environmental monitoring, and other environmental issues are inimical to the interests of many global corporations. These corporations have tremendous power over their governments, and this means that the United Nations is always under attack because it champions people's rights and manages to bring the media on board in understanding what needs to be done. Thus the United Nations is very much a threat to both national governments and major multinational corporations.

Henderson sketches out all the policies that need to be changed at the national level. They include changing the gross national product toward quality of life so that it properly accounts for environmental resources and unpaid work of women and men, which the United Nations finally put a number on in 1995 and said that it was $16 trillion, measured in money terms. This amount was simply missing from the global gross domestic product that year, which was $24 trillion. That is, two-fifths (and probably more) of the productive work going on in the world is not getting into the numbers.

Governments also have to stop subsidizing unsustainability, which they still do to the tune of about a trillion dollars a year. Heavily subsidized activity by most governments does not fully account for the cost of energy to pay for the roads and harbors and airports through tax dollars. Henderson believes that if all of these subsidies were actually accounted for and added into the price of goods that are in world trade, most goods would not be in world trade. The only thing that is really efficient in world trade is services. This is the vision that Henderson talks about. "You shouldn't ship the cookies around, it's better to ship the recipes!" This is how trade can be harmonized with the environment.

Shipping goods around the world, using nonrenewable energy sources to do it, and then selling goods below cost prices do not make sense. These practices use up natural resources for unnecessary transport, put local goods at a disadvantage, and destroy local economies. Instead, we should be exchanging the recipes and the licenses for cleaner, greener technologies. As an example, most countries in the world have a car industry, and it is pure entropy to have a ship full of cars coming from Tokyo to San Francisco passing another ship of cars going from San Francisco to Tokyo. Henderson comments, "That's entropy, that's not world trade. That's not making anyone wealthier. In fact, it depletes resources and makes everybody poorer. But at full cost prices, all of those things would disappear and you would find that when economists understand efficiency in the real world way that engineers do, thermo-dynamic efficiency, you would find that local goods and services are the cheapest and most efficient. So the way world trade can be harmonized with the needs of local economies and nature is to get the prices right. Once they are full cost prices, most of this trading of goods I think will disappear."

At the national level, governments have to shift their taxes off of payrolls and incomes and onto pollution and waste and resource depletion. Then the corporate sector has to be brought under a regime of global standards, many of which they are promulgating themselves: codes of conduct, best practices, submitting to environmental and social audits, and, again, getting the prices right, so they are not undercutting local producers by these subsidies. Corporate charters have to be changed so that we go back to the original corporate ones, which had a social contract where the social purpose of the corporation was built in. We need to shift corporations away from simply servicing shareholders to balancing with all the stakeholders, including employees, customers, communities, and the environment.

There are some things that local governments can do best, such as creating pedestrian malls, dealing with zoning to make solar energy more efficient, and helping with local venture capital for green technologies and local producers. Also, local governments can encourage local currencies, because now the idea of everybody having to use the national currency puts local communities at the mercy of however good or bad the monetary policy is at the central bank.

Henderson has promoted local currencies in all her books. Henderson gives an example of a local currency that she found in Ithaca, New York, called Ithaca Hours. If there are unemployed people in a community like Ithaca and there is a local currency, people can match needs and resources with each other without having to earn the coin of the realm. The day Henderson was in Ithaca, she ate at a restaurant where she paid in Ithaca Hours. The restaurant owner could use the Ithaca Hours to buy groceries at the supermarket. The supermarket owner could use them to work out at the health club. The owner of the health club could use them to pay the points to generate a mortgage, and so on. The idea is that having an alternative currency means that one has many more options to use local markets and to match needs and resources.

Henderson also likes the use of barter and is now an advisor to a new Internet barter company. For example, there a hundreds of barter clubs, many computerized, in North America and Europe. Bartering worldwide is estimated at $2 trillion annually. In Australia Barter Card did about $2 billion worth of business in 1998. The barter card looks just like an American Express card. In Australia, there are many tiny local businesses that are separated by enormous distances, and it is much more efficient to use a barter network to exchange goods and services than it is to go through a marketplace in Sydney, which is 2,000 miles away. Bartering will be the next wave of dot.com companies.

Part of Henderson's mission for the rest of her life is to help the two billion poor people on this planet realize that they can barter from local to global. She believes that this is one of the great gifts of the information age. It is like the home shopping club of the world. Artisan-based goods that are made by people in villages or indigenous areas are put up on television or the Web in a catalog with a toll-free number. People can buy them all over the world. Henderson's idea is that they need to be auctioned as works of art in order not to overload these poor people to crank out unnecessary and excessive amounts of goods. She is working on this with WETV, a global public access television network in Ottawa. The "We" means we the people and the Whole Earth. WETV is already operating in thirty-one countries as a public-private, civic partnership with 38 independent stations, bartering air-time and programming alternatives to commercial, consumerist mass media.

Henderson believes that this is a good way of closing the gap between the rich and the poor. She says that we have the money system and we have the information system and both are equally good for trading. But we must also go beyond economics and work for human values beyond economics: an "earth ethics" that recognizes our human interdependence within nature's tolerances.

BIBLIOGRAPHY

Caracas Report on Alternative Development Indicators. (1990). *Redefining Wealth and Progress. New Ways to Measure Economic, Social, and Environmental Change.* New York: Bootstrap Press.

Cleveland, H., Henderson, H., & Kaul, I. (Eds.). (1995). *The United Nations: Policy and Financing Alternatives.* Washington, DC: Global Commission to Fund the United Nations.

Hazel Henderson Web site. *http://www.hazelhenderson.com*

Henderson, H. (1988). *The Politics of the Solar Age: Alternatives to Economics.* Indianapolis: Knowledge Systems, Inc.

Henderson, H. (1995). *Paradigms in Progress: Life Beyond Economics.* San Francisco: Berrett-Koehler.

Henderson, H. (1996). *Building a Win-Win World: Life Beyond Global Economic Warfare.* San Francisco: Berrett-Koehler.

Henderson, H. (1996). *Creating Alternative Futures. The End of Economics.* West Hartford, CT: Kumarian Press.

Henderson, H. (1999). *Beyond Globalization: Shaping a Sustainable Global Economy.* West Hartford, CT: Kumarian Press.

Henderson, H., Lickerman, J., & Flynn, P. (Eds.). (2000). *Calvert-Henderson Quality of Life Indicators*[SM]: Desk Reference Manual. Bethesda, MD: Calvert Group.

—*John Mongillo*

HUEY D. JOHNSON
(1933–)

Huey D. Johnson is the founder and president of the Resource Renewal Institute (RRI), a nonprofit organization founded in 1985 and headquartered in San Francisco. The organization supports innovative environmental management in the United States and worldwide. Its goal is to promote the implementation of green plans, which are long-term, comprehensive strategies designed to achieve sustainability. Johnson's belief is that the implementation of green plans is the path that countries should take to respond to environmental decline. Unlike conventional approaches that address environmental problems in isolation from each other, green plans treat the environment as it really exists—a single, interconnected ecosystem that can only be safeguarded for future generations through a systemic, long-range plan of action.

Johnson believes that the accelerating environmental decline in the United States and worldwide demands a powerful response. He character-

Photo courtesy of Huey D. Johnson.

izes our present response to solving environmental problems as putting "Band-Aids on broken legs." We attack the problem one part at a time. This year we will concentrate on air quality, the next year on toxic wastes, and the year after on soil conservation. Johnson's green plans work on the problems at the same time. As he states, "You need to link the environmental problems and recreate the whole picture."

Green plans are designed to solve environmental problems through a systemic, multisector process that mandates proactive management strategies and establishes measurable goals for recovery. Green plans are also comprehensive; no one popular issue dominates the overall strategy. Each issue is important as a whole. Green plans create a basis for cooperation among industry, government, and nongovernmental organizations (NGOs). Cooperation of this sort is a radical concept for sectors whose mutual antagonism has led to virtual gridlock in the environmental policy arena.

The green plans of the Netherlands and New Zealand provide excellent models of effective cooperation among the sectors. Before these countries developed green plans, their environmental policies were as conflict ridden as they are now in the United States. Through the process of developing green plans, industry, government, and NGOs came together to work toward ambitious, long-term environmental quality goals.

Johnson's favorite books are *A Sand County Almanac* by Aldo Leopold and *The Lorax* by Dr. Seuss. Both books provided inspiration to him and are sources of many of his ideals and beliefs. Indeed, today, one of his criteria for hiring young staff is whether or not they are familiar with the wisdom contained within *A Sand County Almanac*, a book that provides an intimate portrait of the natural world and the symbiotic relationship with it that man must adopt. Johnson became interested in the environment as a young child and credits his mother's encouragement as the beginning of this love of nature. From a young age, he took long walks outdoors and explored flowers, forests, and waterways.

Before founding Resource Renewal Institute, Johnson served as the Nature Conservancy's western regional director in the 1960s, building that organization's operations in the western half of the United States. In 1972, he started the Trust for Public Land, now the fifth-largest environmental organization in the United States. From 1978 to 1982, Johnson served as California's secretary for resources under Governor Jerry Brown. During this time, he was responsible for creating and implementing one of the first comprehensive approaches to resource management, the hundred-year Investing for Prosperity (IFP) program. IFP has saved billions of dollars and has created jobs by investing in the restoration and maintenance of California's renewable resources.

As resources secretary, Johnson was responsible for the conservation, enhancement, and management of the state's natural and cultural resources, including wildlife, land, water, parks, minerals, and historic sites. The agency has about 14,000 employees and a budget of over $1 billion a year. As a cabinet member, Johnson was responsible for creating and implementing one of the first comprehensive approaches to resource management. At that time, some of California's resource problems included soil erosion of farmlands, declines in salmon runs, low levels of timber production, and other problems caused by the energy crisis. Investing for Prosperity became the predecessor for Johnson's green plans. The idea behind IFP was to establish funding that would set up programs to restore the state's natural resources and to show that the returns to the state would be worth the investment. Under the program, funds made available by the legislation were invested in resource areas that included forestry and wildlands, fish and wildlife, water, soils, coastal resources, parks and recreation, and energy. The success of the IFP was due to the support of a diverse coalition of corporations, government agencies, labor unions, environmental organizations, and individuals. In March 1996, Johnson and RRI received an award from the President's Council on Sustainable Development for initiating IFP and for their ongoing work with green plans.

The Resource Renewal Institute's mission is to catalyze the development and implementation of green plans—working models of sustainability. RRI evaluates the effectiveness of existing and emerging green plans, educates government, business, and public-interest leaders about the attributes and implementation of green plans, and inspires state and community leaders to spearhead green planning processes throughout the United States. RRI's long-term goal is to have green plans implemented in each of the fifty states and, eventually, worldwide. RRI also maintains operations in Washington, D.C., New York, and Europe.

—John Mongillo

DANNY KENNEDY
(1971–)

One needs to inform oneself about issues by getting involved with community groups.

—Danny Kennedy

Danny Kennedy is the director of Project Underground (PU), an environmental and human rights organization based in Berkeley, California. Its mission is to support communities and environmental organizations that are subject to environmental threats by mining and oil industries and those who are oppressed in terms of their civil or political rights.

Kennedy was born in 1971 in the United States of Australian parents. When he was six years old, he moved back to Australia. He came back to the United States for a couple of years and also lived in England for a few years as well. As a youngster, he enjoyed reading a variety of books about world wildlife, and the stories piqued his interest in conservation issues. Kennedy calls himself a wilderness freak. He continues to enjoy bush walking and birdwatching, which inspire him and fill him with purpose.

When Kennedy was twelve years old, he became very active with the Australian Conservation Foundation, which was similar to the Sierra Club in the United States. The young Kennedy worked on an election campaign that was organizing the "green" votes for a federal election in Australia. At that time, the government in power was supporting a proposal for the Franklin Dam in Tasmania. The opposition Labor party was also running in the election. The Labor party sought the votes of the "greens" and would block the dam if it were elected. As a result, the environmentalists put their votes behind the Labor government, which was elected due to the large support of the green vote. Kennedy's experience going door to door in his neighborhood and passing out leaflets at the ballot box gave him his first taste of what good organizing skills can do to support key objectives. It was a successful experience for him.

Kennedy went to high school in Australia and then on to Macquarie University there. While Kennedy was at the university, he worked with an international student network that called for solidarity, equality, environment, and people-centered development. He was also the founder of a youth organization in Australia that was under the umbrella of the Australian Conservation Foundation called Green Youth Action (GYA). He also spent three or four months in the field in Papua New Guinea while working for Greenpeace around the oil industry. He graduated in science, with a degree in resource management. He studied for a while to be a lawyer but decided to quit so he could become more involved in activism.

During Kennedy's young career, mentors have played an important part in his life. One he admired was Jack Mundy, an Australian environmentalist who was also a labor-union leader. Mundy was well known in Australia for being a leader of the Green Bans, a movement of unions in Australia that do not work on sites that are culturally and environmentally sensitive. Another mentor was Peter Garrett, who was lead singer for Midnight Oil and president of the Australian Conservation Foundation. Kennedy also speaks highly of others he has been exposed to especially indigenous communities fighting unfair practices of forestry and mining interests.

Kennedy came back to the United States and founded Project Underground in 1996. At that time, there was no organization in the United States with the technological, legal, and scientific capacity to work on environmental issues related to mines and oil-drilling operations. Project Underground is currently funded about 30 percent by private individuals and about 70 percent from foundations. The project works in a number of ways. It has partnerships with different communities and conducts campaigns against certain corporations at different times. It uses a very sophisticated information program that includes producing a newsletter every two weeks explaining struggles with companies around the world. Its Web site includes a database of mining and oil companies. Project Underground is supported by students, particularly those attending the University of California at Berkeley. Some of the students organize protests at shareholder meetings of different companies around the world.

In 1996, Project Underground was contacted by a community on the Caribbean island of Dominica, a single nation of 70,000 people. It reported that an Australian copper mining company wanted to develop a copper mine on the island. Although Kennedy knew nothing about the Caribbean, he did know a lot about the Australian copper-mine operators. Kennedy had campaigned with the Australian Conservation Foundation against the mining conditions of this company when it was operating in Papua New Guinea. Kennedy sent a friend of his, an ecologist and mining expert, to the island to look at the plans the company had written for the government of Dominica. Kennedy's expert evaluated the plan and reported that it was going to be another unmitigated environmental disaster. The company plans proposed a pit that would have covered 10 percent of the surface area of the country of Dominica. All tailings (waste rock) would be dumped into the coral reefs adjacent to the island. Project Underground informed the community of these plans and gave it some information about the company's track record in Papua New Guinea. The community group, the Dominican Conservation Association, started to organize heavily and did grassroots work, running a petition drive that got approximately 15,000 of the entire population of 70,000 signed on to oppose the mine. Project Underground members also attended the company's shareholder meeting in Sydney, Australia. There they confronted the CEO, who stated that he would not proceed with the operation without the prior informed consent of the community. The island politicians were also getting cold feet about the plan, and within six months the company announced that it was pulling out of Dominica. There would be no copper mining on the island of Dominica.

Kennedy also recalls a tragic story of people being repressed by their government in collusion with oil companies. In Nigeria in 1995, about 1,000 Ogoni were forced to flee to neighboring countries as refugees because of their opposition to oil development in their country. About 800 of them ended up in a refugee camp in Benin, which is next door to Nigeria. In 1998, Kennedy visited the camp in Benin and was appalled by what he saw. The camp consisted of young and old people as well as leaders from some of the villages affected by Shell Oil. They were not receiving enough food and medicine, and there were no educational opportunities. The camp was poorly organized. Project Underground hired an investigator to document the conditions of the camp, her account was published in a report that was circulated to many sympathetic community groups around the world. Project Underground also conducted a postcard campaign targeting the United Nations high commissioner for refugees in Geneva, who was responsible for administering the camp. In April 1999, Project Underground visited with the United Nations high commissioner's representative in New York at the United Nations headquarters to discuss the conditions of the refugees. By July, Project Underground got word that 600 of the refugees were being resettled in the United States. It was another success story for the work of Project Underground.

Kennedy's advice for young people is to get active locally at first, learn all the rules of the political power games, and then commit to one of the local organizations that is fighting to save the local greenbelt, stopping urban sprawl, or challenging various social injustices. Kennedy believes that one cannot learn all about these issues in high school. One needs to inform oneself about issues by getting involved with community groups. By doing so, one gets engaged and starts to learn the ropes of how change is made. In this way, one is involved in a social movement, not just an environmental issue or human rights.

BIBLIOGRAPHY

Drilling Papua New Guinea: Chevron comes to Lake Kutubu. (1996, March). *Multinational Monitor, 17* (3).

Kennedy, D. (1996). Kutubu: A case-study of unsustainable development. In J. Connell & R. Howitt (Eds.), *Resources, Nations, and Indigenous Peoples: Case Studies from the Asia-Pacific.* New York: Oxford University Press.

Kennedy, D., Chatterjee, P., & Moody, R. (1998). *Risky Business: An Independent Annual Report on PT Freeport Indonesia.* Project Underground.

Loosening the golden handcuffs. (2000, July/ August). *Washington Monthly, 32* (7/8).

Project Underground Web site. *http://www.moles.org*

—John Mongillo

HELEN SUNHEE KIM
(1963–   )

Helen Sunhee Kim is a community organizer of Asian Immigrant Women Advocates (AIWA). Kim was born on July 10, 1963, in Seoul, Korea, the oldest of four children. Her family immigrated to the United States when she was twelve years old and settled outside Chicago. There were many challenges for her family when they immigrated, especially for her parents, whose education and work experience meant nothing in the United States. For many years, they worked at several minimum-wage jobs to support the family. Sometimes, because Kim did not speak English and because she was a person of color, people sneered at her or made racist remarks. Her background as an immigrant, as a person of color, and as a woman has profoundly shaped her world view and politics and has also impacted her work as an environmental justice activist and as a community organizer.

After attending secondary schools in Des Plaines and Park Ridge, Illinois, Kim went to Carleton College in Northfield, Minnesota. She majored in history with a concentration in East Asian studies. During college, she began to understand that racism in America is not just personal; it is deeply institutional. From the genocide of Native Americans and the enslavement of African Americans to the incarceration of Japanese Americans during World War II, U.S. history was and continues to be marred by racism. During one of her college summer breaks, she interned at the Chicago Mayor's Office Asian American Liaison Desk, whose role was to facilitate better understanding between Asian Pacific Islander and other communities as well as to advocate for the rights of Asian immigrants.

Following a brief period of working in the media in Seoul, Korea, and finishing law school at the University of Minnesota, Kim moved to northern California in 1991. She wanted to work in the immigrant community and

Photo courtesy of Helen Kim.

among more Asians and other people of color. Early on, her parents had emphasized the importance of giving back to the community. Growing up as an immigrant, as a person of color, and as a woman, Kim had personally experienced and observed many small and big injustices around her, so she wanted to find work that was politically meaningful for her. She believes that getting involved in the community through volunteer work or internship is a great way to find out what one wants to do. She began volunteering at an organization called Asian Immigrant Women Advocates (AIWA; email: aiwa@igc.org) and a few months later was hired as its community organizer.

Founded in 1983, AIWA is a grassroots, community-based organization whose mission is to empower disenfranchised low-income Asian immigrant women who work as garment workers, electronics assemblers, hotel-room cleaners, and janitors in the greater San Francisco, Oakland, and Santa Clara County Bay Area. Asian immigrant women face discrimination as immigrants, people of color, women, and workers, as well as from language and cultural barriers and lack of knowledge of U.S. institutions.

AIWA's struggle for environmental justice stems from its belief that people have the right to live and work in a safe and healthy environment. It identifies where workplace environmental and economic injustices exist and fights to improve these conditions through education, advocacy, and collective action. As a part of her work with AIWA, Kim conducted training workshops on community organizing as well as on environmental and economic justice issues at many college campuses across the country. She recruited many young people to work as volunteers and interns at AIWA and other community organizations so they could understand that in their own communities, the issues of justice and the environment intersect, and that there is much to be done.

From 1992 to 1996, Kim coordinated a national support network for the Garment Workers Justice Campaign. In 1996, the leadership of Asian immigrant garment workers in Oakland, California, made a historic impact on garment workers throughout the San Francisco Bay Area and globally when they signed an agreement with manufacturer Jessica McClintock, Inc., who had been the workers' target in a campaign lasting three and a half years. AIWA's Garment Workers Justice Campaign was launched in 1992 to demand corporate responsibility from million-dollar garment manufacturers, who had built their fashion empires off the backs of immigrant and Third World women workers. The campaign started in 1992, when a group of garment workers came to AIWA with bounced paychecks. The workers' employer was a contractor for Jessica McClintock, Inc., but went out of business without paying the workers. This is not an uncommon occurrence in the garment industry, where workers are spending six to seven days a week, ten to twelve hours a day sewing and being paid less than the minimum wage. Working conditions are unsafe—blocked fire exits, fire hazards, exposure to dyes and bleaches, and repetitive stress injury from sewing.

Over the next three and a half years, the Garment Workers Justice Campaign mobilized people across the nation. Supporters from labor, environmental justice groups, religious communities, and high schools and colleges walked picket lines in thirteen cities, sent delegations, wrote letters, and donated time and money to support the garment workers. In March 1996, AIWA and Jessica McClintock, Inc. signed a "historic agreement" to uphold the working conditions of garment workers. In a press release, then U.S. Labor Secretary Robert Reich announced that (1) the Garment Workers Education Fund would be set up for workers and their families, and (2) workers would have a toll-free confidential hotline to call and report workplace problems, holding the manufacturers accountable. The hotline is a significant tool to educate more immigrant workers about their rights.

The domino effect of the agreement followed the next year in February 1997, when AIWA and garment workers demanded meetings and successfully negotiated with three other major San Francisco Bay Area garment manufacturers, Esprit de Corp, Fritzi California, and Byer California. All three companies agreed to take corporate responsibility for the garment workers who sew their labels by establishing toll-free, confidential hotlines for workers. In total, the four agreements affect an estimated 2,000 garment workers in the San Francisco Bay Area.

According to Kim, just south of San Francisco, in Santa Clara County, better known to outsiders as Silicon Valley, there is another kind of sweatshop industry. In the electronics industry, many low-income immigrant women who speak limited English work in small to medium-sized contractor shops, often set up in warehouses without adequate ventilation. The women spend eight to fourteen hours a day handling hazardous chemicals and inhaling toxic fumes as they assemble, clean, and test printed circuit boards for everything from watches to computers. According to a 1990 research study from the U.S. Department of Commerce, the high-tech industry has become one of the most hazardous for both workers and the communities: (1) the California Division of Industrial Relations receives three times as many reports of worker illness from this industry as the average for the manufacturing industry as a whole; (2) the rate of systemic poisoning is twice the state rate; (3) a growing number of workers are permanently disabled due to exposure to toxins; and (4) there are higher incidences of birth defects and miscarriages among workers in this industry and among families in neighborhoods where toxins have contaminated the groundwater. In fact, these workers are on the front line of toxic poisoning, before the toxins spill out to the surrounding community.

AIWA's Environmental Justice Project seeks to counter this trend. AIWA has educated and organized Asian immigrant women electronics assemblers about toxic chemicals and hazards in the industry. AIWA seeks to promote active participation of immigrant workers in the electronics industry in the struggle for safe working conditions and in the demand for policy and institutional changes that will protect the health and welfare of workers, families, and the surrounding communities. To this end, AIWA has successfully completed the first year of a two-year project called Peer Health Promoter Network (PHPN). This first phase was focused on advocating immigrant women workers on health and safety issues and training them to become peer trainers so they can train their coworkers. An initial group of women was trained by public and occupational health specialists on the various injuries workers in the electronics assembly industry commonly experience. Additionally, these women learned about effective prevention techniques and the responsibilities of the employers for their health and well-being. Participants in the first training went on to train their peers, making use of the same techniques to pass on the same information. The ripple effect continues in workplaces throughout the Bay Area.

AIWA has also been conducting research on the chemicals used in the electronics assembly plants in Silicon Valley. Its reports indicate that the workers have long complained of persistent headaches, dizziness, and memory loss associated with the chemicals they use. Yet the women often do not know what the chemicals are. In order to solve some of these problems, in the second year of the project, AIWA plans to meet with employers in the electronics industries to discuss and recommend plans that can be instituted to establish occupational safety and health in the workplace. They hope to press for a complete and open list of chemicals used in the factories and the substitution of nontoxic chemicals in place of ones that are harmful.

AIWA is also an affiliate organization of the Southwest Network for Environmental and Economic Justice (SNEEJ), which is a network of grassroots organizations of people of color working on environmental and economic justice issues. Kim served as cochairs of the Workers Justice Taskforce and the Sustainable Communities Campaign of SNEEJ. Through SNEEJ, she has had the opportunity to work with many Chicano/Latino, African-American, and Native American environmental justice activists. Kim believes that there is an immeasurable value in learning from our elders and in learning about the history of people's struggles that are too often omitted in our history books. From them and from her work with AIWA, she has also learned how important it is to organize people and give them the voice to demand their rights. For too long, people of color and poor people have taken the brunt of toxic industries and dumping in their communities and hazardous conditions in their workplaces. One of the basic principles of the environmental justice movement is "We speak for ourselves." Through her work with AIWA and in the environmental justice movement, Kim has found a place for herself to channel her anger at injustice and work toward building the power of the people.

The experience of AIWA's Garment Workers Justice Campaign, in which leadership of immigrant garment workers was forged with the overwhelming support and solidarity of their allies, showed Kim how a traditionally disenfranchised population can not only gain power, but have a wide-ranging impact locally and nationally. Members of AIWA will continue to build on their past experience as they demand environmental and economic justice for workers and their community. But Kim knows that this will take the effort and resources of many people. This is why she wants to emphasize the importance of a long-term vision and commitment to any cause one takes up. It is easy to get excited about some issue or one event, but real changes do not happen overnight. Working for social change is very gratifying, but also is not easy. She emphasizes that change does not just happen; people have to fight for it. Kim reiterates that it takes a clear vision, a lot of support and hard work, and a belief in oneself and in the people. We have a responsibility to ourselves and to one another to strive toward a world that respects the life and dignity of all people and beings. On the front of her desk is a quote from the Talmud that says, "Look ahead. You are not expected to complete the task. Neither are you permitted to lay it down."

BIBLIOGRAPHY

Kim, H. (1994). Justice for workers in sweatshops. In R. D. Bullard (Ed.), *People of Color Environmental Groups Directory 1994–95*. Environmental Justice Resource Center at Clarke Atlanta University.

Omatsu, G. (1994). "The four prisons" and the movements of liberation: Asian American activism from the 1960s to the 1990s. In K. Aguilar-San Juan (Ed.), *The State of Asian America: Activism and Resistance in the 1990s*. Boston: South End Press.

Shaffer, G. (1994, Winter). Asian Americans organize for justice. *Environmental Action Magazine*, p. 30.

Shin, Y. (1992). Environmental concerns of immigrant Asian American women. In C. Lee (Ed.), *Proceedings: The First National People of Color Environmental Leadership Summit*. United Church of Christ Commission for Racial Justice.

—Helen Sunhee Kim

WINONA LaDUKE
(1959–)

> [Americans should] consider an amendment to the U.S. Constitution in which all decisions made today will be considered in light of their impact on the seventh generation from now. That is . . . what sustainability is all about.
>
> —Winona LaDuke

Winona LaDuke is one of the most prominent Native American environmental activists today. Her work has been devoted to turning society "from the synthetic reality of consumption and expendability to the natural reality of conservation and harmony." LaDuke believes that science can and should incorporate traditional aboriginal knowledge and respect for the land and all living things. LaDuke is an enrolled member of, and activist for, the Mississippi Band of Anishinabeg (also known as the Ojibwe or Chippewa) of northern Minnesota. She serves on the board of directors of Greenpeace USA and is board cochair of the Indigenous Women's Network (IWN), an international organization that works in rural and urban communities to apply indigenous values to the resolution of contemporary problems. LaDuke also publishes the journal *Indigenous Woman* and serves as a program officer for Honor the Earth, a foundation controlled by Native Americans. Both throughout the United States and internationally, LaDuke is celebrated as a voice for American Indian economic and environmental concerns.

Winona LaDuke was raised by radical activist parents in a Native American neighborhood of East Los Angeles. Her father, Vincent LaDuke, was a Hollywood supporting actor as well as an Anishinabeg spiritual philosopher and writer known as "Sun Bear." When LaDuke was born in 1959, her father enrolled her in his tribe, as was customary for his people. From the time she was a young child, she attended local powwows with him. LaDuke's mother, Betty, a Jewish art professor, painter, and activist, encouraged her to identify with her paternal Native American heritage. "She was very political and very determined," says LaDuke of her mother. "She's very supportive of what I do" (Nelson). LaDuke credits both her parents for passing on their spirit of activism to her. In 1964, when her parents separated, LaDuke moved with her mother to Ashland, Oregon, where she increasingly perceived herself to be isolated by her heritage. "I grew up thinking of myself as Indian," she explains. "I was the darkest person in my school. You just know you don't fit in" (Nelson).

In 1976, Harvard University, impressed with her outstanding test scores and high-school grades, recruited LaDuke. She says that at that time, "all the good schools were looking for Indians. It was kind of the thing" (Nelson). When she joined a small Native American group on the Harvard campus, to her great relief, LaDuke felt that she finally "belonged." Early in her college career, she performed some research on issues related to mining on American Indian reservations for the International Indian Treaty Council, the first native peoples' nongovernmental organization to be part of the United Nations. The organization asked her to present the results of her research to the United Nations when she was only eighteen years old.

LaDuke became further involved in Native American environmental issues after meeting Jimmy Durham, a well-known Cherokee activist and artist, at a Harvard lecture. LaDuke recalls, "Basically, he said, 'There's no such thing as an Indian problem—it's a government problem.' When I heard that, it shook something loose in me. It changed my life." She immediately began working for Durham, investigating the health impacts of uranium mining on Navajo reservations. LaDuke spent time in South Dakota, New Mexico, and Arizona on additional projects with grassroots Native American organizations.

After graduating from Harvard in 1982 with a degree in native economic development (she later also obtained a master's degree in rural development), LaDuke moved to the White Earth Reservation (the Anishinabegs' ancestral lands) in northern Minnesota, one of the poorest areas of the country. She accepted the position of principal of the local reservation school. "I am my father's daughter in terms of identity and culture," LaDuke commented at the time. "This was his land, and now it is mine. I've always wanted to come home to my dad's reservation. But I also have a lot of my mother in me, too—her values, her spirit" (Nelson). She began to learn the tribal language, Ojibwe, and stayed on after the job ended. LaDuke eventually became involved in a lawsuit to recover lands that had been illegally taken from the Anishinabeg people by the federal government and land speculators. As some of the lands are contained in what is now the Tamarac National Wildlife Refuge, the Sierra Club and other main-stream environmental groups opposed the tribe's efforts.

The lands in question had been reserved for LaDuke's tribe in an 1867 treaty; however, questionable sales and foreclosures eventually reduced Native American ownership of the reserved lands to less than one-tenth of the intended acreage. In 1988, after losing the four-year suit and exhausting nearly all available legal remedies, LaDuke founded the White Earth Land Recovery Project and began the difficult work of raising funds to recover the 800,000 acres that had been taken from the reservation. The first 1,000 acres, which are now held in trust, were purchased by the Land Recovery Project with a $200,000 human rights award granted to LaDuke by the Reebok Corporation. Project leaders expect to acquire an additional 30,000 acres over the next ten to fifteen years through direct purchases, bequests, and legislation.

LaDuke maintains that if the members of a people do not have control of their land, they do not control their destiny, and that "decision-making should not be the exclusive right of the privileged. . . . [T]hose who are affected by policy—not those who, by default, often stand above it—should be heard in the debate" (LaDuke 1996a). The White Earth Reservation is using its lands for the Native Harvest project, which encourages a diverse economic base through such enterprises as forestry, maple-sugar production, and the cultivation of corn, raspberries, and wild rice. Tribe members are also strengthening their native culture by teaching the Ojibwe language, culture, and history and by preserving the natural environment that lies at that culture's spiritual core. Though LaDuke's forcefulness and frankness have won her widespread admiration, they have also drawn criticism. Even on the White Earth Reservation, some regard her with suspicion because she works outside the tribe's traditionally all-male council.

In 1986, while attending a Native American conference in Toronto, Canada, LaDuke met Randy Kapashesit, a representative of the Cree tribe from Moose Factory, Ontario. They married in 1988 and had two children. Their long-distance, commuter relationship eventually deteriorated, however. "I couldn't give up my life, because in a little while I'd be angry," LaDuke says. "I couldn't give up who I am and what I believe for my husband" (Nelson). The couple separated in 1992.

LaDuke later founded the Indigenous Women's Network (IWN), which she led to 1995's Fourth World Conference on Women in Beijing, China. The IWN works on a variety of projects, including research into small-scale, sustainable industries and alternative energy resources for adoption on reservations. At the official governmental forum in Beijing, the IWN pressed two major issues. First, members wanted the United Nations to legally acknowledge indigenous "peoples," as opposed to "populations" or "people." Such acknowledgment would guarantee them an enhanced degree of international recognition and legal rights. Second, they urged world governments to adopt the Declaration of the Rights of Indigenous Peoples, a comprehensive legal document that had been drafted by indigenous organizations in Geneva over the preceding several years.

In October 1994, LaDuke was arrested with four other activists in Los Angeles while protesting the clearcutting of forests near Clayoquot Sound, British Columbia, for pulp to manufacture telephone books. The demonstrators nevertheless succeeded in shutting down the GTE telephone company's printing plant for five hours, and LaDuke spent four hours in jail. LaDuke has also lashed out at the general consumer culture that she believes currently permeates America and contributes to the "corporatization of our democracy" (Schierenbeck).

In the spring of 1995, LaDuke organized and hosted a national tour of the popular singing act the Indigo Girls. Known as the "Honor the Earth Tour," the series of concert events was organized to raise money for grassroots organizations. The tour raised $250,000, and it was estimated that over 40,000 individuals around the country had participated. In March 1995, LaDuke was named by *Time* magazine as one of 1994's fifty most promising leaders under the age of forty.

In the 1996 and 2000 American presidential campaigns, LaDuke served as Ralph Nader's running mate in the Green party, which in 1996, excluding the Reform party, garnered the most support of all independent parties—over a half million votes. The Green party presidential campaigns had three main goals: to use the tools of democracy to extend the power of citizens and facilitate the banding together of people, to help galvanize a new generation of activists who wanted to reform politics, and to inform Democrats and Republicans that they could no longer tell millions of disaffected Americans that they had nowhere to go politically. In announcing her 1996 candidacy, LaDuke said that Americans should "consider an amendment to the U.S. Constitution in which all decisions made today will be considered in light of their impact on the seventh generation from now. That is . . . what sustainability is all about" (LaDuke 1996b). LaDuke was named a *Ms.* magazine "Woman of the Year" in 1997.

LaDuke has written two books and numerous articles about Native American economic, environmental, and land-use issues. In her 1997 historical novel, *Last Standing Woman*, LaDuke traced the life and times of seven generations of Anishinabeg. The story begins around 1862 with the Sioux uprising in Minnesota and chronicles its impact on the Anishinabeg community. The story ends in 2018, about seven generations later. It is notable that parts of the novel depict an optimistically harmonious future, a purposeful strategy on LaDuke's part. "A lot of Indian writing is history," LaDuke has said. "I think Indian people need to be in the future, too" (Cronin).

In 1998, LaDuke was one of five recipients of Global Green USA's Third

Annual Millennium Awards, which recognize those who foster a global value shift toward the environment. LaDuke received the organization's Award for Individual Leadership. In honoring LaDuke, the group's director stated, "LaDuke has worked across the world and in her own backyard to empower others to reconnect humankind to the earth" (Business Wire).

Currently, LaDuke still lives in Fergus Falls, Minnesota, on the White Earth Reservation. Her latest work, *All Our Relations: Native Struggles for Land and Life*, an in-depth account of Native American resistance to environmental and cultural degradation, was published in late 1999. In this volume, LaDuke speaks forcefully for native self-determination and community and describes her vision of political, spiritual, and ecological transformation.

"Things take a long time to fix," LaDuke says. "If it takes 100 years to take back the land, then that is the way it is. Just don't let them have the power that they make you feel useless, disgusted, powerless. I just turn my back and do my own thing" (Cronin).

BIBLIOGRAPHY

Barsamian, D. (1998). Being left: Activism on and off the reservation; David Barsamian interviews Winona LaDuke, Z Magazine. Available: *http://www.lol.shareworld.com/zmag/articles/being_leftja98.htm*

Business Wire. (1998, October 6). Five win Global Green USA's prestigious Millennium Awards for environmental activism. Available: *http://www.elibrary.com*

Cronin, M. E. (1998, April 23). Activist/author looks to the future. *Seattle Times*. Available: *http://www.seattletimes.com/news/lifestyles/html98/altduke_042398.html*

Green Party, USA, National Green Program. (n.d.). Indigenous People. Available: *http://www.greens.org/gpusa/p_indig.html*

Greenpeace. (1994, October 18). Four arrested in L.A. for concern over Clayoquot. Available: *http://nativenet.uthscsa.edu/archive/nl/9410/0272.html*

Honor the Earth online. *http://www.honorearth.com/newwinona.html*

LaDuke, W. (1994). Native Americans and the environment: Native environmentalism. Rice University Center for Conservation Biology *Cultural Survival Quarterly, 17*. Available: *http://conbio.rice.edu/nae/docs/nat_env.html*

LaDuke, W. (1995, August 31). The Indigenous Women's Network: Our future, our responsibility [plenary speech at the Non-Governmental Organization Forum of the United Nations, Fourth World Conference on Women, Beijing, China]. Available: *http://www.igc.org/beijing/plenary/laduke.html*

LaDuke, W. (1996a, August 29). Address at the press conference on her vice-presidential candidacy in St. Paul, Minnesota, on August 29, 1996. Available: *http://ic.net/~harvey/greens/laduke.txt*

LaDuke, W. (1996b, November/December). The growing strength of native environmentalism. *Sierra, 81*, pp. 38–45.

LaDuke, W. (1999). *All Our Relations: Native Struggles for Land and Life*. Cambridge, MA: South End Press.

LaDuke, W. (1997). *Last Standing Woman*. Stillwater, MN: Voyageur Press.

LaDuke, W. (n.d.). Women writers of color. Available: *http://voices.cla.umn.edu/authors/WinonaLaduke.html*

Mazza, P. (n.d.). Native activists call on environmentalists to surrender legacy of conquest. Available: *http://www.tnews.com/text/laduke.html*

Muldrew, F. (1996, June 1). Winona LaDuke: A Mother Earth revolt. *Contemporary Women's Issues Database*, p. 23.

Natural Resources Defense Council online. Winona LaDuke. *http://www.nrdc.org/sitings/lookup/biola.html*

Nelson, M. R. M. (1994, November 28). Indian activist Winona LaDuke fights to return tribal lands to her people. *People*, p. 165.

Ordower, G. (1999, April 23). Environmentalist tells Northwestern U. students to follow "green path." *Daily Northwestern*/University Wire. Available: *http://www.elibrary.com*

Owen, R. (1999, March 1). Put up your dukes [interview with Winona LaDuke]. *Contemporary Women's Issues Database*, p. 28.

Paul, S., & Perkinson, R. (1995, October 1). Winona LaDuke [interview]. *Progressive, 59* (10), pp. 36–39.

Schierenbeck, T. (1997, April 10). Speaker targets Native American issues; hopes for change, control of decision making. University of Wisconsin, River Falls, *Student Voice*. Available: *http://www.uwrf.edu/student-voice/issues/1997/1997apr10/972207.html*

South End Press online. *All Our Relations: Native Struggles for Land and Life*; Winona LaDuke. *http://www.lol.shareworld.com/SEP/AllOurRe.htm*

Walljasper, J. (1996, January/February). Celebrating hellraisers: Winona LaDuke. *Mother Jones*. Available: *http://www.mojones.com/mother-jones/JF96/anniversary/laduke.html*

Washington State Campaign for Democracy online. Winona LaDuke biography. *http://www.nader96.org/winona.htm*

—*Bibi Booth*

CHARLES LEE
(1950–   )

Charles Lee is an environmental justice activist and advocate. Lee, a Chinese American, was born in Taiwan and came to the United States when he was eight years old. He grew up in New Jersey and New York and attended Harvard as an undergraduate. He later finished his bachelor's degree at the State University of New York.

During Lee's early years, he wanted to be an architect and planner. But part of Lee's generation became very socially conscious in the 1970s. Lee believes that living during the 1970s helped him become an activist. He began to think about the issues related to race, poverty, and the environment.

Lee was further motivated by an incident that occurred in Warren County, North Carolina, in 1982. The county was planning to build a landfill for PCB (polychlorinated biphenyl) wastes. Many groups felt that the proposed site was racially motivated and posed an environmental health problem for the residents. As a result, there were widespread protests, including marches. More than 500 demonstrators were arrested. Although the protestors were unsuccessful in blocking the PCB landfill, they brought national attention to inequities in siting of waste facilities and galvanized African-American church and civil rights leaders' support for environmental justice. Lee believes that the incident was observed as a juncture that transformed local environmental issues in poor minority communities into a national movement of environmental justice.

When Lee was in his late twenties, he went to work in New York for the United Church of Christ Commission for Racial Justice (UCC/CRJ). The UCC is a 1.6-million-member denomination headquartered in Cleveland, Ohio, that is socially active and socially conscious on a national level. It assists other grassroots organizations of people of color on issues related to air and water pollution, toxic-waste sites, incinerators, and landfills. Lee was hired as the research director for the civil rights agency of the UCC/CRJ.

At the United Church of Christ, Lee directed the environmental justice program that addresses minority issues. He was the principal author of the first national study on the demographic patterns associated with location of hazardous-waste sites, *Toxic Wastes and Race in the United States: A National Report on the Racial and Socioeconomic Characteristics of Communities with Hazardous Waste Sites*, published in 1987. It is considered a landmark study in the field, a seminal work that gave rise to much national attention. Lee also pulled together and organized the first environmental leadership of people of color, which gave rise to and crystallized the movement around environmental justice.

Lee did much work with communities, doing technical assistance advocacy on the local, regional, and national levels. He served on numerous committees for bodies including the National Academy of Sciences, the Institute of Medicine, the Environmental Protection Agency (EPA), other federal agency panels, and numerous organizations. Many see Lee as one of the pioneers of the whole environmental justice movement.

Lee's observations include the need to enforce the present environmental laws that address the kinds of issues that arise in environmental justice issues, some of which are regulated, while others are not. As he states, certainly, equal protection under the law is a major concern and central issue in the environmental justice movement. Environmental justice is also a land-use and economic development issue, according to Lee. One needs to deal with the conflict between environmental protection and economic development, as both are needed within environmentally degraded and economically depressed communities, which are the environmental justice communities. It is essential to find a way toward developing a balance between the two, addressing these concerns within communities so they are truly healthy and sustainable. Lee thinks that economic development and environmental protection are not antithetical to each other. It is his feeling that the mainstream environmentalists forget about the needs of the working poor or communities of color. That does not mean that one needs to develop practices that are environmentally destructive. However, it does require a certain amount of thinking, planning, education, and assistance. One must achieve a balance between them. Often, in a mainstream environmental view, these issues are forgotten. One must also understand that in many of those communities where environmental neglect is the greatest, people are taken advantage of. One cannot talk about the environment without talking about the economy.

Environmental justice also addresses issues of disease prevention and health improvement in communities. Lee further states that U.S. environmental issues in communities of color have an intrinsic tie to communities in other parts of the world. Lee is confident that over the next decade there will be many more relationships and many more common issues on an international level for people of color.

Lee is currently the associate director for policy and interagency liaison with the Environmental Protection Agency Office of Environmental Justice. He is primarily responsible for developing the policy agenda and other types of efforts that will integrate environmental justice and the policies, programs, and activities of the EPA and other federal agencies. Lee states that environmental justice is very broadly defined because it includes not just the physical environment but also issues of the social environment, including economic concerns. Environmental justice involves issues like transportation, housing, and health care, all of which are interrelated. That is why Lee believes that one has to develop a very comprehensive, holistic, integrated approach toward addressing these issues of these communities.

BIBLIOGRAPHY

United Church of Christ Commission for Racial Justice. (1987). *Toxic Wastes and Race in the United States: A National Report on the Racial and Socioeconomic Characteristics of Communities with Hazardous Waste Sites.* National United Church of Christ.

—*John Mongillo*

ALDO LEOPOLD
(1887–1948)

> That land is a community is the basic concept of ecology, but that land
> is to be loved and respected is an extension of ethics.
>
> —Aldo Leopold

Aldo Leopold was a biologist and conservationist who cofounded the Wilderness Society and is considered to be the father of wildlife ecology. His book of nature sketches and philosophical essays, *A Sand County Almanac*, published posthumously in 1949, has influenced the way generations of hikers and environmentalists have looked at the world. Leopold won over other ecologists to his holistic view of nature, and the book became the bible of environmental activists of the 1960s and 1970s. It is often acclaimed as this century's literary landmark in conservation. With the precision of a scientist and the sensitivity of a poet, Leopold documented the emotional connections between humans and the natural world.

Leopold was one of the first scientists to promote the importance of wildlife diversity for ecosystem health. His *Almanac* writings on the "Land Ethic" and his theories on humanity's place in the complex web of nature have served as the foundation for much of the ecological movement. According to Leopold, "All ethics so far evolved rest upon a single premise that the individual is a member of a community of interdependent parts. . . . The land ethic simply enlarges the boundaries of the community to include soils, waters, plants, and animals, or collectively: the land" (Leopold 1968, 203–4). Furthermore, he said, "That land is a community is the basic concept of ecology, but that land is to be loved and respected is an extension of ethics" (viii–ix). Throughout his life, Leopold persevered in his personal intellectual quest to better understand the land community and his own participation in it. Rather than viewing nature as an object to be manipulated for utilitarian purposes, Leopold made a case for granting moral status to the "land community" as a whole.

Aldo Leopold was the oldest of four children of Clara Starker and Carl Leopold. The well-off Leopold family lived in Burlington, Iowa, in a house located on top of a limestone bluff overlooking the Mississippi River. The big river was a migratory pathway for a quarter of the continent's waterfowl and provided a year-round natural playground for a young boy. From childhood on, Leopold kept a journal, detailing his emotions and his encounters with wildlife, plants, and the weather. He grew up as both a sportsman and a naturalist. In those days, there were no restrictions on hunting methods, seasons, or bag limits, except for those dictated by the personal code of an individual sportsman. Carl Leopold was among those who noticed the trend of declining wildlife populations, and he adjusted his approach to hunting accordingly, even going so far as to avoid some species altogether. By the time his sons began to hunt, they, too, had a well-developed personal code of sportsmanship.

In 1904, at the age of sixteen, Leopold was sent to the Lawrenceville Academy in New Jersey, where, besides his formal studies, he pursued his love of ornithology and natural history. Typically, he would set aside a few hours each day to set off with a small notebook and field glasses, recording his observations in a journal. Written at a rate that sometimes reached four and five letters a week, Leopold's correspondence with his family served as his reprieve from schoolwork, his literary training ground, and his naturalist's notebook; the letters allowed him to explore and express his absorbing relationship with nature. The letters became a systematic, chronological record of the natural events of the seasons.

This early attachment to the natural world, coupled with an unusual skill for both observation and description, led Leopold to pursue a degree in forestry at Yale University. In 1909, he received his master's degree after a year at the Yale School of Forestry. He then headed to Albuquerque, New Mexico, and accepted a job with the U.S. Forest Service, which was then headed by Gifford Pinchot, a strong conservationist. Assigned to the Arizona–New Mexico (Arizona Territories) district, Leopold spent fifteen years in the field, eventually rising to the position of chief of the district. He made meticulous surveys of the farthest reaches of the district and discovered that the greatest threat to the survival of most animal populations was not overhunting, but the loss of suitable habitat. To a keen observer such as Leopold, it was obvious that he was witnessing the end of an era. Because of human-caused activities such as grazing, fencing, logging, the construction of new roads, and increased tourism, the land was losing its essential wildness, and it appeared that no one was counting the cost.

Leopold began considering the concept of wilderness as a management tool as early as 1913. Almost single-handedly, he pushed for creation of the nation's first designated wilderness area, beginning the movement toward establishment of today's National Wilderness Preservation System. Throughout his life, he defined and redefined his thinking on the subject, spelling out the importance of wilderness to science and recreation and to the preservation of beleaguered species such as wolves and grizzlies. He also solidified his belief that wilderness was vital to human culture. "Raw wilderness," he wrote, "gives definition and meaning to the human enterprise" (Rennicke). He also posed the question, "Of what avail are forty freedoms without a blank spot on the map?" (Rennicke).

In 1921, Leopold wrote an article published in the *Journal of Forestry* in which he urged that representative portions of some forests be preserved as wilderness. In that article, he defined wilderness as "a continuous stretch of country preserved in its natural state, open to lawful hunting and fishing, big enough to absorb a two weeks' pack trip, and kept devoid of roads, artificial trails, cottages, or other works of man" (Leopold 1921, 718). He also proposed some limits on the extent of such a system, which he suggested should include "only a small fraction of the total National Forest area—probably not to exceed one in each State" (719). He even proposed a candidate area: the headwaters of the Gila River in New Mexico.

There was little opposition to Leopold's proposal, since by that time, he was a respected forester and an insider with the Forest Service. Also, as was his nature, Leopold had done much work beforehand, laying out a detailed proposal that could deflect every foreseeable argument with the combined forces of fact and eloquence. All that remained, Leopold said, was "for the Government to draw a line around each one and say, 'This is wilderness, and wilderness it shall remain' " (Rennicke). On June 3, 1924, the Forest Service's first officially designated wilderness, the 500,000-acre Gila Primitive Area, was born within the Gila National Forest. In 1931, the area was bisected by a road that would enable hunters to reach deer more easily. The larger, western section became the Gila Wilderness, while the smaller eastern piece (which would become the Aldo Leopold Wilderness in 1987) was renamed the Black Range Primitive Area.

Leopold came to Madison, Wisconsin, in 1924 as the associate director of the U.S. Forest Products Laboratory, the principal research institution of the Forest Service at that time. He worked at the laboratory from 1924 to 1928 and then spent several years surveying game populations as a forestry consultant in the north central United States. Leopold began teaching at the University of Wisconsin in 1928, and in 1933 he became professor of game management, a position created specifically for him and that he held until his death. His 1933 textbook *Game Management* laid the foundation for the modern science of wildlife management by defining fundamental skills and techniques for managing and restoring wildlife populations. Known simply as "the Professor" to many of his students, Leopold trained hundreds of ecologists. Nevertheless, throughout his career, he argued that scientific knowledge alone was not enough, and that people must also learn to live in harmony with nature.

In Germany in 1935, Leopold studied that country's game management and forestry, which provided him with a fresh perspective. He was appalled by the Germans' highly artificial system of natural-area management and the resulting loss of biodiversity; he came to realize that what was lacking in the clean German forests was not wilderness per se, but wildness. He wrote, "The forest landscape is deprived of a certain exuberance which arises from a rich variety of plants fighting with each other for a place in the sun" (Leopold Bradley, 1998). His experience in Germany enabled him to begin looking on America's land in a new light, and with new dismay.

Another key field experience that profoundly influenced Leopold's philosophy was a trip in the 1930s to the Rio Gavilan in northern Mexico. "It is here," Leopold reflected years later, "that I first clearly realized that land is an organism, that all my life I had seen only sick land, whereas here was a biota still in perfect aboriginal health" (Leopold Bradley, 1998). This vital new idea was the concept of biotic health. Though the term "biodiversity" had not yet been coined, Leopold was moving toward the realization that conserving interconnections and wildlife diversity required broader measures than just conserving wildlife. In 1935, with the rapid loss of wilderness in America weighing heavily on his mind, Leopold joined seven other leading conservationists to form the Wilderness Society.

In 1935, Leopold acquired his own land on the Wisconsin River in Sauk County, Wisconsin (part of an area known as the "sand counties"), which ultimately greatly influenced his later thinking. The purchase comprised an abandoned farm and chicken coop, which he later rebuilt into the family cabin known as "the Shack." The land, once used for growing corn, was worn out and eroded. Over the next thirteen years, the entire family, including Leopold's wife, Estella, and their five children, spent weekends and vacations working to restore life to the depleted acres of river bottom. As his daughter, Nina, later recalled, "During family weekends at The Shack, we became participants in the drama of the land's inner workings. In the very process of restoration, of planting, of successes and failures, of animals and birds responding to changes, we grew increasingly to appreciate and admire the interconnectedness of the living systems. As we transformed the land, it transformed us" (Eisele 1998). As an experimental scientist, Leopold carried out some of the first experiments in restoration ecology there. In the course of these studies, he discovered that the native prairie ecosystem would not return without human management. Leopold then dedicated himself to the study, protection, and restoration of natural habitats; his nature observations and insights into land reclamation were recorded in *A Sand County Almanac*. In 1948, he wrote the last chapters for *A Sand County Almanac* just a few months before he died of a heart attack while fighting a grass fire on a neighbor's property. The book was published the following year and earned the longtime professor of wildlife ecology worldwide acclaim. In subsequent years, all five of his children became distinguished natural scientists in their own rights. Since 1968, his daughter, Nina, has been conducting ecological research at the 1,500-acre Aldo Leopold Memorial Reserve in Wisconsin. The Shack, now listed on the National Register of Historic Places, is available for special tours and serves as a symbol of the land ethic espoused by Leopold.

In 1999, the U.S. Fish and Wildlife Service announced that as much as 8,120 acres straddling the Baraboo River in south central Wisconsin will become the Aldo Leopold National Wildlife Refuge in honor of the late conservationist. The land, in Columbia and Sauk counties, could become an important breeding area for sandhill cranes, waterfowl, and shorebirds, and its designation is also expected to benefit water quality and decrease downstream flooding.

Leopold's ecological understanding led to the development of a more deeply rooted appreciation of the complexity and interconnectedness of ecosystem processes. Leopold is credited with having instilled an "ecological conscience," the concept that human beings are not "conquerors" of the land, but "plain members" of it, "biotic citizens," without any type of privileged status. He has been named to the National Wildlife Federation's Conservation Hall of Fame, and in 1978, the John Burroughs Memorial Association awarded him the John Burroughs Medal for his lifework and, in particular for *A Sand County Almanac*.

BIBLIOGRAPHY

Ambelang, J. (1995, May 29). Leopold's shack just that. *Capital Times* (Madison, WI), p. 3A.

Biography online. Leopold, (Rand) Aldo. *http://www.biography.com/cgi-bin/biomain.cgi*

Earth Explorer online. (1995). Aldo Leopold: (1887–1948): Champion of wildlife diversity. Available: *http://www.elibrary.com*

Ecology Hall of Fame online. *http://ecotopia.org/ehof/leopold/index.html*

Eisele, T. (1995, December 8). Leopold's lessons still hold true. *Capital Times* (Madison, WI), p. 4B.

Eisele, T. (1998, May 29). Leopold's legacy, teachings have stood the test of time. *Capital Times* (Madison, WI), p. 5B.

Hall, D. J. (1999, January 27). Leopold refuge plan: Wildlife and views—meetings, environmental assessment planned. *Wisconsin State Journal*, p. 1A.

Jackson, D. D. (1998, September). Aldo Leopold: A sage for all seasons. *Smithsonian*. Available: *http://www.smithsonianmag.si.edu/smithsonian/issues98/sep98/leopold.html*

Leopold, A. (1921, November). The wilderness and its place in forest recreational policy. *Journal of Forestry, 10* (7), pp. 718–721.

Leopold, A. (1968). *A Sand County Almanac and Sketches Here and There*. New York: Oxford University Press.

Leopold, A. (1990). *Game Management*. Madison: University of Wisconsin Press.

Leopold Bradley, N. (1996, June 23). A legacy filled with love of outdoors. *Minneapolis Star Tribune*, p. 20C.

Leopold Bradley, N. (1998, April/May). A man for all seasons: The lasting environmental impact of Aldo Leopold. *National Wildlife, 36* (3), pp. 30–35.

Leopold Bradley, N. (n.d.). How hunting affected Aldo Leopold's thinking and his "Commitment to a Land Ethic." Fourth Annual Governor's Symposium on North America's Hunting Heritage, hosted by the Wisconsin Department of Natural Resources. Available: *http://www.ool.com/wildlife-forever/symposium/bradley.html*

Leopold Education Project online. *http://www.lep.org/about.html*

The Leopold Foundation online. *http://www.aldoleopold.org*

Leopold refuge a good idea. (1999, June 3). *Capital Times* (Madison, WI), p. 14A.

McCann, D. (1998, February 22). Naturalist Leopold left lasting legacy: Author's writings while on Baraboo refuge helped foster conservationism. *Journal Sentinel* (Milwaukee, WI). Available: *http://www.antigonews.com/wis/50/stories/02225esq.htm*

Naturenet Online. *http://www.naturenet.com/alnc/aldo.html*

Nix, S. (1998, January 4, March 1, March 8). A Leopold biography: Interview with Marybeth Lorbiecki, Parts 1–3. Available: *http://forestry.miningco.com/education/forestry/library/weekly/aa030898.htm*

Purser, R., Park, C., & Montuori, A. (1995, October 1). Limits to anthropocentrism: Toward an ecocentric organization paradigm? Special Topic, Forum on Ecologically Sustainable Organizations, Academy of Management Review.

Rennicke, J. (1998). A blank spot on the map: Conservationist Aldo Leopold. *Backpacker, 26*, p. 88. Available: *http://www.elibrary.com*

Seely, R. (1995, June 17). In his footsteps: Leopold's lessons can still be learned. *Wisconsin State Journal*, p. 1B.

Seely, R. (1999, September 26). Leopold's words still ring out, environmental trails he blazed continue to be followed. *Wisconsin State Journal*, p. 1C.

Van Putten, M. (1998, April/May). Leopold's enduring legacy. *National Wildlife, 36* (3) p. 7.

The Wilderness Society online. *http://wilderness.org/ethic/index.htm*

—Bibi Booth

ROBERT MacARTHUR
(1930–1972)

> Unraveling the history of a phenomenon has always appealed to some people and describing the machinery of the phenomenon to others. In both processes generalizations can be made and tested against new information so both are scientific, but the same person seldom excels at both.
>
> —Robert MacArthur (*Geographical Ecology*)

Robert MacArthur was a pioneering ecologist, population biologist, and biogeographer who developed influential theories about the process of species distribution and natural selection in animal populations. MacArthur believed that ecosystem science should venture beyond qualitative description, but at the same time should not limit itself to the mere collection and cataloging of facts. His view was that the discipline should find broader patterns in the natural world and from those patterns extract general principles.

Few ecologists of MacArthur's time could match his mathematical skill and sophistication. But for MacArthur, mathematics was only a means toward the larger goal of understanding how evolutionary processes and ecological pressures combined to shape biological communities. He did much to change ecology from a descriptive discipline to a quantitative, predictive science. One theme underlying most of MacArthur's work was the search for patterns, and he is credited by many with having advanced the field of ecology into the predictive, theoretical stage.

Born in Toronto, Canada, MacArthur moved to the United States at the age of seventeen. He studied mathematics at Vermont's Marlboro College, from which he graduated in 1951, and at Brown University, where he earned his master's degree in mathematics in 1953. MacArthur then went to Yale University to obtain his doctorate in mathematics, but switched his field to zoology at the end of his second year. During this time, he studied ecology under G. Evelyn Hutchinson, one of the leading American ecologists of the time, who had started to blend a moderately "mathematized" variety of ecology with more traditional elements, such as taxonomy and genetics.

MacArthur's studies were interrupted by a two-year period of military service; he then returned to Yale in 1957 to complete his doctoral dissertation, which won him the Mercer Award for the best ecology paper of 1957–1958. For this doctoral thesis, MacArthur studied the relationships among several coexisting species of warbler in New England forests; these species (now known as "MacArthur's Warblers") were ecologically very similar, and it was suspected that their coexistence violated biology's "competitive exclusion principle." If the niches were truly identical, MacArthur believed, one of the species should have had a competitive advantage over the others, which would eventually drive the less fit species to extinction or force them to take over other niches. MacArthur was led to a tentative hypothesis: where such a situation was found, careful observation should reveal subtle niche specialization of the apparently competing species. MacArthur discovered that there were, in fact, extremely slight differences in the foraging strategies used by each of the warbler species. MacArthur's dissertation work yielded an article for *Ecology* that appeared in 1958 and became recognized as a minor classic.

Following his work at Yale, MacArthur spent one year as a postdoctoral student at Oxford University. From 1958 to 1964, he taught zoology at the University of Pennsylvania, first as an assistant professor and then as an associate professor. In 1965, MacArthur was appointed professor of biology at Princeton University, where his responsibilities included overall editing of Princeton University Press's Monographs in Population Biology, which offered a forum for the modern, theoretical ecology he preferred.

MacArthur devoted the rest of his life to investigating population biology. He examined how the diversity and relative abundance of species fluctuate over time, and the ways in which species and communities evolve. He also investigated the strategies developed by coexisting species under the pressures of competition and natural selection. In these studies, Mac-

Arthur attempted to correlate several important factors, including the structure of the environment, the morphology of the species, and the dynamics of population change. MacArthur concluded that in the evolution of a natural community, two somewhat-opposing processes were occurring. More efficient species replaced less efficient species, but more stable communities outlasted less stable communities. In the process of community formation, the entry of a new species might involve one of three possibilities: complete displacement of an old species, occupancy of an unfilled niche, or partitioning of a niche with a preexisting species.

Perhaps MacArthur's most important contribution was to quantify some of the many factors involved in the ecological relationships between species. It was already well known, for example, that complex habitats, such as forests, supported more species of birds than did simpler ones, such as grasslands. But it was only after MacArthur had devised an index of vegetational complexity (which he called "foliage height diversity") in 1961 that it became possible for biologists to employ their observations about bird species' diversity in a definite mathematical equation. MacArthur's formula allowed them to compare habitat structures and to predict the diversity of resident bird species for a given habitat.

In the early 1960s, MacArthur and entomologist Edward O. Wilson, Jr., developed a theory describing the distribution of species on islands and the "dynamic stability" of the number of species over time. They found that the number of resident species on islands remained steady while the identities of those species changed continually, with one species replacing another over time. They also discovered that area and distance of islands combined their effects to regulate the balance between species immigration and extinction. They created an intricate mathematical model that could forecast the details of such equilibration. The model offered ecologists a radically new way to understand ecosystems that were becoming increasingly fragmented and islandlike as a result of humans' activities. In 1967, MacArthur and Wilson published their findings as *The Theory of Island Biogeography*. The book has provided the scientific foundation for all subsequent discussions of the decline of ecosystems.

MacArthur continued to teach at Princeton University, wrote fifteen more papers, and also helped to found the journal *Theoretical Population Biology*. In 1972, he was diagnosed with renal cancer. He left Princeton, returned to his house in Vermont, and, working against time and with no library resources, produced a brief volume entitled *Geographical Ecology: Patterns in the Distribution of Species*, which summarized his own final view of what was most important in ecological science. MacArthur gave special thanks in the preface to Edward Wilson, who "showed me how interesting biogeography could be and wrote the lion's share of a joint book on islands." His illness progressed quickly, and he died later the same year.

BIBLIOGRAPHY

Biography online. MacArthur, Robert Helmer. *http://www.biography.com/cgi-bin/biomain.cgi*

Cairns, J., Jr. (1994). Restoration ecology: Protecting our national and global life support systems. In J. Cairns, Jr., *Rehabilitating Damaged Ecosystems*. Boca Raton, FL: CRC Press.

Hutchinson, G. E. (1959, May/June). Homage to Santa Rosalia; or Why are there so many kinds of animals? *American Naturalist 93*, pp. 145–159.

The Hutchinson Dictionary of Scientific Biography. (1998). MacArthur, Robert Helmer. Available: *http://www.elibrary.com*

MacArthur, R. H. (1958). Population ecology of some warblers of northeastern coniferous forests. *Ecology 39* (4), pp. 599–619.

MacArthur, R. H. (1972). *Geographical Ecology: Patterns in the Distribution of Species*. Princeton, NJ: Princeton University Press.

MacArthur, R. H., & Wilson, E. O. (1967). *The Theory of Island Biogeography*. Princeton, NJ: Princeton University Press.

National Academy of Sciences, Working Group on Teaching Evolution. (1998). *Teaching about Evolution and the Nature of Science*. Washington, DC: National Academy Press.

Quammen, D. (1996). Life in equilibrium: Edward O. Wilson and Robert H. MacArthur's 1967 book "The Theory of Island Biogeography" became a cornerstone of new science of population biology. *Discover, 17* (3), pp. 66–77.

—*Bibi Booth*

ROBERT MARSHALL
(1901–1939)

"How many wilderness areas do we need?" How many Brahms symphonies do we need?

—Robert Marshall

Robert Marshall was a forester, a conservationist, and one of America's most eloquent spokesmen on behalf of wilderness. Marshall documented the value of wilderness, asserting that it had intrinsic beauty and allowed people space in which they could regain peace of mind and heart. It also gave city dwellers the opportunity for adventure, by which Marshall meant "venturing beyond the boundary of normal aptitude, extending oneself to the limit of capacity, [and] courageously facing peril" (Kerasote). A visionary in the truest sense of the word, Marshall set an unprecedented course for wilderness preservation in the United States that few have surpassed. He was among the first to suggest that large tracts of Alaska be preserved, shaped the U.S. Forest Service's policy on wilderness designation and management, and wrote passionately on all aspects of conversation and preservation. Marshall's writings not only detailed the aesthetic value of wilderness to humankind, but also pushed for public ownership. A devoted socialist, Marshall believed that private interests would certainly destroy America's forests.

Robert Marshall was born in 1901 in New York City, the third of four children of German immigrants Louis and Florence Marshall. His father was a prominent constitutional lawyer, an active conversationist, and a leader in the Jewish community. Louis Marshall fought for the conservation of the Adirondack and Catskill mountains of upstate New York. He realized their importance in the water supply of New York City and relished their beauty. Young Robert was educated in the city, but spent the summers of his youth at "Knollwood," his family's summer home on Lower Saranac Lake in the Adirondacks. Here he and his brothers, George and James, learned to use a compass and map, and between 1918 and 1924, Robert and George climbed forty-two of the forty-six Adirondack peaks with elevations greater than 4,000 feet; Robert eventually also climbed the remaining four. Marshall unknowingly started a trend with his accomplishment: today, members of the Adirondack Forty-Sixers Club are also dedicated to climbing all forty-six of these peaks. In July 1932, Marshall set a record of a different sort by climbing fourteen Adirondack peaks in less than a day, a feat that comprised a total ascent of 13,600 feet.

Marshall decided in his teens that he wanted to be a forester. After he graduated in 1919 from the Ethical Culture School in New York, he spent a year at Columbia University and then in 1920 enrolled in the New York State College of Forestry at Syracuse University. As part of his experience there, he attended a three-month intensive-study camp in the Adirondacks, where he took it upon himself to inventory and "rate" all of the ponds and lakes in the range's Cranberry Lake Region. He was able to document over ninety-four lakes and ponds. While Marshall was engaged in his pet project, he began to note that the Adirondack wilderness was in a state of demise. He joined the Adirondack Mountain Club and contributed to *The High Peaks of the Adirondacks*, the club's first publication, which included a rating from Marshall for each peak. In 1924, Marshall graduated fourth out of fifty-nine in his class and then scored first in the nation on the civil service test for foresters.

Marshall received his master's degree in forestry from Harvard University in the spring of 1925. While he was working on his master's thesis, he developed a theory about the effect of weather on political history. He demonstrated that poor growth in trees could be detected from the trees' rings, and that there was a correlation between poor growth and presidential election years. He later reported his data in a 1927 article for the *Nation* entitled "Precipitation and the Presidents."

In 1925, at the age of twenty-four, he traveled to western Montana to accept a position with the U.S. Forest Service. He served as the assistant silviculturist at the Northern Rocky Mountain Experiment Station from 1925 to 1928, where he was in charge of research pertaining to forest propagation after fires. On the job, Marshall spent six days a week hiking through the back country of Montana and northern Idaho, and he still spent nearly every Sunday—his only day off—hiking in some of the wildest country of the northern Rocky Mountains.

In 1928, Marshall left Montana to attend graduate school at Johns Hopkins University, where he earned a doctorate in plant physiology in 1930, one of three doctorates he achieved in his lifetime. He then took the first of several extended trips to the small, remote town of Wiseman, Alaska, beginning a long romance with the wildlands of the central Brooks Range. His first book, *Arctic Village*, chronicled his experiences living with both Koyukuk and white people in Wiseman between 1930 and 1931; it was a 1933 best-seller. The total amount of money he received for the book was $3,600, of which he kept half. He then sent a check for $18, along with a copy of the book, to each of the one hundred Koyukuk he had known during his trip. Marshall was one of the first persons to extensively explore this area of the world, especially the headwaters of the North Fork of the Koyukuk River. Much of this acreage would later be protected within the Gates of the Arctic National Park, whose name derived from Marshall's description of two peaks, Frigid Crags and Boreal Mountain, as "the gates from Alaska's central Brooks Range into the Arctic regions of the far north" (Great Outdoors Recreation).

Marshall was a tireless hiker, regularly covering thirty or forty miles a day over steep, rocky terrain. He greatly enjoyed exploring undeveloped country, simply attired in his trademark high-topped sneakers and jeans, carrying a pocket full of raisins and cheese. Marshall also carried a stopwatch and kept highly detailed records of each hike. As he visited the rapidly dwindling wildlands of the United States, he began to formulate a plan to preserve them. In the February 1930 issue of *Scientific Monthly*, he published "The Problem of the Wilderness," which provided a definition of the term. "The dominant attributes of such an area," Marshall wrote, "are: first, that it requires anyone who exists in it to depend exclusively on his own effort for survival; and second, that it preserves as nearly as possible the primitive environment. This means that all roads, power transportation and settlements are barred" (142).

With his doctorate in forestry, Marshall was well educated in the logic of scientific argument and the economic basis for federal forest policies. As chief forester of the Bureau of Indian Affairs from 1933 to 1937, he worked to involve Native Americans more fully in the management of their tribal forests and ranges. "The philosophy that progress is proportional to the amount of alteration imposed upon nature never seemed to have occurred to the Indians," he once commented (National Park Service). In 1934, he personally subsidized a new 12,000-square-mile exploration of forty-six previously uncharted Alaskan roadless areas and then surveyed many of these areas himself; the resultant map was published by the U.S. Geological Survey. In January 1935, with Benton MacKaye, Aldo Leopold (a Forest Service colleague of Marshall's), and six other conservationists and avowed socialists, he also helped found the Wilderness Society, which dedicated itself to protecting roadless areas from commercialization. He was also a member of the Society of American Foresters, the Ecological Society, the Society of Plant Physiologists, the Society of Anthropologists, and the Explorers and Cosmos clubs.

By 1936, Marshall had identified forty-eight unspoiled forested areas larger than 300,000 acres in size and twenty-nine desert areas over 500,000 acres in size within the United States. In 1937, he returned to the Forest Service as chief of the Recreation and Lands Division, where he had tremendous influence over public lands policy. In this position, Marshall helped to set aside 18.8 million acres for study as roadless areas. Marshall became one of the first Forest Service officials to research carrying capacities of wilderness areas and to manage the areas in a way that encouraged visitors but did not upset the natural conditions. Marshall anticipated the position of the motor-vehicle industry, which would assert that wilderness areas would exclude the majority of recreationists. He responded to this argument by noting that "there are certain things that cannot be enjoyed by everybody. If everybody tries to enjoy them, nobody gets any pleasure out of them" (Kerasote). Marshall's goal was to set aside areas where a person could "spend at least a week or two of travel . . . without crossing his own tracks" (Bob Marshall/Great Bear Wilderness Areas). He ultimately succeeded in setting aside at least 5 million acres of such lands. In 1939, Marshall again traveled to Alaska for the summer. While he was returning home by train, he died suddenly of heart failure en route from Washington, D.C., to New York. He was only thirty-eight years old. Independently wealthy, Marshall left one-quarter of his $1.5-million estate to the Wilderness Society, assuring its existence and commitment to wilderness preservation for years to come. Today, the society's highest honor is its Robert Marshall Award.

Marshall had always been active in social causes, actively combating racial and religious discrimination. He left the remainder of his estate for use in the "promotion and advancement of an economic system based upon the theory of production for use and not for profit," and for the establishment of the Robert Marshall Civil Liberties Trust. This fund was established in Marshall's will for the "safeguarding and advancement of the cause of civil liberties in the United States of America." The Robert Marshall Papers, which describe both the life of Robert Marshall and the activities of the Robert Marshall Civil Liberties Trust, were donated to

Hebrew Union College's American Jewish Archives in Cincinnati, Ohio, by Marshall's brother, James, between 1979 and 1983.

Marshall's early definition of wilderness was partially incorporated into the Wilderness Act of 1964. It enabled Congress to set aside selected areas in the national forests, national parks, national wildlife refuges, and other federal lands as units to be kept permanently unchanged by humans: no roads, no structures, no vehicles, no significant impacts of any kind. In 1941, the U.S. Forest Service had set aside the South Fork, Pentagon, and Sun River primitive areas within the Flathead and Lewis and Clark national forests in Montana. After passage of the Wilderness Act, the agency designated these collective territories as the Bob Marshall Wilderness Area in "recognition of his distinguished work in development of [the nation's] system of wilderness areas." The establishment of the contiguous Great Bear and Scapegoat wildernesses in the 1970s created a 1.5-million-acre wilderness complex that is one of the most completely preserved mountain ecosystems in the world. It is often referred to as the "crown jewel" of America's wilderness system and is widely known as "Bob Marshall Country" or simply "The Bob." The Bob Marshall Foundation is dedicated to the restoration of the trail system and wilderness values in the Bob Marshall Wilderness Complex through volunteer labor and the establishment of public and private partnerships. There are also three other geographic points in the United States named in Marshall's honor: Marshall Lake in the Brooks Range, Alaska; Mount Marshall in the Adirondack Mountains, New York; and the Bob Marshall Recreation Camp in the Black Hills National Forest, South Dakota. "How many wilderness areas do we need?" Marshall was once asked, to which he replied, "How many Brahms symphonies do we need?" In 1956, Marshall's Alaskan journals, maps, and photographs were collected in the volume *Arctic Wilderness* (entitled *Alaska Wilderness: Exploring the Central Brooks Range* in later editions), which was edited by his brother, George.

BIBLIOGRAPHY

Audubon Magazine. Champions of Conservation. (1998). Available: *http://magazine.audubon.org/century/champion.html#MARSHALL*

Biography online. Marshall, Robert. *http://www.biography.com/cgi-bin/biomain.cgi*

The Bob Marshall Foundation online. *http://www.bobmarshall.org*

Bob Marshall/Great Bear Wilderness Areas. *http://members.aol.com/CMorHiker4/backpack/Marshall.html*

Glover, J. M. 1986. *A Wilderness Original: The Life of Bob Marshall*. Seattle: Mountaineers Books.

Goldstein, E. (1998, November). Defending the wild places. Reed College (Portland, OR) *Reed* Magazine. Available: *http://web.reed.edu/community/newsandpub/nov1998/defending/index.html*

Great Outdoors Recreation Pages. Bob Marshall Wilderness Area. *http://www.
gorp.com/gorp/resource/US_Wilderness_Area/MT_bob_m.HTM*

Great Outdoors Recreation Pages. Gates of the Arctic National Park and Preserve.
http://gorp.com/gorp/resource/US_National_Park/ak_gates.HTM

Hebrew Union College, American Jewish Archives online. An Inventory to the Robert Marshall Papers, 1919–1973. *http://huc.edu/aja/R-Marsha.htm*

Kerasote, T. (1998, August 1). What we talk about when we talk about wilderness.
Sports Afield, p. 70.

Marshall, R. (1922). *The High Peaks of the Adirondacks*. Lake George, NY: Adirondack Mountain Club.

Marshall, R. (1930, February). The problem of the wilderness. *Scientific Monthly
30* (2), pp. 141–148.

Marshall, R. (1991). *Arctic Village: A Nineteen Thirties Portrait of Wiseman,
Alaska*. Fairbanks, AK: University of Alaska Press.

Marshall, R., & Marshall, G. (1974). *Alaska Wilderness: Exploring the Central
Brooks Range*. Berkeley, CA: University of California Press.

National Park Service online. Nature Notes from Crater Lake, Volume XIII—October 1, 1947. Crater Lake Natural History Association. *http://www.nps.
gov/crla/nn-vol13.htm*

National Wilderness Preservation System. Bob Marshall Wilderness. *http://www.
wilderness.net/nwps/wilderness.cfm/Bob%20Marshall*

Pacific Biodiversity Institute. (1998). History of wild land and roadless area mapping. Available: *http://www.pacificbio.org/roadless-mapping/Washington-
roadless/History.html*

Robert Marshall: Principal founder of the Wilderness Society. The Wilderness Society online. *http://www.tws.org/profiles/marshall.htm*

Southern Illinois University, Edwardsville. Professor Alan Burns: Nature writing
online. Some major figures in American nature writing. *http://www.siue.edu/
~alburns/naturewriters.html*

Spencer, L. T. (1997). A wilderness original: Summaries from the book by Glover.
Plymouth State College (Plymouth, NH), Biology Faculty online. *http://oz.
plymouth.edu/~lts/wilderness/marshall.html*

University of Texas, Austin, College of Engineering, Clint R. Murchison Sr. Chair
of Free Enterprise. (1999, April 19). The environmental movement (1970s).
In Global Governance: Why? How? When? Available: *http://www.engr.
utexas.edu/cofe/governance/g_part05.html*

—*Bibi Booth*

YNES MEXIA
(1870–1938)

Ynes Enriquetta Julietta Mexia was a Mexican-American botanical explorer who provided research academic institutions with extensive collections of specimens. Mexia also supplemented her collections with extensive and accurate research data.

Mexia was born on May 24, 1870, in Washington, D.C., where her father was on a diplomatic mission for the Mexican government. Her paternal grandfather, Jose Antonio Mexia, was an important contributor in the early history of Texas. When she was sixteen, her family moved to Philadelphia, where Mexia attended private schools. Later, she enrolled in a Quaker-sponsored school in Ontario, Canada, and then studied at St. Joseph's Academy in Maryland. Although she spent her early years living in Texas and Maryland, during her late teens and into her late thirties, she lived in Mexico City on her father's hacienda.

In 1921, at the age of fifty-one, Mexia gained admittance as a special student to the University of California at Berkeley. She soon developed an interest in the natural sciences, particularly botany. Mexia spent time in Mexico, Alaska, and Brazil studying and collecting hundreds of plant specimens. In 1925, Mexia took a course on flowering plants at the Hopkins Marine Station in Pacific Grove, California. She then joined Roxana Stinchfield Ferris on an expedition to Mexico sponsored by Stanford University. Mexia returned with a large collection of almost 500 species. After this expedition, Mexia established a working relationship with Nina Floy Bracelin, an assistant at the University of California Herbarium. Bracelin served as Mexia's curator and agent and handled detailed arrangements of equipping Mexia's expeditions. Bracelin kept records of the sale of specimens to various institutions and also made arrangements to have academic experts examine Mexia's specimens.

In 1926, Mexia set off on her own through western Mexico. When she returned to California in April 1927, she brought back with her more than 30,000 specimens, including a new plant genus and almost fifty new species of plants. During the summer of 1928, Mexia traveled to Mount McKinley Park in Alaska. She returned with 6,100 specimens.

In November of the following year, Mexia set off on an expedition to Brazil and Peru, returning in March 1932 with 65,000 specimens. She spent one season working for the Department of Agriculture's Bureau of Plant Industry and Exploration in Ecuador, where she searched for the cinchona, the wax palm, and soil-binding herbs.

In the United States, she spent some time taking photographs and drawing plants in their natural habitat. By October 1935, Mexia was back in the Andes as a member of the University of California Botanical Expedition in search of a variety of items, including wild tobacco. There she recorded valuable information about primitive life in the Andes. Mexia financed her expeditions through the sale of the specimens she collected for the institutions.

Mexia traveled in Peru, Bolivia, Argentina, and Chile before reaching the Straits of Magellan. On this expedition, she collected approximately 15,000 specimens, including plants and animals. During the winter of 1937 and the spring of 1938, Mexia embarked on her final botanical expedition. On this trip, she concentrated her efforts in Guerrero and Oaxaca, Mexico. Due to a stomach ailment, Mexia was forced to return to San Francisco ahead of schedule with some 13,000 specimens. A few months later, she died of lung cancer in Berkeley, California. During her lifetime, Mexia collected more than 100,000 plant specimens and discovered hundreds of new species. Mexia related details of her expeditions in articles that appeared in a number of journals, including "Botanical Trails in Old Mexico," *Madrono* (September 1929); "Three Thousand Miles up the Amazon," *Sierra Club Bulletin* (February 1933); and "Camping on the Equator," *Sierra Club Bulletin* (February 1937).

—*John Mongillo*

STEPHANIE MILLS
(1948–   )

Stephanie Mills is an author, editor, lecturer, and bioregional activist. She has concerned herself with the fate of the earth and humanity since 1969, when her commencement address at Mills College gained national and international attention for its dramatic call to personal engagement with the crises of overpopulation and the degradation of the life of the land.

Mills was born in Berkeley, California, in 1948 and was raised in Phoenix, Arizona. She was the only child of Robert and Edith Mills. Her mother, Edith Garrison Mills, was raised in Mississippi. After serving in the Women's Army Corps, Garrison married and became a homemaker. Robert Mills was raised in various mining towns in the intermountain West. He studied mechanical engineering and had a long career as a salesman of mining equipment throughout the western United States, Mexico, and South America.

Stephanie Mills attended a public elementary school, a private preparatory school for junior high, and a public high school in Phoenix. By the time she was halfway through high school, she had a controversial column in the school paper. In 1965, she entered Mills College in Oakland, Cali-

Photo by Peter M. Mann.

fornia. (The fact that she and her college share the same name is pure coincidence.) A small liberal arts college for women, Mills occupies an idyllic campus within easy reach of both San Francisco and Berkeley.

At college, Mills pursued her interests in modern literature, history, philosophy, art history, and drawing. Despite the earnest efforts of the biology faculty, she barely passed the one course required in that subject. She was involved with the campus radio station and a variety of other student activities. Again she was a columnist for the school paper and, as a senior, edited the college literary magazine. The writings of James Joyce, Thomas Mann, Joseph Campbell, and Carl Jung became her consuming passion in college, thanks in great measure to Hunter Hannum and Diana O'Hehir, the professors who taught these authors.

It was not until near the end of her senior year that Mills developed an interest in the environment. She organized a campus symposium on ecology that led her to read a short essay on the enormity of the population problem by the biologist Paul Ehrlich. That night she drafted a commencement speech and soon after offered it to the senior who seemed the likeliest choice for class speaker. However, the class of 1969 elected Mills to give the valedictory speech, which she had titled "The Future Is a Cruel Hoax." The speech dealt with the peril of ignoring overpopulation and stated the danger in no uncertain terms. *Chemical and Engineering News* called it "strident" and "near-hysterical" but conceded that "she may be symptomatic of a growing trend." According to the *San Francisco Chronicle*, "The remarks left her 1000 listeners and . . . the main speaker visibly shaken . . . She received an enthusiastic ovation that lasted nearly a minute from the 181 graduates and their families when she concluded." The following week the *Los Angeles Times* published the speech in its entirety. *Life* ran an excerpt of it in a feature on graduations. Since then, "The Future Is a Cruel Hoax" has been anthologized in psychology books, rhetoric texts, and collections of writings on population. The stories about the shocking graduation speech hit the wires and went around the world. There were features on Mills in *Look*, the *St. Louis Post-Dispatch*, the *San Francisco Examiner*, and the *Oakland Tribune*.

Consequently, Mills spent the next couple of years being a sought-after young spokesperson for the burgeoning ecology movement. After her graduation, she worked for Planned Parenthood of Alameda and San Francisco counties as a campus organizer. This involved a lot of speechmaking. In the course of the year, she made some eighty talks at places like Yale, Stanford, the University of California at Davis, the University of California at Irvine, the University of Idaho, the opening of the Oakland Museum, and numerous Earth Day celebrations and Planned Parenthood meetings. Perhaps the most notable of these occasions was Mills's sermon at San Francisco's Grace Cathedral, which made her the first woman ever to have preached from that pulpit. Mills also participated, with Gary Snyder, Stewart Brand, Richard Brautigan, and a number of other beaux esprits of the Bay area, in generating the ecological manifesto "Four Changes."

From Planned Parenthood, Mills moved on to edit *Earth Times*, a muckraking monthly environmental tabloid spun off by *Rolling Stone*. *Earth Times* lasted a short four issues and then stopped publication for lack of revenue. After *Earth Times* folded, Mills continued to give talks and to write on a freelance basis.

In 1971, Mills received a "Cookenheim"—a grant from the Point Foundation to throw dinner parties. The grant originated with a salon in Stockholm. Mills hosted small gatherings of NGO participants at the United Nations Conference on the Human Environment. The purpose of these gatherings was to give people a low-pressure, gracious situation where they might exchange ideas and information in an atmosphere of trust generated by the act of breaking bread together. This worked so well in Stockholm that the foundation extended the grant for a year's worth of dinners in Berkeley. At that time, Stewart Brand was quoted in the *Village Voice* as saying, "Some of my best ideas have come out of her parties. . . . Wonderful things would happen in her salons. People would meet people they had been hearing about for years. . . . There was no business purpose implicit in the salons. Having nothing else to do, they became friends."

Mills was encouraged to write an article on the salons for the *Co-Evolution Quarterly*, published by *The Whole Earth Catalog*. The piece, "Salons and Their Keepers," retained its interest and usefulness on into the 1990s when the *Utne Reader* commissioned Mills to write "Salons and Beyond," an article that helped to launch the Utne Salons and to renew an interest in focused but convivial conversation. In 1996, Mills was named one of the *Utne Reader*'s visionaries.

Mills lived and worked in the San Francisco Bay area for fifteen years, freelancing for and editing various publications—*CoEvolution Quarterly, Not Man Apart* (Friends of the Earth's fortnightly news magazine), Sierra Club Legal Defense Fund, the Trust for Public Lands, and California Tomorrow among them. In 1984, she moved to her present home in northwestern lower Michigan and began her first book, *Whatever Happened to Ecology?* (1989). *Kirkus Reviews* called the book "an important personal memoir portraying the human face behind the headlines of the environmental movement." *Library Journal* termed Mills's second book, the anthology *In Praise of Nature* (1990), a "stimulating and useful volume." *Kirkus Reviews* praised Mills's third book, *In Service of the Wild: Restoring and Reinhabiting Damaged Land* (1995), as "an uncommonly clear, commonsensical argument for rehabilitating damaged landscapes and moving toward what the author calls the 'future primitive.'" Mills also edited *Turning away from Technology: A New Vision for the 21st Century* (1997).

The publication of *Whatever Happened to Ecology?* led Mills back to the podium to speak at such venues as the Land Institute's Prairie Festival, the Sitka Summer Writers' Symposium, and Michigan's Backyard Eco-Conference and to deliver one of the E. F. Schumacher Society Lectures in 1991. Mills also has lectured at the Chicago Academy of Sciences and universities and colleges in the heartland and has participated in the Orion Society's Forgotten Language Tours. She continues giving talks and lectures at numerous conferences and symposia.

Mills's staunch insistence, despite criticism from both the left and the right, on the seriousness of human overpopulation as one of the driving forces of environmental degradation and human misery, and her advocacy of taking individual action to do something about overpopulation, namely, refraining from having children, are two of her notable contributions in the ecology movement. More complex, but no less important, in her writing and public speaking is her holistic view and her radical critique of industrial civilization. Mills does not believe that our present way of life can be made to be ecologically sustainable, nor does she think it would be worth trying. Rather, following the thought of Peter Berg, a founder of bioregionalism, Mills sees ecological problems as systemic. Bioregionalism looks at the causes of environmental crisis in industrial civilization and the nation-states that have supported and exported it. It proposes a diverse, decentralized, place-located, ecologically literate array of human societies as a viable alternative. Mills was a leading participant in many of the North American Bioregional Congresses and was one of the organizers of the Great Lakes Bioregional Congress. Her most recent bioregional work has consisted of lectures and seminars on bioregionalism at colleges, universities, and conferences. Above all, Mills hopes that her written work will serve as an enduring source of inspiration for readers who care about and work for the flourishing of life's diversity on earth.

Being female, the greatest single obstacle Mills has faced is male chauvinism, pervasive throughout human societies. The American environmental movement is patriarchal in its power structures and approaches, although at the grassroots level many of the leanest, most innovative, inclusive, and effective organizations are led by women.

Every person seeking to defend life on the planet from civilization's ravages confronts an enormous obstacle in the psychological mechanism of denial. Denial is operative within each individual as well as in the collective. Mills thinks that breaking through is challenging, painful, and liberating open-ended work.

Mills has the following advice for young people: Keep faith with biology. Don't be seduced by reductionist science. Spend only the barest minimum of time at a computer. Read Aldo Leopold's *A Sand County Almanac* and Rachel Carson's *Silent Spring*. Never miss an opportunity to go outdoors with a naturalist. Learn the plants. Appreciate the insects. Watch the birds. Get to know the natural history of your home place. Keep a diary about it. Keep your appetites modest and avoid debt—that way you won't need a lucrative career. You can be free and have a true vocation instead. Think seriously about not having children. Do everything you decently can to preserve wild nature wherever it exists. Nurture the relationships you make in the work. Practice respect. Savor your life to the utmost.

BIBLIOGRAPHY

Mills, S. (1969, June 8). The future is a cruel hoax. *Los Angeles Times*.

Mills, S. (1970, June). We've got to end the war on the people and environment of Indochina. Editorial. *Earth Times*.

Mills, S. (1972, June). Paul Ehrlich: DeFusing commoner. Interview. *Clear Creek*.

Mills, S. (1978, November 9). Ecology and the mind—Two futurists tune in on the cosmic wavelengths. *San Francisco Bay Guardian*.

Mills, S. (1986, February). Thoughts from the Round River rendezvous. *Earth First!*

Mills, S. (1989). *Whatever Happened to Ecology?* San Francisco: Sierra Club Books.

Mills, S. (1989, November/December). Vision of our ecological future. *Utne Reader.*

Mills, S. (1991, March/April). Salons and beyond. *Utne Reader.*

Mills, S. (1995). Alternatives to monoculture: A bioregional perspective. In H. Norberg-Hodge, P. Goering, & S. Gorelick (Eds.), *The Future of Progressing: Reflections on Environment and Development.* Green Books, Ltd.

Mills, S. (1995). *In Service of the Wild: Restoring and Reinhabiting Damaged Land.* Boston: Beacon Press.

Mills, S. (1997, September/October). The journey home: A sense of place, and a place for senses. *Sierra 82* (5).

Mills, S. (1998, May). Could you live with less? *Glamour.*

Mills, S. (1999). Can't get that extinction crisis out of my mind. In R. Rosenblatt (Ed.)., *Consuming Our Future.* Washington, DC: Island Press.

Mills, S. (Ed.). (1990). *In Praise of Nature.* Washington, DC: Island Press.

Mills, S. (Ed.). (1997). *Turning Away from Technology.* San Francisco: Sierra Club Books.

Mills, S., & Nelson, J. (1977, December). AgriBusiness vs. AgriCulture. Editorial. *Not Man Apart.*

—*Stephanie Mills*

RICHARD MOORE
(1946–   )

Richard Moore is the executive director of the Southwest Network located in Albuquerque, New Mexico. He was born in 1946 in Harrisburg, Pennsylvania. His mother, who came from Pennsylvania, and his father, who came from Puerto Rico, did migrant work throughout Pennsylvania picking apples and cherries. Moore described his family life and parents as being very good, but he found himself involved in gangs around Harrisburg, so he got himself into some legal situations as a teenager. At the age of sixteen, he dropped out of school in the tenth grade. Because of his love for horses, especially walk and trot horses and Tennessee walking horses, Moore traveled throughout the east coast working for different horse stables.

However, a few years after dropping out of school, he and his mother met with a school counselor, and they encouraged him to join the Job Corps Program in 1966. Moore was in the Job Corps for about a year and a half being trained as a repairer of office machines. At the same time, he worked toward getting his high-school diploma. Later, Moore also joined the Volunteers in Service to America (VISTA) program after graduating from the Job Corps center. During this time, Moore worked in a street-gang program with the Catholic Christian Brothers who were running the program in Santa Fe.

Moore started to become interested in doing more community organizing

Photo courtesy of Richard Moore.

and social justice work. In 1968, he participated in the New Mexico Delegation to the Poor People's Campaign held in Washington, D.C. The march was called by Martin Luther King, Jr., and was organized to express concerns over social justice, civil rights, and human rights issues.

In 1980, Moore became executive director of the Southwest Organizing Project of New Mexico, a citywide and statewide grassroots organizing effort in Albuquerque and throughout New Mexico. He worked on many issues with grassroots organizations; the tasks included health care, childcare issues, water rights issues, land issues, housing, and unemployment—whatever the needs of the communities were.

In 1990, Moore became the coordinator of the Southwest Network. Initiated by the Southwest Organizing Project of New Mexico, the network was formed in Albuquerque, New Mexico, at the People of Color Regional Activist Dialogue on Environmental Justice when over eighty representatives from thirty-two organizations came together to regionalize their issues. The issues included lethal pesticides used in farming, contaminated water, obnoxious odors from waste plants and incinerators, toxic materials used in industry, lead poisoning, and how these issues affected the communities of color. As an example, when Moore arrived in Albuquerque, the local sewage plant was located in the middle of a primarily Chicano-Latino community. The smell from that facility was so bad that people could not use their backyards for recreation. Now the plant has been re-located.

Southwest Network is also interested in the health of children. Other issues include nuclear facilities, industrial sites, military contamination of communities, and landfills. Southwest Network has also been involved with Native American concerns. It supported the Western Shoshone in efforts to stop missile testing in Nevada and other Native American groups seeking to end the mining of coal and uranium on their lands.

Incorporating youth into the leadership of the Southwest Network has been one of the goals from the day the network was founded. A youth component is called the Youth Leadership Development Campaign. The program prepares young people not only to take on leadership roles within their local organizations of the Southwest Network, but to be involved nationally and internationally in the environmental and economic justice movement. The program prepares them in several areas of training. As an example, they learn how to conduct a press conference. Other skills include public speaking and constructing an analysis of a power structure, which involves looking at a problem and then determining what kinds of solutions would be needed to solve the problem.

Moore believes that policy issues will be a major environmental issue in the next decade. These include policy regulations already in the books but not being enforced for people of color or low-income communities. Enforcing these regulations would prevent some of the devastating problems that are located in many of these communities and workplaces. However, Southwest Network has won major battles in getting the EPA to enforce regulations. Moore believes that the Presidential Executive Order on Environmental Justice signed by President Bill Clinton in 1994 was a very important benchmark for the community and the environmental justice movement. The order mandates all federal agencies to prepare plans for integrating environmental justice into their work. Moore would also like to see the strengthening of Title 6 of the Civil Rights Act implemented in 1964 as it relates to environmental justice issues.

There are over eighty organizations in six states that are affiliated with the Southwest Network. The states include Colorado, Arizona, New Mexico, Nevada, Texas, and California. In 1994, Southwest Network opened up its membership to grassroots organizations in the country of Mexico who were working on economic and environmental issues. Now a very large constituency of the Southwest Network is made up of organizations in four Mexican states in northern Mexico. The network has provided a model for other environmental and economic justice networks throughout the United States and Mexico. However, Moore insists on calling Southwest Network "a developing binational organization."

—*John Mongillo*

Photo courtesy of The Bancroft Library,
University of California, Berkeley.

JOHN MUIR
(1838–1914)

John Muir was a farmer, inventor, naturalist, explorer, writer, and conservationist who was born on April 21, 1838, in Dunbar, Scotland. Muir is
often referred to as the "father of our national park system." At a very
young age, he was fascinated by natural habitats in hills and valleys. He
attended grammar school and studied Latin, French, English, mathematics,
and geography. He loved natural history and was particularly interested in
books about North American animals written and illustrated by John James
Audubon and Alexander Wilson.

When he was eleven years old, the Muir family emigrated to the United
States and settled in Wisconsin. Muir spent much of his youth helping his
father on the family farm. As a teenager, he did not have any formal education but found time to teach himself geometry, mathematics, and philosophy. He also liked to invent gadgets. Some of his inventions included

a clock that kept accurate time and a machine that tipped him out of bed in the morning. He built barometers, table saws, and hydrometers.

In 1860, Muir left the farm for good and took his inventions to exhibit them at the state fair at Madison, where he won several prizes. He also decided to enroll at the University of Wisconsin. In college, he studied botany and geology and earned good grades, but after three years, he left Madison to travel the northern United States and Canada.

In 1867, while working at a factory in Indianapolis, Muir suffered a blinding eye injury. When he regained his sight one month later, he planned a thousand-mile trip that would take him from Indianapolis to the Gulf of Mexico. He sailed to Cuba and to Panama, where he crossed the isthmus. He then sailed up the west coast, landing in San Francisco in March 1868. In California, he walked across the San Joaquin Valley through waist-high wildflowers and into the high country for the first time. Later he would write, "Then it seemed to me the Sierra should be called not the Nevada, or Snowy Range, but the Range of Light . . . the most divinely beautiful of all the mountain chains I have ever seen" (Sierra Club).

Muir herded sheep through that first summer and made his home in Yosemite. From that moment on, California became his home. He hiked, climbed, and studied the Sierra ranges and glaciers and made the first ascent of Cathedral Peak.

During the 1870s, Muir wrote a series of articles entitled "Studies in the Sierra" for San Francisco's *Overland Monthly*, which launched his career as a writer. In 1871, the *New York Tribune* published an article titled "Yosemite Glaciers." Muir hypothesized that the Yosemite valley was carved up by glaciers. At the time, his theory was controversial. However, years later, Muir's theory was proven correct.

During his writing days, Muir still had time to take many trips, including his first to Alaska in 1879, where he discovered Glacier Bay. He climbed Mount Whitney (14,494 feet), the first recorded ascent by an eastern route, and Mount Shasta (14,162 feet) and explored the Modoc Lava Beds (now a national monument) just south of the Oregon border in northern California.

In 1880, Muir married Louie Wanda Strentzel and moved to Martinez, California, where they raised their two daughters, Wanda and Helen. Muir went into partnership with his father-in-law and managed the family fruit ranch with great success, but the activity of ranching did not quell Muir's wanderlust. He continued to travel back to Alaska many more times and to Australia, South America, Africa, Europe, China, Japan, and, of course, the Sierra Nevada. Throughout the 1880s, Muir worked as an editor and author on *Picturesque California*. He visited many sites in the Pacific Northwest, including Portland, Oregon, Columbia Gorge, Mount Rainier, Snoqualmie Falls, Spokane Falls, and Crater Lake. This trip provided the material for articles about Oregon and Washington published in *Pictur-esque California*, a two-volume series. In 1885, Muir visited and wrote about Yellowstone National Park.

Muir published 300 articles and ten major books that recounted his travels, expounded his naturalist philosophy, and beckoned everyone to "climb the mountains and get their good tidings" (Sierra Club). Muir's love of the high country gave his writings a spiritual quality. His readers, whether they were presidents, congressmen, or plain folks, were inspired and often moved to action by the enthusiasm of Muir's own love of nature.

Through a series of articles appearing in *Century* magazine, Muir drew attention to the devastation of mountain meadows and forests by sheep and cattle. With the help of *Century*'s associate editor, Robert Underwood Johnson, Muir worked on plans to stop this destruction. In 1890, due in large part to the efforts of Muir and Johnson, an act of Congress created Yosemite National Park.

Johnson and others suggested to Muir that an association be formed to protect the newly created Yosemite National Park from the assaults of stockmen and others who would diminish its boundaries. In 1892, Muir and a number of his supporters founded the Sierra Club to, in Muir's words, "do something for wildness and make the mountains glad" (Sierra Club). Muir served as the club's president until his death in 1914. Also in 1892, Muir's brother David and his family moved to the ranch, relieving Muir of the burden of ranch management and allowing him more time to write and travel.

In one of his visits east, Muir first met Gifford Pinchot in New York in 1893. Muir traveled with Pinchot and others with the National Forestry Commission surveying the problems of the western forest reserves. At that time, they became good friends. The friendship was not to last, however. While Muir was returning from Alaska in 1897, he read a newspaper account attributing an opinion to Pinchot that sheep grazing in forest reserves did no harm. Muir had observed sheep firsthand as "hoofed locusts" literally destroying the flowery mountain meadows, so when Muir found that Pinchot was staying at the same hotel in Seattle, he confronted him in the lobby. Surrounded by a pack of reporters, clutching the newspaper article in his hand, Muir asked Pinchot, "Are you correctly quoted here?" "Yes," Pinchot admitted. "Then," replied Muir, "I don't want anything more to do with you" (Wolfe).

Muir went on to crusade for the preservation of wilderness, while Pinchot campaigned for utilitarian use of public lands. The deepening schism between Muir and Pinchot eventually grew into the great split between the preservation wing and the utilitarian wing of the conservation movement. Pinchot and Muir became major antagonists on the issue of damming Hetch Hetchy Valley in Yosemite National Park.

In 1901, Muir published *Our National Parks*, the book that brought him to the attention of President Theodore Roosevelt. In 1903, Roosevelt visited Muir in Yosemite. There, together, beneath the trees, they laid the foundation of Roosevelt's innovative and notable conservation programs. Years later, Roosevelt recalled his time and friendship with Muir. In an article written in the *Outlook* in 1915, Roosevelt spoke of Muir: "His was a dauntless soul. . . . Not only are his books delightful, not only is he the author to whom all men turn when they think of the Sierras and Northern glaciers, and the giant trees of the California slope, but he was also—what few nature-lovers are—a man able to influence contemporary thought and action on the subjects to which he had devoted his life. He was a great factor in influencing the thought of California and the thought of the entire country so as to secure the preservation of those great natural phenomena—wonderful canyons, giant trees, slopes of flower-spangled hillsides—which make California a veritable Garden of the Lord. Our generation owes much to John Muir."

In 1903, Muir studied the Petrified Forest and campaigned for its protection. In 1906, the Petrified Forest was proclaimed a national monument by President Theodore Roosevelt. In 1908, Roosevelt proclaimed Muir Woods National Monument, named after Muir by request of the donor of the land, William Kent, who later became a U.S. congressman.

Muir and the Sierra Club fought many battles to protect Yosemite and the Sierra Nevada, the most dramatic being the campaign to prevent the damming of Hetch Hetchy Valley. In 1913, after years of effort, the battle was lost, and the valley that Muir likened to Yosemite itself was doomed to become a reservoir to supply the water needs of a growing San Francisco. The following year, after a short illness, Muir died in a Los Angeles hospital after visiting his daughter Wanda.

John Muir was perhaps this country's most famous and influential naturalist and conservationist. He taught the people of his time and ours the importance of experiencing and protecting our natural heritage. His words have heightened our perception of nature. His personal and determined involvement in the great conservation questions of the day was and remains an inspiration for environmental activists everywhere. He played a role in the creation of several national parks and wrote many articles and several books expounding the virtues of conservation and the natural world. The Muir house and historic Martinez adobe became part of the National Park Service in 1964. In 1992, Mt. Wanda was added to the site. The 325-acre tract of oak woodland and grassland was historically owned by the Muir family.

BIBLIOGRAPHY

Cutright, P. R. (1985). *Theodore Roosevelt: The Making of a Conservationist*. Urbana: University of Illinois Press.

The John Muir Homepage. *http://oz.plymouth.edu/~lts/muir.html*

Muir, J. (1888/1889). *Picturesque California*. 2 vol. New York: J. Dewing.

Muir, J. (1894). *The Mountains of California*. New York: The Century Co. Muir's first book conveys with words and pictures the wonder and delight Muir found on his travels in the Sierra Nevada. This book was an immediate success and served to rally and solidify the conservation sentiment of the entire nation.

Muir, J. (1901). *Our National Parks*. Boston, New York: Houghton Mifflin. This is a supreme guidebook, an exciting introduction to Yosemite and several other magnificent parks. It is part reminiscence, part philosophy, and mostly enticing description, all vintage Muir.

Muir, J. (1911). *My First Summer in the Sierra*. Boston, New York: Houghton Mifflin. Muir's journal of his first explorations of the Sierra Nevada. He colorfully describes the wildlife, the plant life, and the geology of the area.

Muir, J. (1913). *The Story of My Boyhood and Youth*. Boston, New York: Houghton Mifflin.

Muir, J. (1918). *Steep Trails*. Boston: Houghton Mifflin. Collection of letters and magazine articles that highlight numerous areas from the Great Salt Lake to Mt. Rainier to the Grand Canyon.

Muir, J. (1992). *John Muir: The Eight Wilderness Discovery Books*. Seattle: Mountaineers Books. A collection of Muir's eight most influential books arranged in order of his life.

Muir, J. (1996). *John Muir: His Life and Letters and Other Writings*. London: Bâton Wicks. Seattle: Mountaineers. The second volume of a two-part omnibus set, this book is a diverse collection with a broader, more intimate reflection of Muir himself.

Sierra Club. John Muir. *http://www.sierraclub.org/history/Muir/*

Sterling, K. B., Harmond, R. P., Cevasco, G. A., & Hammond, L. F. (Eds.). (1997). *Biographical Dictionary of American and Canadian Naturalists and Environmentalists*. Westport, CT: Greenwood Press.

Wolfe, L. M. (1946). *Son of the Wilderness: The Life of John Muir*. New York: Alfred A. Knopf. Availible: *http://earthinstitute.org/uap3354/notes/history.htm*

—*John Mongillo*

MARGARET E. MURIE AND OLAUS JOHAN MURIE
(1902–   ) and (1889–1963)

Margaret and Olaus Murie were members of the Wilderness Society and contributed greatly to the appreciation and protection of Alaska's wilderness. Margaret E. ("Mardy") Thomas was born in Seattle in 1902, the daughter of a sea captain, and was raised in Fairbanks, Alaska. Her husband, Olaus Murie, was born in Moorhead, Minnesota, in March 1889, the son of Joachim D. and Marie Frimanslund Murie. Olaus and Mardy married in 1924 and had three children. In 1924, Margaret Murie was the second person to graduate from the University of Alaska at Fairbanks. Olaus first attended Fargo College in North Dakota and then received an A.B. from Pacific University in 1912. Later, he obtained a graduate degree from the University of Michigan in 1927 and an honorary doctorate of sciences from Pacific University in 1949.

Mardy Murie has been called the "grandmother of the conservation movement," but her love of the land began early. After a honeymoon spent researching caribou on a 500-mile dogsled journey through Alaska's Brooks Range, Mardy grew accustomed to extended family camping in the Alaska wilds, accompanying Olaus on expeditions for the U.S. Biological Survey (later the U.S. Fish and Wildlife Service). Mardy and Olaus moved to near Jackson Hole, Wyoming, to study declining elk populations in 1926.

Brought up in a rural setting, Olaus developed a deep appreciation for wilderness preservation while he was young. He explored the Hudson Bay region (1914–1915) and later the Labrador Peninsula (1917) while he was with the Carnegie Museum, studying birds and mammals in natural habitats. During 1920 to 1927, he conducted similar research in British Columbia and in northern Canada and Alaska with the Biological Survey. Olaus also served as a fur warden in Alaska. The Biological Survey tolerated Olaus Murie's publicly acknowledged disagreement with its policy of eradicating predators in the late 1920s and early 1930s. In 1945, however, the Fish and Wildlife Service declined to publish Murie's major study on coyotes because it reflected his continuing criticism of this policy.

The Muries published, traveled, and lectured extensively, devoting much of their time to the promotion of legislation that would protect remaining American wilderness from irresponsible developers. Olaus was director of the Wilderness Society from 1945 to 1962. Important accomplishments of the Muries included the establishment of major wilderness-protection legislation. Margaret and Olaus Murie worked with other conservation leaders to bring about the passage of the 1964 Wilderness Act. Through their activities with the Wilderness Society, the Muries were instrumental in the preservation of two major wilderness areas, Grand Teton National Park and the Arctic National Wildlife Refuge. As authorities on wildlife and wilderness preservation, the Muries believed that the ultimate success of the wilderness movement depended on the extent to which it received the support of the general public. Toward this end, they worked to expand the support base for wilderness protection and enhance the effectiveness of citizen advocacy organizations.

Olaus Murie died in Moose, Wyoming, on October 21, 1963, just months before passage of the Wilderness Act. After her husband's death, Mardy Murie continued the work that she and Olaus had begun, joining the Governing Council of the Wilderness Society and working for the protection of wild Alaska. In 1980, President Jimmy Carter signed the Alaska

Lands Act, which added 54 million acres to the National Wildlife Refuge System and 56 million acres to the National Wilderness Preservation System. In congressional testimony about the Alaska Lands Act, Mardy Murie said: "I am testifying as an emotional woman and I would like to ask you, gentlemen, what's wrong with emotion? Beauty is a resource in and of itself. Alaska must be allowed to be Alaska; that is her greatest economy. I hope the United States of America is not so rich that she can afford to let these wildernesses pass by, or so poor she cannot afford to keep them." In 1984, Mardy wrote in *Wilderness* magazine: "The shining, comforting thought now is that the parks, the forests, the refuges, the wild rivers, are there . . . and my feeling about it all is that when the oil and the minerals have all been found and taken away, the one hundred million acres of national parks and refuges and wild rivers and forests will be the most beneficent treasure in the whole state. I would plead with all administrators: Please allow Alaska to be different, to be herself, to nourish our souls." For her lifelong commitment to conservation, President William Clinton awarded Margaret Murie the nation's highest civilian honor, the Medal of Freedom, in January 1998. As of 2000, Mardy Murie was living in Moose, Wyoming, and continuing her work for conservation causes.

During his career, Olaus Murie authored major publications on American wildlife, won a number of awards from wilderness and conservation organizations, and served as a trustee of the Robert Marshall Wilderness Fund. In presenting him the Audubon Medal, its highest award (1959), the National Audubon Society described Olaus Murie as a "personification of the spirit of wilderness." In addition to being the recipient of several awards, Margaret Murie was the author of *Journeys to the Far North, Island Between, Two in the Far North*, and *Wapiti Wilderness*, which was written with her late husband.

BIBLIOGRAPHY

Bryant, J. (1995). *Margaret Murie: A Wilderness Life*. New York: Twenty-First Century Books.

Murie, M. (1977). *Island Between*. Fairbanks: University of Alaska Press.

Murie, M. E. (1997). *Two in the Far North: 35th Anniversary Edition*. 5th ed. Portland, OR: Graphic Arts Center Publishing Company.

Murie, M., & Murie, O. (1966). *Wapati Wilderness*. New York: Knopf.

Sterling, K. B., Harmond, R. P., Cevasco, G. A., & Hammond, L. F. (Eds.). (1997). *Biographical Dictionary of American and Canadian Naturalists and Environmentalists*. Westport, CT: Greenwood Press.

—*John Mongillo*

RALPH NADER
(1934–   )

The most important office in America for anyone to achieve is full-time citizen. . . . You've got to keep the pressure on, even if you lose.
—Ralph Nader

Ralph Nader is a lawyer, a public-interest activist, and America's best-known consumer advocate. He has often been rated in national pools as "the most respected person in America." Honored by *Time* magazine as one of the one hundred most influential Americans of the twentieth century, Nader has devoted his life to defending the American public against corporate negligence and government indifference.

Nader has been credited with direct or indirect responsibility for laws that have improved environmental protection and provided for safer workplaces. Among the reforms Nader has had a hand in initiating are the Environmental Protection Agency, the Safe Drinking Water Act, and the Occupational Health and Safety Administration. Although Nader is widely known as a consumer advocate, his overriding concern is actually representative democracy: he has a high regard for the individual's ability to understand and impact politics.

Nader was born in Winsted, Connecticut, the youngest of four children of Lebanese immigrants Nathra and Rose Nader. His father, a reformer with strong democratic beliefs, encouraged activism as well as critical thinking, and both parents believed that a citizen owed a debt to society. In later life, the senior Nader told his son that if he ever ran for president, it should not be as a Democrat or Republican, "because both parties sell the country very cheap." Indeed, Ralph Nader has consistently been very critical of the two major American political parties, especially in their handling of health, environmental, and ecological issues. Over twenty years ago, for example, he complained, "The Democrats have got to get rid of the idea that because a couple of them have staked out claims, ecology is their exclusive issue. Everybody has got to get involved" (Sheasby).

Nader's interest in civic activism developed at an early age when his older brother, Shafik, encouraged him to read the *Congressional Record*. By the age of fourteen, Nader had read the works of many early activist authors, who inspired his later thinking about the distribution of power in America and the potential of citizenship. One of Nader's favorite boyhood novels was Upton Sinclair's *The Jungle*, a 1906 exposé of the horrors of the American meat industry.

In 1951, Nader graduated from the Gilbert School in Winsted and entered Princeton University, where he majored in Far East politics and languages. He conducted his first protest action during his sophomore year after he noticed dead birds around the trunks of trees treated with DDT. "Shouldn't people know that DDT kills birds and it can harm people too?" he asked at the news office of the *Daily Princetonian*, waving a bag full of dead birds.

In 1955, Nader enrolled in Harvard Law School, which he quickly dismissed as morally complacent. "It was a high-priced tool factory, only instead of tools and dies, they were producing hired advocates for corporate law firms and corporations," Nader later said (Sheasby). It was at Harvard that Nader first explored the engineering design of automobiles and began to talk hopefully about "offensive lawyers, lawyers without clients, lawyers who are primary human beings" (McCarry).

In 1959, Nader set up a law practice in Hartford, Connecticut, and published his first article on auto safety, "The Safe Car You Can't Buy," in the *Nation*. Convinced that he could make little progress against automobile manufacturers while working from Hartford, in 1963 he abandoned his practice and hitchhiked to Washington, D.C., to begin work as a public activist. He became a researcher on auto safety for the U.S. Department of Labor the following year. Nader also worked as a freelance writer for the *Nation* and the *Christian Science Monitor* and acted as an unpaid advisor to a Senate subcommittee that was exploring the federal government's potential role in auto safety.

Nader first gained national attention in 1965 with the publication of his first book, *Unsafe at Any Speed: The Designed-in Dangers of the American Automobile*. The book both revealed automobile design flaws and condemned an industry that would knowingly market and sell products that caused preventable deaths. He also criticized the General Motors Corporation, the world's largest corporation, for spending so little of its profits on objective safety research. Nader's belief that blame often rests not with consumers but instead with irresponsible manufacturers became a recurring theme in many of his later investigations.

In March 1966, Nader testified before a congressional committee that General Motors had hired a private investigator to search for information to discredit him. The company's attempts to ruin Nader backfired, and he instead gained a national reputation for honesty and integrity. He successfully sued for invasion of privacy and was awarded $280,000 in damages. The public furor Nader created yielded results when the National Traffic and Vehicle Safety Act was signed later that year. The incident served as a proof of an essential Nader belief: one person, acting with intelligence and persistence, can make a difference. The U.S. Jaycees honored Nader as one of the nation's Ten Outstanding Young Men of 1966.

In 1969, Nader established the Center for the Study of Responsive Law in Washington and staffed it with young lawyers and students, whom the media quickly tagged "Nader's Raiders." As Nader says of his approach to activism, "The function of leadership is to produce more leaders." Nader's associates published reports on many subjects, including the environment, insecticides, radiation dangers, freedom of information in government, and public health. The Raiders discovered inadequate government regulations that were weakly enforced and inappropriate relationships between regulators and those they were charged with regulating. Some of the earliest Nader's Raiders books helped to structure public debate on major environmental issues and shape several landmark laws.

In 1971, Nader helped to found the organization Public Citizen, Inc., a federation of advocacy groups that would act as the consumer's "eyes and ears" in Washington. What distinguished Public Citizen from the single-issue public-interest groups that emerged around the same time was its simultaneous activism on multiple fronts: Congress, federal agencies, the courts, and the media. Public Citizen soon spawned Congress Watch, a champion of consumer and citizen interests before Congress. Since its establishment, Congress Watch has been the driving force behind new regulatory standards in such areas as pesticide and food safety, auto fuel efficiency, toxic-chemical control, and toxic-waste cleanup.

The consumer movement has often played a crucial role in environmental crusades. In 1970, for example, Nader's skepticism of nuclear power reached a peak when he read about the findings of an Atomic Energy Commission (AEC) scientific study on nuclear power plant emissions. To help publicize the sobering findings and expose AEC retribution against the study's two scientists, Nader helped to provoke one of the first national inquiries into nuclear power safety. Nader and Public Citizen then convened Critical Mass '74, the first national antinuclear conference, which led to the creation of the Public Citizen Critical Mass Energy Project. Over the next decade, Critical Mass became one of the leading national voices for the safe energy movement.

What has distinguished the consumer movement's environmental advocacy is its linking of environmental problems to corporate power. "The modern corporation's structure, impact and public accountability," Nader wrote in 1970, "are the central issues in any program designed to curb or forestall the contamination of air, water and soil by industrial activity" (Bollier). In the early 1970s, Nader sponsored several reports designed to spark citizen action against specific industry-caused environmental problems and weaknesses in existing legislation. For example, John C. Esposito's Nader-supported book *The Vanishing Air* was a devastating critique of the historic failures of federal programs to control air pollution, particularly the Clean Air Act of 1967. The book prodded the Senate Subcommittee on Air and Water Pollution into taking leadership on environmental issues and ultimately resulted in the inclusion of tough provisions in the Clean Air Act of 1970. In a similar manner, a second Nader-sponsored book led the way for passage of the Clean Water Act, another cornerstone of modern environmental law.

During the 1980s, Nader's critics asserted that his activism had run its course, and America's best-known public advocate was out of the head-lines. Of the many causes of Nader's decline, success was, ironically, one of the most significant. Several Nader loyalists had moved to high-level positions within the Carter administration. At one point, for example, there were at least fifty Federal Trade Commission staff members who had come out of the consumer movement, many from Nader's groups. Nader's behind-the-scenes influence actually increased as a result, but he had lost his most experienced people. In addition, business leaders and the public had become exasperated with what they viewed as excessive regulation and its hidden costs.

The public's regard for Nader remained strong, however. For example, when Mark Green, the primary author of the Nader-sponsored book *Who Runs Congress?*, ran for a U.S. Senate seat from New York in 1986, his pollsters found that 80 percent of New York State residents knew who Nader was, and more than 70 percent had a favorable opinion of him. But 1986 was a devastating year for Nader personally. Not only did he contract Bell's palsy, which paralyzed his facial nerves for months, but his brother, Shafik, was diagnosed with prostate cancer. Shafik's death later that year crushed Nader, and he withdrew for months afterwards. During the period of his brother's illness, Nader had also been battling a California initiative that restricted consumers' rights to sue for damages. Nader ardently op-posed any such restrictions, including no-fault insurance; his anger at the unsuccessful outcome finally snapped him out of his depression.

By the early 1990s, the political winds had shifted again as Americans seemed to be again receptive to the consumer movement. In 1996 and 2000, Nader accepted the Green party's invitation to be its candidate for president, a gesture that he intended primarily as a means of raising public awareness of the power that corporate America exerted over decision mak-ers in Washington. His presidential campaigns focused on this corporate power, asserting that special interests bought and sold our president and members of Congress and were moving to restrict citizen access to the courts as well. He focused his campaigns on removing special-interest money from elections, adding that he would neither seek nor accept cam-paign contributions.

In 1997, Nader took on the Microsoft Corporation, alleging that it was a ruthless, fearsome monopoly with an alarming potential to undermine democracy and sabotage market freedom. At Nader's conference, "Ap-praising Microsoft and Its Global Strategy," speakers blasted Microsoft's stranglehold on the computer software industry as the bane of innovation and the death of choice.

The issue of new technologies is also a major concern for Nader. In his view, "We have all kinds of historic examples where the leading scientists in the world were absolutely wrong in the way they evaluated the consequences of a new technology. I remember when they told us nuclear power would be too cheap to meter. Nuclear power would replace all fossil fuels and we'd have a totally clean environment. They never told us about radioactive waste . . . [or that] this was the only fuel system producing electricity that had to have an elaborate evacuation plan. . . . [T]he broader democracy must develop the consequences, the probabilities, and the value system" (Sheasby).

Today, Nader works out of the Center for the Study of Responsive Law in Washington, D.C. From there, he acts as an advisor to a consortium of public-interest groups, the most visible of which is Public Citizen. Equally formidable is Nader's network of Public Interest Research Groups, which lobby in two dozen states.

Nader urges a reinvigoration of civic life before society swings too far toward a corporate state and the redefinition of the individual American as a citizen. His teams of Raiders are exploring multimedia approaches to bring this information to millions of Americans, including an online information service called Essential Information. "The most important office in America for anyone to achieve," Nader says, "is full-time citizen. . . . You've got to keep the pressure on, even if you lose. The essence of the citizens' movement is persistence" (Reader's Companion to American History).

BIBLIOGRAPHY

Biography online. Nader, Ralph. *http://www.biography.com/cgi-bin/biomain.cgi*

Bollier, D. (n.d.) Citizen action and other big ideas: A history of Ralph Nader and the modern consumer movement. Available: *http://www.nader.org/history*

Carlin, J. (1997, November 18). Bill Gates beware: America's consumer crusader is on. *Independent*, p. N2.

Compton's Encyclopedia online. Nader, Ralph. *http://www.comptons.com*

Esposito, J. C. (1970). *The Vanishing Air; Ralph Nader's Study Report on Air Pollution*. New York: Grossman Publishers.

Funk and Wagnalls Encyclopedia online. Nader, Ralph. *http://www.fwkc.com/encyclopedia/low/articles/n/n017000009f.html*

Green, M. J., Rosenthal, B., & Darling, L. (1979). *Who Runs Congress?* New York: Bantam Books.

Judis, J. B. (1998, August 31). The spirit of '68: What really caused the Sixties. *The New Republic*, pp. 20–27.

McCarry, C. (1972). *Citizen Nader*. New York: Saturday Review Press.

Nader, R. (1959). The safe car you can't buy. *The Nation, 188* (15), pp. 310–313.

Nader, R. (1991). *Unsafe at Any Speed: The Designed-in Dangers of the American Automobile* (Rev. ed.). New York: Knightsbridge Publishing Company.

Public Citizen online. More about Public Citizen. *http://www.citizen.org/newweb/more_about.htm*

Ralph Nader for President online. *http://nader96.org/aboutrn.htm*

Reader's Companion to American History online. (1991). Nader, Ralph. Available: *http://www.elibrary.com*

Sheasby, W. C. (n.d.). Ralph Nader: The ecological crisis: "Everybody has got to get involved." Available: *http://petra.greens.org/~cls/nader/rnbio-e.html*

Stewart, T. A. (1989, May 22). The resurrection of Ralph Nader: After years in the shadows, America's most famous consumer activist shines again; why the comeback? it's not that he's changed: it's that society has. *Fortune*, pp. 106–108.

Suarez, R. (1998, April 16). Ralph Nader [interview]. National Public Radio, Talk of the Nation. Available: *http://www.elibrary.com*

—*Bibi Booth*

WILLIAM ALBERT NIERING
(1924–1999)

William Albert Niering was born on August 28, 1924, in Scotrun, Pennsylvania, to George and Emma Everitt Niering. His grandfather, William, ran the local grocery, and his father was a butcher.

The Poconos of the 1920s were very different from those of today. The area was a favorite spot for honeymooners and vacationers from nearby urban areas who came to the big hotels that were very popular at that time. However, there were still many wild spots untouched by the so-called development that has engulfed the area since those days. Young Bill Niering grew up knowing these natural areas. He came under the influence of his grandmother and mother, who saw to his religious education. His grandfather, a man very like the man Bill would become, was his link to the land. His father taught him about the world of the butcher; Bill and his buddies were employed to pluck chickens at a penny a chicken. He also raised pigs and helped his father on his round, which involved selling meat to individual housewives and the chefs of the big hotels. This experience led to a lifelong knowledge of meat and, with his mother's help, the proper way to cook it.

His life was that of an average boy: chores, school, church, and play. His sister Annabelle was born in 1928, and they remained close throughout adulthood. After graduation from the local high school, he went on to Pennsylvania State University in 1942 where he became engrossed in the agricultural program. World War II interrupted his education, and he served in the army from December 1942 until November 1945, attaining the rank of staff sergeant. He was an instructor in jungle living. His active service took place in the South Pacific where he and other young men occupied the coral islands that had been wrested from the Japanese. During this time, he received word of the birth of his sister Bettina. On the way to the islands he had a stopover in Hawaii, where the young botanist became enthralled by the tropical vegetation and spent many happy hours exploring the islands and visiting the Bishop Museum.

Photo courtesy of Catherine Niering.

After his demobilization, he returned to Penn State and graduated in 1948 with a B.S. in biology. In 1950, he received his M.S. in botany. His thesis was titled "The Flora of Monroe and Pike Counties, Pennsylvania." From there, he went to Rutgers University, where his future was set. Murray Buell, one of the early ecologists, became his mentor and set him on the path he trod for the rest of his life. He was introduced to the science of ecology, which sees the holism of life. He concentrated on botany and received his Ph.D. in 1952. His thesis was titled "The Past and Present Vegetation of High Point State Park, New Jersey." The Buells became his friends, and he learned a great deal from them.

After graduation, the young botanist went to the Audubon Center in Greenwich, Connecticut, as a counselor. It was here that he met Richard Goodwin, chairman of the Botany Department at Connecticut College, New London. Niering arrived at the college in September 1952 as an assistant professor in the Botany Department. He spent his entire academic life, forty-seven years, at this beautiful college that he loved. He taught, as part of a team, the introductory biology course, taxonomy, and, later, human ecology. In 1964 he became a professor and in 1965 took on the additional job of director of the Connecticut Arboretum (now the Connecticut College Arboretum). He was named Katharine Blunt Professor of Botany and the Lucretia L. Allyn Professor of Botany chair.

Niering had many areas of interest. In the summer of 1954, Niering was land ecologist with a team of other scientists on Kapingamarangi, one of the South Pacific's Caroline Islands atolls. With Richard Goodwin, he challenged the state of Connecticut's treatment of roadside vegetation. With Robert Whittaker, he studied saguaro in the Sonoran Desert of Arizona, and in 1967 their paper on saguaro received the Mercer Award from the Ecological Society of America. During the years that followed, additional research was done to check on the development of the giant cacti. At home, he became interested in wetlands, both salt and fresh, and became a well-known authority on these "liquid assets." Several Arboretum bulletins dealt with this subject and were instrumental in the passage of Connecticut's Tidal Wetlands Act (1969). As a consultant, Niering testified in numerous cases involving the altering of wetlands. Another interest lay in natural areas that he considered vital to the preservation of a healthy environment. He advocated the use of native plants in landscaping and started the SALT (Smaller American Lawns Today) movement. His home in Gales Ferry is where he practiced what he preached. His lawn through the years diminished to the point where it can be mowed in about ten minutes with a hand mower.

His many accomplishments are too numerous to all be mentioned here. In 1995, the Garden Club of America awarded him the Margaret Douglas Award for Conservation Education. In 1998, he received the Connecticut College medal for accomplishments and services that enhanced the reputation of the college and nourished its growth. That year he also received the Governor's Environmental Award. With Richard Goodwin, he was honored when the Center for Conservation Biology and Environmental Studies was named the Goodwin-Niering Center. In honor of his years as director of the Arboretum, there is a Niering tract. In 1993, Niering assumed the editorship of a new journal, *Restoration Ecology*, which deals with the problems of altered and degraded landscapes and ways to restore them to functioning, self-sustaining ecosystems.

Perhaps Niering's greatest contribution lay in his teaching. During his forty-seven years at the college, he inspired students with his enthusiasm and energy. It was apparent that he loved teaching, and many students were inspired to follow in his footsteps. Today his students have achieved positions of the highest order and carry on the work to which he devoted his life, trying to save the world for future generations. His lectures were legendary. He illustrated them with plants, animals, soda cans collected during his numerous walks on campus, the clothes he rescued from those cast off by students at the end of the year, and his shoes, more than forty years old, that he still wore. He was known for his walks, leaving behind students or others puffing to keep up. In his taxonomy class, the students enjoyed his homebaked goods that were appropriate to the subject matter.

Niering was the author of over 200 scientific articles and five books. He belonged to numerous organizations, including the Ecological Society of America, the Botanical Society of America, Sigma Xi, the Connecticut Botanical Society, the American Institute of Biological Sciences, the American Association for the Advancement of Science, the Society for Ecological Restoration, the Society of Wetland Scientists, the Connecticut Forest and Park Association, and the Explorers Club. He died suddenly on August 30, 1999, after giving a talk to the incoming freshmen. At his memorial service, he was remembered as a gentle, kind, patient man who almost never said no and was ever humble.

BIBLIOGRAPHY

Niering, W. (1966) *The Life of the Marsh: The North American Wetlands.* New York: McGraw-Hill Book Company.

Niering, W. (1984). *A Book of Wildflowers.* New York: William Morrow and Company.

Niering, W. (1985). *Wetlands: The Audubon Society Nature Guides.* New York: Alfred A. Knopf.

Niering, W., & Littlehales, B. (1991). *Wetlands of North America.* Charlottesville, VA: Thomasson-Grant.

Niering, W., & Olmstead, N. (1979). *The Audubon Society Field Guide to North American Wildflowers; Eastern Region.* New York: Alfred A. Knopf.

A Tribute to William A. Niering. *http://arboretum.conncoll.edu/niering/memorial.html*

—*Catherine Niering*

EUGENE ODUM
(1913–)

> Without healthy, natural systems to support and buffer industrial, urban, and agricultural activities, there can be no healthy economy or high quality of life.
>
> —Eugene Odum

Eugene Odum is a zoologist who is one of the most influential figures in the history of ecology. For more than half a century, Odum has challenged scientific assumptions about the world around us. His groundbreaking book *Fundamentals of Ecology*, which was published in 1953, influenced an entire generation of ecologists, and he is often referred to as the "father of modern ecosystem ecology" in both popular and scientific circles. At first, his ideas had little support in the scientific community, but over the decades they have become accepted throughout the world. For an incredible ten years, his book was the only textbook available worldwide on ecosystem ecology. As a result, his influence was enormous. "The work of Dr. Odum changed the way we look at the natural world and our place in it," former President Jimmy Carter said in 1996. "We cannot overestimate the value of [his] work in making spaceship Earth a better place for us all" (University of Georgia).

Before Odum, ecology had been studied on a small scale within individual disciplines. Many scientists doubted that it could be studied on a large scale, which entailed examining how one natural system interacts with another. Odum began with the system as a whole and showed the power of that concept, advocating a holistic or ecosystem approach to ecology. Instead of breaking things down and starting with individual pieces—birds, trees, or soil types—Odum starts at the top, studying a whole forest, sand dune, or geographic region, and then works down to individual components. It was he who made "ecosystem" a household word.

The son of Howard Washington Odum, a prominent sociologist, Odum was born in 1913 in Lake Sunapee, New Hampshire, and was raised in Chapel Hill, North Carolina. He was an avid birdwatcher, and his earliest studies began informally in the 1920s, when, as teenagers, he and a friend parlayed their hobby into a column for the local Chapel Hill newspaper. Odum received a bachelor's degree from the University of North Carolina in 1934 and his doctorate in zoology from the University of Illinois in 1939. He then worked as the resident biologist at a New York State nature preserve. Accompanied by his wife, Martha, he came to the University of Georgia in Athens as a zoology instructor in 1940.

Odum continued to study birds and other animals. His first research findings about the effects of DDT on fish were published in the journal *Science* in 1946, nearly a generation before Rachel Carson wrote *Silent Spring*, America's environmental wakeup call. His early studies also helped shape his theory that an entire ecosystem should be studied and not just its individual animal populations or other components.

Odum honed his pioneering research methods at such locations as the Atomic Energy Commission's Savannah River Site, a nuclear production facility near Aiken, South Carolina. In 1951, he and three University of Georgia graduate students began surveying fields, streams, and forests near the facility as it was under construction. Their research became the foundation for ecosystem-level studies and led to many additional radiation ecology studies.

In 1953, Odum began an ecosystem study of the estuaries and salt marshes on Sapelo Island, an ecologically pristine coastal barrier island five miles off the Georgia coast. His team studied how salt-marsh organisms affect and contribute to the ocean and its life-support systems by their recycling of essential elements such as carbon and nitrogen. Odum and his students helped to establish the importance of coastal areas as nurseries for shrimp, oysters, and many other ocean organisms.

When Odum began teaching ecology at the University of Georgia in the 1940s, there were no textbooks on the subject. First published in 1953,

Fundamentals of Ecology became the world's most widely read ecology text. Translated into more than a dozen languages as well as braille, the book has helped teach generations of ecologists to think in terms of whole ecosystems. Decades later, it is still a definitive textbook on ecosystem ecology. A fifth edition of the landmark text will soon be published once Odum and Gary Barrett, a former Odum doctoral student and now the University of Georgia Eugene Odum Professor of Ecology, finish their revisions.

By 1954, Odum had switched from terrestrial and coastal studies to coral-reef studies in the Eniwetok Atoll in the southern Pacific Ocean. He was joined in his research by his colleague and brother, Howard T. Odum, a world-renowned ecologist and now a University of Florida professor emeritus. Their study, criticized by some scientists who claimed that the brothers were venturing too far afield from their expertise, showed how important the delicately balanced nutrient cycle is to coral-reef survival. They won the Ecological Society of America's Mercer Award for their outstanding research.

In 1954, Odum was responsible for the founding of the University of Georgia's Marine Institute on Sapelo Island off the Georgia coast. Tobacco magnate R. J. Reynolds had donated the use of the southern part of the island for the study of ecosystems in Georgia's coastal marshes. Odum was also chiefly responsible for the founding of the Savannah River Ecology Laboratory, operated by the university under contract with the U.S. Department of Energy. This 300-square-mile area has become one of the largest outdoor science classrooms on earth. The laboratory continues to perform ecosystem-level research in such areas as ecological stewardship, radioecology, ecotoxicology, wetlands monitoring, and environmental assessment.

Odum's belief in his vision led to the founding of the University of Georgia Institute of Ecology in 1961. Odum took a broad view of the environment, pioneering ecosystem ecology as a new "integrative science." He explored connections between pristine watersheds, plants, animals, climate, and weather. In addition, Odum was among the first scientists to use radioactive tracers to study food chains, nutrient cycles, and animal populations; recognize energy flow as an important ecological principle; recognize the importance of wetlands to the health of rivers and streams; and look at reduction of pollution sources, including ways to reduce toxic materials and other nonpoint pollution that goes into a system.

Although it did not happen immediately, thanks to Odum's persistence, ecology courses were added to the University of Georgia curriculum, and eventually the school began to award bachelor's, master's, and doctoral degrees in ecology. Today, in fact, environmental literacy is a requirement for University of Georgia undergraduates.

Odum remained on the University of Georgia faculty until his retirement in 1984, and he still serves the university's Institute of Ecology as its di-

rector emeritus. He is also Callaway Professor Emeritus of Ecology and
Alumni Foundation Distinguished Professor Emeritus of Zoology. Since re-
tiring, Odum has devoted his time to writing and lecturing, promoting
awareness of ecology nationwide.

Odum was elected a member of the National Academy of Sciences in
1970 and has received three international awards: the Crafoord Prize from
the Royal Swedish Academy of Sciences in 1987, France's Institut de la Vie
Environmental Prize in 1975, and the University of Southern California's
Tyler Prize for Environmental Achievement, which was presented to Odum
by President Jimmy Carter in ceremonies at the White House in 1977.
Odum shared the Crafoord and Institut prizes with his brother and donated
his portion of the prize monies to advance ecological research. His donations included an endowment for the Institute of Ecology. Odum has also
won the National Wildlife Federation's Educator of the Year Award, the
Ecological Society of America's Eminent Ecologist Award, and the U.S.
Department of the Interior's Conservation Service Award. His latest book
is *Ecological Vignettes: Ecological Approaches to Dealing with Human
Predicaments* (1998), which presents basic ecological principles in a series
of chapters on such subjects as population growth, consumption of natural
resources, and resource management.

Today, Odum notes that ecosystem theory provides a common denominator for humans and nature and that the goods and services of both are
coupled. "Without healthy, natural systems to support and buffer industrial, urban, and agricultural activities," he says, "there can be no healthy
economy or high quality of life" (University of Georgia). Among his
thought-provoking recommendations to achieve healthy systems is the assignment of fair market values to nature's free services whenever decisions
are made about protecting or developing large expanses of natural areas,
such as forests or beaches. "Just about everything that concerns us as humans has parallels in nature," he says. "By presenting ecological concepts
this way I hope to help us avoid overpopulation, overconsumption, damage
to life-support systems and other suicidal behaviors" (Purdy).

BIBLIOGRAPHY

Biography online. Odum, Eugene Pleasants. *http://www.biography.com/cgi-bin/
biomain.cgi*
Gordon and Beach Publishers online. Ecological Vignettes. *http://www.gbhap-us.
com/Features/Ecological_Vignettes/top.htm*
Odum, E. P. (1953). *Fundamentals of Ecology.* Philadelphia: Saunders.
Odum, E. P. (1998). *Ecological Vignettes: Ecological Approaches to Dealing with
Human Predicaments.* New York: Harwood Academic Publishers.
Purdy, J. B. (1998, Fall). A voice for the wild. University of Georgia *Research Re-
porter* online. *http://www.ovpr.uga.edu/rcd/researchreporter/fall98/voice.
html*

University of Georgia, *Georgia Center Quarterly* (Spring 1997) online. Eugene Odum: An ecologist's life. *http://www.gactr.uga.edu/gcq/gcqspr97/odum.html*

—*Bibi Booth*

FREDERICK LAW OLMSTED
(1822–1903)

> In its influence as an educator, as a source of scientific interest, and its effect upon the health, happiness, and comfort of our people may be found the chief value of landscape.
>
> —Frederick Law Olmsted, Sr., *Report to the Commissioners of Central Park*, 1870

Frederick Law Olmsted, author, environmentalist, and abolitionist, was widely recognized as the founder of American landscape architecture and the nation's foremost park maker. He is also referred to as the "father of landscape architecture in the United States." Olmsted was born in Hartford, Connecticut, on April 26, 1822, the son of John and Charlotte Law Hull Olmsted. His father ran a successful dry-goods store. He attended local schools and studied for one semester at Yale. His early jobs included being a sailor and farmer. In the 1850s, he began his career as an author and journalist for a monthly magazine. He also had time to travel to England and study British landscape gardening.

In 1857, due in part to his knowledge of British landscape gardening and his background in scientific farming, he became superintendent of Central Park in New York City. He and his partner, Calvert Vaux, won the competition for the landscape plan for the new park in 1858. Olmsted spent much of the next three years implementing his and Vaux's plans for the park. To start, Olmsted employed 900 men to clear 770 acres of waste ground.

In 1861, Olmsted left the park work to lead the U.S. Sanitary Commission in an effort to help the suffering of ill and wounded Northern soldiers in the Civil War. While he was traveling in the South, he wrote a series of letters to the *New York Daily Times*, later reprinted in three volumes as *The Cotton Kingdom* in 1861. The letters were considered strong antislavery statements.

In 1863, Olmsted resigned his position in the Sanitary Commission to become superintendent of the Mariposa Company, a mining company with headquarters near Yosemite Valley, California. In California, he became involved in proposing that the Mariposa Big Trees and Yosemite Valley become state parks. As a member of California's state Yosemite Commission, he produced a report on preserving Yosemite as a park. He warned as early as 1865 that the goal of management of Yosemite should be "the preservation and maintenance as exactly as is possible of the natural scenery." However, other members of the Yosemite park commission saw to it that the report never reached the state legislature. Not until 1952 was Olmsted's document finally discovered and published for the first time.

In 1865, Olmsted returned to New York, where he rejoined his partner, Calvert Vaux, and the two continued working as landscape architects for Central Park and Brooklyn's Prospect Park. Olmsted continued in landscape architecture for the rest of his life. Olmsted was influenced by his observations and studies of European parks, gardens, and estates. He was impressed with the landscape design of Capability Brown in England. Olmsted also visited Paxton's Birkenhead Park in Liverpool, England, as well as other gardens. As a result of his landscape observations in Europe, the design Olmsted and Vaux implemented at Central Park had many elements of English landscape design. Central Park has long, sweeping vistas and scenery, and even the outcroppings of rocks were left in place to remind visitors of the original landscape. Central Park was designed as a place of peace, a refuge from the stress of the city.

Olmsted also incorporated in his landscape designs detailed plans for integrating traffic flow. His urban parks, including Central Park, integrated the city and landscape by taking the radical step of routing traffic through the park, not away from it. He did this without disrupting the landscape by physically separating roads by either sinking them out of sight or hiding them behind landforms. In Central Park, he included transverses at 66th, 81st, 86th, and 96th streets. These transverses were sunk about eight feet below the level of the park and today help minimize the visual and noise pollution of the cars. He created a loop around the park for strolling, running, and parading.

Although the design of Central Park is associated with Olmsted's name, it was just the first of his many important landscape architecture designs. The construction of Central Park brought Olmsted and Vaux fame and recognition, and as a result the two men attracted clients across the United States. In 1868, Olmsted, Vaux, and other partners began their work in Buffalo, New York. The city of Buffalo is home to America's oldest co-ordinated system of public parks and parkways. Olmsted's design for Buffalo consisted of three public grounds: a very large park featuring a naturalistic landscape, a public ceremonial space, and a military drill ground, all of which were connected by broad "parkways" that excluded all commercial traffic and extended the park throughout the city. Today the majority of Olmsted's designs in Buffalo are substantially intact and represent one of the largest bodies of work by the master landscape architect. The portions of the Buffalo park system designed by Olmsted are listed on the National Register of Historic Places.

From 1874 to 1892, Olmsted planned the expansion and landscaping of the U.S. Capitol grounds. In describing his plan for the Capitol grounds, Olmsted noted that "the ground is in design part of the Capitol, but in all respects subsidiary to the central structure" (United States Capitol Web site). Therefore, he was careful not to group trees or other landscape features in any way that would distract the viewer from the Capitol. The use of sculpture and other ornaments was kept to a minimum. Many of the trees on the Capitol grounds have historic or memorial associations. Among the oldest is the Cameron Elm near the House entrance. This tree was named in honor of the Pennsylvania senator who ensured its preservation during Olmsted's landscaping project. Other trees commemorate members of Congress and other notable citizens, national organizations, and special events. In addition, over thirty states have made symbolic gifts of their state trees to the Capitol grounds. Many of the trees on the grounds bear plaques that identify their species and their historic significance. The eastern part of the grounds contains the greatest number of historic and commemorative trees.

Olmsted's most ambitious plan, however, was for the "Emerald Necklace" series of parks in Boston. The Emerald Necklace is one of the oldest series of public parks and parkways in the country. The "necklace" stretches from the Boston Common downtown to the Arnold Arboretum and Franklin Park in Roslindale and Roxbury. This involved 2,000 acres for five large parks that interconnect riverfront, parkways, and playgrounds. Once again, Olmsted was integrating the city with the parks. For this project, he consulted Charles Sprague Sargent, a scientist and Harvard professor, who believed in the scientific study of trees. In 1872, their separate visions came together on a former farm in the Jamaica Plain section of Boston to create the Arnold Arboretum, a part of the necklace. The Arnold Arboretum is a 265-acre public park and center for the scientific study of North American trees and shrubs. The arboretum landscape remains one of the best-preserved Olmsted landscapes in existence today.

When Olmsted returned to New England to plan the Boston park system, he worked out of his Brookline office from 1883 until his retirement in 1895. From 1888 to 1895, Olmsted worked on the Biltmore estate in North Carolina owned by G. W. Vanderbilt. Olmsted hired a young forester, Gifford Pinchot, to manage the forests on the estate. The grounds of Biltmore, a total of almost 8,000 acres, were unparalleled in scale and diversity when they were conceived a century ago and remain one of the premier achievements of Olmsted. Many of the original plants have been preserved.

Olmsted's other famous parks include the Belle Isle Park in Detroit, Mount Royal Park in Montreal, and Roger Williams Park in Providence. He also planned city parks for Seattle, Baltimore, Atlanta, Rochester, and Louisville, among other cities. As a city planner, Olmsted designed suburban developments such as Riverside in Illinois and Parkside in Buffalo. He designed university campuses such as the University of California at Berkeley and Stanford University.

Olmsted's office in Brookline, Massachusetts, remains much as it was in his day. Designated a national historic landmark in 1963 and a national park in 1979, the office contains close to one million original documents and plans of nearly five thousand projects, making it one of the most researched collections in the national park system. Today, researchers have the opportunity to survey and study historic records related to parks, gardens, communities, estates, and campuses designed by Frederick Law Olmsted and the Olmsted firm, including Acadia National Park in Maine, Atlanta Parks in Georgia, the Baltimore park system in Maryland, the Boston park system in Massachusetts, the Buffalo park system in New York, Central Park in New York City, the Chicago park system in Illinois, the Denver Park Commission in Colorado, the Louisville Park System in Kentucky, Prospect Park in Brooklyn, New York, the Seattle park system in Washington, Stanford University in California, the U.S. Capitol grounds in Washington, D.C., the U.S. Military Academy in West Point, New York, the White House grounds in Washington, D.C., and Yosemite National Park in California. The National Park Service is currently cataloging and conserving the collections, which contain landscape photographs, initial surveys, field sketches, general plans, planting lists, presentation drawings, business records, and scale models.

In giving advice to young professionals, Olmsted said, "The best thoughts come to us unaware: not by study; that is, not directly by study. But if entirely without study, you will not have knowledge enough or strength enough to pick up the gold you stumble upon. But don't let study or what you have arrived at or hope to arrive at by study be the end, only the means. What you want is unconsciously and incidentally to cultivate your eye and the eye of your mind and heart. It will come when you do not know it—This appreciation of excellency. 'Never fear' " (Fredrick Law Olmsted National Historic Site). Olmsted died in Waverley, Massachusetts, in August 1903.

BIBLIOGRAPHY

Beveridge, C. E., & Rocheleau, P. (1995). *Frederick Law Olmsted: Designing the American Landscape*. New York: Rizzoli Books.
Frederick Law Olmsted Homepage. *http://www.fredericklawolmsted.com/*
Frederick Law Olmsted National Historic Site. *http://www.nps.gov/frla*
Olmsted, F. L. (1861). *The Cotton Kingdom*. New York: Mason Brothers.
Olmsted, F. L. (1865). Yosemite and Mariposa Grove: A Priliminary Report, 1865. Available: *http://www.yosemite.ca.us/history/olmsted/report.html*
Sterling, K. B., Harmond, R. P., Cevasco, G. A., & Hammond, L. F. (Eds.) (1997). *Biographical Dictionary of American and Canadian Naturalists and Environmentalists*. Westport, CT: Greenwood Press.
United States Capitol Web site. Frederick Law Olmsted's 1874 plan for capitol grounds. *http://www.aoc.gov/grounds/olmsplan.htm*

                 —*John Mongillo*

SIGURD OLSON
(1899–1982)

I have discovered in a lifetime of traveling in primitive regions, a lifetime of seeing people living in the wilderness and using it, that there is
a hard core of wilderness need in everyone, a core that makes its spiritual values a basic human necessity.

—Sigurd Olson

Sigurd Olson was a writer, conservationist, and environmental activist who
believed that wilderness was essential to the sustenance of the human spirit.
Olson believed that the psychic, as well as physical, needs of humanity were
rooted in the Pleistocene environment that dominated the evolutionary history of our species. This view, combined with his single-minded focus on
spiritual values, distinguished him from other leading philosophers of the
wilderness-preservation movement. Olson was influenced by the naturalist
writers W. H. Hudson and John Burroughs, as well as many other thinkers
and social critics of the nineteenth and twentieth centuries, including Henry
David Thoreau and Aldous Huxley. Olson argued that people could best
come to know their "true selves" by returning to their biological roots.
Olson's wilderness philosophy may well be considered a theology because
it was deeply connected with his beliefs about the nature of God.

Olson was born in Chicago, Illinois, the second of three sons of Ida May
Cederholm and Lawrence J. Olson, a stern Swedish Baptist minister. At the
age of seven, Olson moved with his family to Sister Bay, Wisconsin, to a
house surrounded by woods, farms, and swamplands. Olson lived for three
years on the Door County peninsula, followed by three years in the north
central logging town of Prentice. He then spent his teenage years in Ashland, on the shore of Lake Superior. During his childhood in northern
Wisconsin, he formed his lifelong attachment to nature and outdoor recreation. Olson's early explorations of the woods, fields, and shorelines of
Door County and his later excursions at Prentice and Ashland were partly
an outgrowth of his inherent passion for natural beauty. However, they
also served as an escape from the oppressive atmosphere of his home and
school and as an outlet to express his identity.

Olson earned a bachelor's degree from the University of Wisconsin in
1920. In 1921, he married Elizabeth Dorothy Uhrenholdt, with whom he
eventually had two sons. He started teaching high-school biology in 1923
in Ely, Minnesota, an iron-mining town at the edge of a huge, lake-studded
wilderness known then as the Quetico-Superior (which is now called the
Boundary Waters Canoe Wilderness Area). During the summers, Olson

earned extra money and satisfied his appetite for the outdoors by working as a guide for a local outfitting company.

One summer evening in the early 1920s, after setting up camp on an island in Robinson Lake, Olson paddled his canoe across the lake, where there was a rock summit that offered a spectacular view of the wilderness to the west. Olson climbed to the crest in time to watch the sunset, but for him the experience amounted to more than admiration of a beautiful red sky. As he recalled in his journal in 1930, "The sun, a round red ball on the horizon, [was] separated from me by leagues and leagues of primitive wilderness. . . . For a brief moment I experienced the sensation of feeling the earth move away from the sun. . . . Never before had I experienced anything which placed me so in harmony with the infinite" (Backes n.d.). Olson described his feelings of peace and contentment as a sense of "union of myself with the plan of creation."

After this profound event, Olson began to seek such epiphanies, or "flashes of insight" as he began to call them, whenever he could take time alone outdoors. During canoe trips, after his clients had bedded down for the night, Olson often would head out by himself, seeking another brush with the cosmos. Olson's observations convinced him that nature in general, and wilderness in particular, could play a crucial spiritual role for modern civilization. The sense of oneness with creation that he had felt at Robinson Peak and many times afterward was apparently similar to a religious awakening. Gradually, Olson came to believe that his life's mission was to spread the joyful news of wilderness salvation. This conviction formed the basis of both his conservation and writing careers.

Ely remained Olson's home for the rest of his life, although in 1922 he returned briefly to the University of Wisconsin for graduate work in geology and in 1932 earned a master's degree in animal ecology from the University of Illinois. Olson was active as a conservationist in the 1920s, battling to keep roads, and later dams, out of the Quetico-Superior area. In 1936, he became dean of Ely Junior College, a position he held until he resigned in 1947 to become a full-time writer and professional conservationist.

One central element of Olson's wilderness theology was his belief that human instincts demand contact with wild nature; it was the earliest precept of his wilderness credo. "We all have a pronounced streak of the primitive set deep in us," he wrote in a statement in *Field & Stream* in 1928, "an instinctive longing that compels us to leave the confines of civilization and bury ourselves periodically in the most inaccessible spots we can penetrate" (29). This was his first description of what he came to call humans' "racial memory." Olson put far more emphasis on the idea of racial memory than did any other leading thinker or nature writer of the wilderness movement; he referred to it repeatedly in his speeches and writings. To Olson, the theory of racial memory provided rational, evolutionary support for his justification of wilderness preservation on spiritual grounds.

In his early life, Olson struggled to develop an effective writing style, sometimes succeeding and frequently failing, but eventually becoming known to many as "the poetic voice of the modern wilderness movement." Robert Marshall, a founder of the Wilderness Society, identified Olson's strongest gift as his skill at describing the effects of wilderness on the human psyche in an unpretentious, lyrical style that commanded people's attention and held their interest. In September 1938, Olson published an article entitled "Why Wilderness?" in *American Forests*, his second article to appear in a national magazine and a piece that helped to shape national debate about the meaning of wilderness. Olson believed that by returning to their evolutionary foundations in the wilderness, humans could more easily open themselves to all spiritual experience and understand themselves more fully.

"Love" was also an important element in Olson's brand of environmentalism, because he believed that "[t]here can be no real, lasting land ethic without love." But Olson's idea of love went beyond the anthropocentrism common to both the Christian and humanistic world views. While he often spoke of stewardship, a concept that is based in love but implies human domination over the rest of nature, his essential philosophy delved more deeply. "What civilization needs today," he wrote, "is a culture of sensitivity and tolerance and an abiding love of all creatures, including mankind" (Backes n.d.).

While Olson was dean of the local junior college, he often felt stalled in his career and sometimes despaired of his chances to achieve his dream of writing full-time. In the 1940s, Olson spearheaded a precedent-setting campaign to prohibit airplanes from flying into the Quetico-Superior area, and the conflict propelled him into the forefront of the conservation movement. Olson served as the wilderness ecologist for the Izaak Walton League of America from 1948 until his death, as vice president and then president of the National Parks Association from 1951 to 1959, as vice president and then president of the Wilderness Society from 1963 to 1971, and as an advisor to the National Park Service and to the secretary of the interior from 1959 to the early 1970s.

Olson helped to draft the Wilderness Act, which became law in 1964 and established the U.S. National Wilderness Preservation System. In 1956, he wrote to Senator Hubert Humphrey, "In view of the mounting pressures of population, commercialization, and industrial expansion . . . the only way to assure future generations that there will be any wilderness left for them to enjoy is to give such areas congressional sanction now. . . . I feel strongly that this is the last chance to preserve the wilderness on this continent for we are on the verge of an era where the pressures to destroy or change it will become greater than anything we have ever experienced" (Olson 1956).

Olson became a living icon to many environmentalists of his day, "the personification of the wilderness defender," according to former Sierra Club president Edgar Wayburn (Profile of Sigurd Olson). As Olson said at a Sierra Club conference in 1965, "I have discovered in a lifetime of traveling in primitive regions, a lifetime of seeing people living in the wilderness and using it, that there is a hard core of wilderness need in everyone, a core that makes its spiritual values a basic human necessity. There is no hiding it. . . . Unless we can preserve places where the endless spiritual needs of man can be fulfilled and nourished, we will destroy our culture and ourselves."

Olson was trusting and sentimental, but he was also a strong leader who could bring together conflicting factions of environmentalists. Despite this widespread support, however, even Olson had his opponents. In his hometown of Ely, for example, many blamed wilderness regulations for the poor local economy. In 1977, as he came forward to testify on the bill that would become the Boundary Waters Canoe Area Wilderness Act of 1978, he was jeered and hanged in effigy at a public field meeting. His position never faltered, and he maintained that "[s]ome places should be preserved from development and exploitation for they satisfy a human need for solace, belonging, and perspective. In the end we turn to nature in a frenzied chaotic world to find silence—oneness—wholeness—spiritual release" (Feingold).

Olson also played a role in the establishment of Alaska's Arctic Wildlife Refuge and helped to identify and recommend other Alaskan lands that were ultimately preserved by the Alaska National Interest Lands Conservation Act of 1980. Among his many other activities, he played key roles in the establishment of Point Reyes National Seashore in California and Voyageurs National Park in Minnesota. In recognition of his many accomplishments, four of the five largest U.S. conservation organizations—the Sierra Club, the Wilderness Society, the National Wildlife Federation, and the Izaak Walton League—each gave Olson their highest awards. In 1974, he received the Burroughs Medal, the nation's highest honor in nature writing.

Today, Olson's large, at times almost worshipful following derives especially from the humanistic philosophy that he professed in nine popular books, including *The Singing Wilderness* (1956), *Listening Point* (1958), *The Lonely Land* (1961), and *Of Time and Place* (1982); in magazine articles; and in myriad speeches and interviews. Olson had a way of writing about the natural world that continues to touch deep emotions in his readers. The Listening Point Foundation was founded in 1998 to preserve and protect Listening Point, a rustic lakeshore cabin on a tree-covered, glaciated point of rock that served as a sanctuary for Olson. The organization also promotes Olson's philosophy and seeks to build upon his legacy in the field of wilderness education.

In an August 1999 congressional tribute, Senator Russ Feingold of Wisconsin commemorated the hundredth anniversary of Olson's birth with a speech that included the following words: "Mr. President, I rise today to pay tribute to one of our nation's most beloved nature writers and dedicated wilderness conservationists, Mr. Sigurd Olson. As an architect of the federal government's protection of wilderness areas, as well as a poetic voice that captured the importance of these pristine sites, Mr. Olson left us and our children a legacy of natural sanctuaries and an ethic by which to better appreciate them. . . . On behalf of myself and the citizens of my state, as well as all Americans, I wish Sigurd Olson a very happy birthday. We are a greater country for his dedication" (Feingold).

BIBLIOGRAPHY

Backes, D. (1997). *A Wilderness Within: The Life of Sigurd F. Olson*. Minneapolis: University of Minnesota Press.

Backes. D. (n.d.). The land beyond the rim: Sigurd Olson's wilderness theology. Available: *http://www.uwm.edu/Dept/JMC/Olson/theology.htm*

Feingold, R. (1999, August 3). U.S. Senate tribute to Sigurd Olson. *Congressional Record, 145*, S10134.

The Listening Point Foundation online. The Listening Point Foundation. *http://www.uwm.edu/Dept/JMC/Olson/lpf/index.htm*

Olson, S. F. (1928, June). Reflections of a guide. *Field & Stream* 33 (2), pp. 28–29, 106–108.

Olson, S. (1938, September). Why wilderness? *American Forests, 100* (11/12) pp. 49–52.

Olson, S. (1956). Letter to Sen. Hubert Humphrey on the need for wilderness preservation. Available: *http://www.uwm.edu/Dept/JMC/Olson/letters/1950s/1956_04_03_humphrey_WildAct.htm*

Olson, S. F. (1997a). [1958]. *Listening Point*. Minneapolis: University of Minnesota Press.

Olson, S. F. (1997b). [1961]. *The Lonely Land*. Minneapolis: University of Minnesota Press.

Olson, S. F. (1997c). [1956]. *The Singing Wilderness*. Minneapolis: University of Minnesota Press.

Olson, S. F. (1998). [1982]. *Of Time and Place*. Minneapolis: University of Minnesota Press.

Profile of Sigurd Olson. Available: *http://www.uwm.edu/Dept/JMC/Olson/profile.htm*

—*Bibi Booth*

MARGARET WENTWORTH OWINGS
(1913–1999)

> Wilderness and wildlife each carry a renewing sense of meanings that were elusive—suddenly become clear. Each should be treasured in the same way.
> —Margaret Wentworth Owings (Friends of the Sea Otter online)

Margaret Wentworth Owings was a conservationist and artist who was the founder and president of Friends of the Sea Otter, and who worked on a broad spectrum of conservation and wildlife issues for nearly forty years. She played crucial roles in successful campaigns to prevent construction of a freeway through California's Redwood National Park, to protect California sea lions from near annihilation, and to protect mountain lions from bounty and trophy hunters. Though Owings sometimes described herself as "the proverbial little old lady in tennis shoes," those familiar with her work knew her as one of the leading conservationists of the late twentieth century. Her resolute will to protect endangered and threatened species left an enduring legacy for future generations: native sea otters swimming in California's coastal waters, mountain lions prowling their woodland habitat, and the splendor of old-growth redwoods.

Margaret Owings was born in Berkeley, California, in 1913 to Frank and Jean Wentworth; her father was a trustee of Mills College in nearby Oakland. "Living in a rock and live-oak garden throughout my childhood, I was aware of the environment from a very early age," Owings said (Valtin). She graduated from Mills College in 1934 and completed graduate work in art history at Harvard University's Fogg Art Museum in 1935. "My formal training was in art, and art and nature just seemed to go hand-in-hand," Owings decided (Valtin). Her environmental activism dates back at least to 1947, when she joined the Point Lobos League, a small group of people who opposed the sand mining of a spectacular beach and the construction of a housing development along the coast south of Carmel, California.

Owings's first marriage, to Malcolm Millard, took her to Chicago for the next ten years, produced one daughter, Lisa, and ended in divorce. Owings then returned to California to live in Carmel, adjacent to beautiful Point Lobos. By the time she and her second husband, renowned Chicago architect Nathaniel Owings, married in 1953, Owings was becoming known as an artist. However, her interest in conservation soon overtook her career in art. With her husband—"not really an activist at first," she said—she campaigned to keep the scenic Big Sur coast in its undeveloped state (Goldberger). Their position was initially attacked as being "anti–free enterprise," but public support gradually grew. Ultimately, with private money that the Owingses raised to match public funds, they were able to buy and preserve a beautiful, 1.5-mile stretch of coastline between the Carmel River and Point Lobos State Reserve. Owings often recalled this early achievement fondly as a major conservation victory.

In 1958, the couple took up residence in the dramatic house they had designed at Big Sur. Straddling a narrow, craggy ledge, Wild Bird, as they named their clifftop home, afforded a magnificent vantage point for studying Big Sur's diverse marine wildlife. At the same time, the couple also shared a love for the landscape and peoples of the southwestern United States—particularly New Mexico's Rio Grande Valley—and spent much of their time there as well. Together, they built an adobe farmhouse, outfitting it with antique woodwork and rustic floors, and the house became a haven for them. Margaret became part of a close-knit artistic community in New Mexico, counting among her close friends the painter Georgia O'Keeffe.

In 1962, after a mountain lion had been treed and killed for a $100 bounty near her home in Big Sur, an outraged Owings fought to abolish both bounty and sport hunting of mountain lions in the state. She approached Senator Fred Farr to draw up a bill, and the effort was rewarded with passage of a temporary moratorium on hunting of mountain lions in the state. In 1985, Owings's "Lion People," with the guidance of Assemblyman Sam Farr and Senator Henry Mello, set out to gather 600,000 signatures for an initiative petition (Proposition 117), called the California Wildlife Protection Act, which they referred to as the "mountain lion bill." The measure passed, and the mountain lions were granted a reprieve from hunting.

As recently as the 1930s, the California (southern) sea otter (*Enhydra lutris nereis*) had been feared to be extinct until a group of them was seen in 1938 at Big Sur, confirming the species' continuing, albeit precarious, existence. In 1968, when Owings and Jim Mattison, a physician and avid outdoorsman, founded the Friends of the Sea Otter (FSO), the entire population of the California sea otter numbered only about 650. As Owings said much later, "I told my husband, 'two or three years to save otters.' I know now it's not only a lifetime, but it will be many a lifetimes" (Friends of the Sea Otter). When FSO first began, it was operated on a volunteer basis; many times, meetings were held at the Owingses' home, and it was through Margaret Owings's fierce determination that the organization continued functioning. She helped to establish environmental policy to benefit the otter, approached legislators in both Sacramento and Washington, D.C., and used her insight and knowledge to unite scientists, conservationists, and educators behind FSO's mission. In 1977, FSO was instrumental in the California sea otter's formal designation as threatened under the

Endangered Species Act. The organization had worded and circulated a petition that sent over 65,000 names from every state to the U.S. Fish and Wildlife Service.

Since then, FSO has worked to protect the sea otter and its habitat from a wide range of threats. In Owings's view, "This little sea mammal, which lived along what is now called the California coast for millions of years . . . these same otters today offer the human race a personal reunion with life in the sea" (Owings 1995). As the world's first advocacy organization for sea otters, FSO has become an important voice in the fight to protect marine communities. It is in large part due to FSO's efforts that the population of the California sea otter has since grown in number and range and now includes about 2,000 individuals along the central California coastline. In 1993, FSO expanded its mission to include not just protection of the California sea otter, but sea otters throughout the North Pacific and all habitat of the sea otter worldwide. FSO, now claiming more than 4,000 members around the world, was a key advocacy group in ensuring that the Monterey Bay National Marine Sanctuary became a reality in 1995.

Margaret Owings also served as a commissioner on the California State Parks Commission from 1963 to 1969 and was on the boards of the National Parks Foundation, Defenders of Wildlife, and the African Wildlife Leadership Foundation. In 1964, in affiliation with the National Audubon Society, she helped to found the Rachel Carson Council to carry on Carson's work, which had inspired Owings early in her life. Owings served on the Environmental Defense Fund's (EDF) Board of Trustees from 1972 to 1982 and was an honorary trustee until her death. She recalled in 1986, "I'm proud, indeed, to have been associated with EDF, . . . because it was EDF that grew into action out of the voiced alarm and studies of Rachel Carson. . . . It's a proud cause that EDF has been serving over the past two decades" (Valtin). Owings also chaired the California Mountain Lion Preservation Fund and was a member of the Big Sur Land Trust.

In 1995, Owings's longtime ally Sam Farr, then a U.S. congressman, paid tribute to her before the House of Representatives, praising her as "one of the nation's most outspoken and respected conservationists" and declaring that she was "perhaps most responsible for the natural beauty that is seen in her community to this day." Farr closed by saying, "I am proud to have people like Margaret Owings in my district. Her unfaltering dedication to maintaining Big Sur's beauty and species variety sets an example that we should all strive to follow" (Farr).

Over her lifetime, Owings was nationally recognized by many organizations for her activism. Among many other honors, she received the National Audubon Society Medal, one of the most prestigious environmental awards, and was honored with the U.S. Department of the Interior's Conservation Service Award, the highest honor possible for a nonemployee.

She also earned the United Nations Environment Programme's Gold Medal Award. In 1998, the Audubon Society included Owings as one of this century's "100 Champions of Conservation." The following year, the Monterey Bay National Marine Sanctuary posthumously awarded its 1998 Organization/Institution Award to Owings for being a "renowned and cherished advocate and defender of wildlife within the Sanctuary."

In January 1999, fifteen years after her husband's death, Margaret Owings passed away of heart failure in her home, Wild Bird, at the age of eighty-five. She had spent the last months of her life preparing a compilation of her writings and artwork, *Voice from the Sea: And Other Reflections on Wildlife and Wilderness*. Published less than a month before her death, the book spans fifty years of Owings's writing and conveys the passion that underlay her lifetime of activism in defense of the California coast and its wildlife. "It's so important for every person to do whatever they can," Owings often urged. "I sometimes think people feel that one individual can't make a difference, and I'm afraid a lot of young people these days don't involve themselves in environmental issues for just that reason" (Valtin).

BIBLIOGRAPHY

Associated Press. (1999). California enviro legend passes away. *http://www.defenders.org/defenders/gline803.html*

Audubon Magazine. Champions of Conservation. *http://magazine.audubon.org/century/ champion.htm1#OWINGS*

Fahys, J. (1997, April 27). California woman honored for efforts to protect environment. *Salt Lake Tribune.* Available: *http://www.sltrib.com/97/apr/042797/utah/13449.htm*

Farr, S. (1995, October 11). Tribute to Margaret Owings in the House of Representatives. *Congressional Record.* Available: *http://www.torhouse.org/thfsfarr.htm*

Friends of the Sea Otter online. *http://www.seaotters.org/index2.html*

Goldberger, P. (1996, June). Revisiting Big Sur's Wild Bird: The California house of Nathaniel and Margaret Owings. *Architectural Digest.* Available: *http://www.benheinrich.com/bigsur_wild_bird_article.htm*

Monterey Bay National Marine Sanctuary online. *http://www.mbnms.nos.noaa.gov/Research/MBNMSawards.html*

Owings, M. W. (1995, October 7). A talk at the annual meeting of the Tor House Foundation, with prophetic words by Robinson Jeffers. Available: *http://www.torhouse.org/thfowing.htm*

Owings, M. W. (1998). *Voice from the Sea: And Other Reflections on Wildlife and Wilderness.* Monterey, CA: Monterey Bay Aquarium Press.

Valtin, T. (1986, October). Margaret Owings. *Environmental Defense Fund Newsletter.* Available: *http://www.edf.org/pubs/EDF-letter/1986/Oct/m_owings.html*

—*Bibi Booth*

Photo courtesy of Academy of Natural Sciences.

RUTH PATRICK
(1907–   )

Ruth Patrick was born on November 27, 1907, in Topeka, Kansas. She spent her childhood, along with her sister, growing up in both Topeka and Kansas City. Her interest in the environment was sparked at an early age by her father, Frank Patrick. Although Frank Patrick was a lawyer and banker by profession, he had a degree in botany from Cornell University. It was this interest in the environment that he instilled in Ruth. When Ruth was seven, he gave Ruth her first microscope. Together they spent countless hours looking at microscopic organisms gathered from local rivers, streams, and ponds. It was under the microscope that Ruth first learned about diatoms, organisms that she would spend the rest of her life studying.

After graduating from high school and after taking a few biology courses at the University of Kansas in Lawrence, Patrick attended Coker College in Hartsville, South Carolina. At that time, Coker was a small college for woman. During summer breaks, Frank Patrick sent Ruth to Woods Hole

Oceanographic Institute in Massachusetts and to Cold Spring Harbor Laboratory on Long Island, New York, to participate in summer courses. Upon graduating from Coker College, Patrick attended the University of Virginia in Charlottesville, where she earned both her M.S. and Ph.D. studying the diatoms of Southeast Asia. She was twenty-six years old when she received her Ph.D. in 1934.

In 1931, while Patrick was still in graduate school, she married Charles Hodge IV. She had met Hodge during one of her summer stints at Cold Spring Harbor Laboratory. At the request of her father, she continued to use her maiden name of Patrick for her research. It was his way of keeping the Patrick name alive and connecting his name to his real love, science. Shortly after her marriage, Patrick moved to Philadelphia with Hodge, where his family had long been established. Hodge was a direct descendant of Benjamin Franklin, and like Franklin and Patrick, he was a scientist in his own right. Hodge's area of expertise was entomology, with a particular interest in beetles and grasshoppers. He taught zoology at Temple University in Philadelphia for over thirty years. He and Patrick had one child, Charles Hodge V, who inherited his parents' interest in science and became a pediatrician in Kansas City, Missouri, conducting research on the digestive tract.

In 1985, after fifty-four years of marriage, Charles Hodge IV died, leaving Patrick to carry on her research. In 1995, at the age of eighty-seven, Patrick married longtime family friend Lewis H. Van Dusen, Jr., a prominent lawyer from Philadelphia. To this day, both Van Dusen and Patrick continue to work at their professions.

From the beginning, Patrick has been an outspoken proponent for the environment, even before the age of *Silent Spring*. From an early age, her father encouraged her to stand by her beliefs. Because she grew up during a time when it was unusual for a woman to go to college, never mind become a scientist, Patrick had to learn to be forthright and plainspoken. Her father told her, "When you enter a boardroom full of men, you can't afford to be feminine. Don't waste time talking about pleasantries. Always speak to the point and always back up your views with data." Although many boardroom men were at first caught off guard by Patrick's manner, her father's advice was well heeded, and she quickly gained the respect of all those who came in contact with her.

It was this outlook that allowed her to get her foot in the door as a researcher at the Academy of Natural Sciences in Philadelphia in 1931. Although she was unpaid at first, her relationship with the academy has lasted to this day, first as a researcher (1931–1937), then as the academy's curator of the Leidy Microscopical Society (1937–1947), chairman (and founder) of the academy's Limnology Department (1947–1973), and, finally, the holder of the Francis Boyer Chair of Limnology (1973–present). In 1940, as a result of a scientific paper she presented, Patrick was hired by an oil-company executive to head a study to determine if diatoms could be used to describe the health of a given body of water. This was the breakthrough that Patrick needed. To this day, Patrick credits William B. Hart, that oil-company executive, as the person who established her credibility as a scientist and a leader in the science community, regardless of her gender.

From the study of diatoms as an indicator species grew Patrick's theory that in order to study the health of an aquatic ecosystem, one must study all major groups of plants and animals within an aquatic system. She established a study of a 475-square-mile region drained by Conestoga Creek located near Lancaster, Pennsylvania. Through this study, her team of scientists determined that healthy aquatic systems have a rich diversity of organisms, while unhealthy (polluted) aquatic systems decreased the diversity to a few hardy organisms that could resist the pollutions. This community approach established Patrick as an expert in freshwater aquatic systems and led to a lifelong relationship with major industries trying to curb their impact on the environment. Her study on the effects of pollution on aquatic communities became well known within the scientific community long before the rest of the world became conscious of the impact of pollution through Rachel Carson's *Silent Spring*.

Patrick's chance meeting with Hart led to many other such opportunities with other corporate CEOs, including Crawford Greenwalt. In 1949, Greenwalt contracted Patrick to begin baseline studies of the Savannah River on the future site of what now is the Department of Energy's Savannah River Site. Greenwalt, at that time president of Du Pont, was concerned with the possible impact his company might have on the health of the Savannah River when it constructed the Savannah River Nuclear Plant, the country's first and only tritium-production facility. Patrick's involvement with the Savannah River Site continued until 1997, when she stepped down from the Savannah River Site's Environmental Committee.

Through the years, Patrick has amassed many recognitions and awards, including twenty-five honorary degrees and over forty other honors and awards such as the National Medal of Science (1996), the American Society of Limnology and Oceanography's Lifetime Achievement Award (1996), the Benjamin Franklin Award for Outstanding Scientific Achievement from the American Philosophical Society (1993), the Golden Medal of the Royal Zoological Society of Antwerp, Belgium (1978), election to the American Academy of Arts and Sciences (1976), and election to the National Academy of Sciences (1970) (only the twelfth woman at that time to be so honored). In addition to these awards and honors, Patrick has been a member of numerous councils, committees, and boards, including President Lyndon Johnson's Science Advisory Committee, Panel on Water Blooms, 1966; chairman of the Panel of the Committee on Pollution for the National

Academy of Sciences, 1966; chairman of the Panel on Ecology of the Executive Advisory Committee, Environmental Protection Agency, 1974–1976; chairman of the National Academy of Sciences, Section on Population Biology, Evolution, and Ecology, 1980–1986; member of the National Council of the World Wildlife Fund, 1986–present; member of President Ronald Reagan's Advisory Commission on Acid Rain, 1982; and member of the Board of Governors of the Nature Conservancy.

Those lucky enough to know Patrick and have an opportunity to have a casual conversation with her are captivated by her energy, her wit, her self-effaciveness, and her charm. When the conversation turns to the environment, one can see her father's advice kick in when she becomes very serious as she discusses the state of the environment and what we should do as citizens to ensure the health of the environment. Because of Patrick's strong reputation as an environmentalist, many would think that she is antiindustry or antibusiness. Patrick is neither. She believes that business and industry play an important role in the world as we know it; however, she also believes that they have an obligation to do everything within their means to ensure that their actions do not impair the environment. It is because of this that during her lifetime, Patrick has been sought after by CEOs of many of the largest corporations to help them devise environmental plans that will enable them to have as minimal an impact on the environment as possible.

Patrick has a soft spot in her heart for children. She believes very strongly that the future of the ecological health of the world is in its children. She encourages youngsters to get involved with the environment. She encourages parents and teachers to take their children outside and experience the environments in which they live. Even today, she is known to put her waders on and lead groups of students into waterways to discuss their ecology. In 1991, the University of South Carolina Aiken named its science education center after Patrick. Today, over 50,000 students, teachers, and members of the general public participate in science and mathematics programs sponsored by the Ruth Patrick Science Education Center.

Although she turned ninety-two in 1999, Patrick has no plans to slow down. She continues to go to her office in Philadelphia at least three days a week, while working at home the remaining days of the week. She is currently working on the final volume of her series *Rivers of the United States*. Volume 5a was published in March 2000.

Associates and colleagues of Patrick use these terms to describe her: intelligent, tireless, devoted, committed, and a person of integrity. Probably the term most used to describe her is friend. Ruth Patrick is not only a friend to those who know her, but a friend of the environment. What better way is there to be remembered for a lifelong devotion to unlocking the doors of knowledge concerning the environment?

BIBLIOGRAPHY

Center for Environmental Literacy. *http://www.mtholyoke.edu/proj/cel/patrick.
htm*

Patrick, R. (1994). *Rivers of the United States*. Vol. 1, *Estuaries*. New York: John
Wiley and Sons.

Patrick, R. (1995). *Rivers of the United States*. Vol. 2, *Chemical and Physical Characteristics*. New York: John Wiley and Sons.

Patrick, R. (1996). *Rivers of the United States*. Vol. 3, *The Eastern and Southeastern States*. New York: John Wiley and Sons.

Patrick, R. (1998). *Rivers of the United States*. Vol. 4, Part A, *The Mississippi River
and Tributaries North of St. Louis*; Part B, *The Mississippi River Tributaries
South of St. Louis*. New York: John Wiley and Sons.

Patrick, R. (2000). *Rivers of the United States, Vol. 5*, Part A, *The Colorado River*.
New York: John Wiley and Sons.

Patrick, R., Douglass, F., Palavage, D., & Stewart, P. (1992). *Surface Water Quality: Have the Laws Been Successful?* Princeton, NJ: Princeton University
Press.

Patrick, R., Ford, E., & Quarles, J. (1987). *Groundwater Contamination in the
United States* (2nd ed.). Philadelphia: University of Pennsylvania Press.

Patrick Center for Environmental Research. *http://www.acnatsci.org/research/pcer*

Ruth Patrick Science Education Center. *http://rpsec.usa.sc.edu*

Steinmann, M. (1985). *Ruth Patrick in Science Year: The World Book Science
Annual, 1985*. pp. 351–362.

—*Jeff Priest*

MARIA PEREZ
(1983–)

At only sixteen years of age, Maria Lidia Perez is a youth environmental
organizer at Concerned Citizens of South Central Los Angeles. She is also
a junior at Thomas Jefferson High School in South Central Los Angeles.
Maria has resided in Los Angeles her whole life. She lives with her mother,
father, and sister.

Perez was born on May 25, 1983, in Los Angeles. She spent six years of
her life in Hollywood, then moved down to South Central Los Angeles,
where she continues to reside. She has moved three times within the South
Central area. The second house was the one where she spent most of her
years, from the time she was seven years old until she was almost sixteen.
She lived there with her sister, Esmeralda Perez, her brother, Hugo Asencio,
her mother, Lidia Perez, and her father, Cruz Perez.

Growing up was not easy for Perez. Things started to change when her
sister Esmeralda was born. Perez was nine at the time. Her parents had just
bought a new house, so when Perez's sister was born, the family experi-

Photo courtesy of James A. Brown.

enced much economic hardship because there was another mouth to feed. Soon her uncle had to come to help the family out. Everything went well for about three years until her uncle passed away. Perez was only eleven years old. Around the same time, her mother had an epileptic seizure, fell, broke her shoulder, and had to have surgery. Only twelve years old, Perez had to grow up, play the role of an adult, and take care of her three-year-old sister. All the opportunities she had in the sixth grade of going on trips with her classmates became only a dream because she had to stay home and help her mother out. Even though Perez was trying to act like an adult and help around the house, she still had dreams of becoming a lawyer or a doctor because her father had always encouraged her to do this so that she could build the family income. As with many young people, Perez's dreams of becoming a doctor changed as she grew up. The only dream left was the one of becoming a lawyer. Even though Perez had to take more responsibilities at home, she still continued to do well at school, getting As and Bs and remaining on the honor roll. Perez never disappointed her parents at school. At parent conferences, her teachers always had something good to say. Perez recalls her sixth-grade English teacher Mrs. Ruiz telling her father, "It is such a pleasure having Maria in my class, she is always helping me. However, sometimes she does talk a little bit too much." That is one of the problems Perez has always had.

During Perez's vacations in the winter of 1997, she started working for Clean and Green, a program that hires youth as young as thirteen to work cleaning up their neighborhood. When Perez started working for Clean and

Green, she made $4.25 an hour. She started working there to help her family out. After the first day at work, Perez remembers telling her mother, "I'm never dropping out of school. I am going to graduate from high school, go to college, and get a good job that will not have me working so hard." Perez had to go back to school, but she continued working for Clean and Green on Saturdays thanks to Steve, her supervisor. Having a job was a tremendous help to Perez and her family. She was now able to buy her own clothes and shoes. She was really happy to be able to buy her own name-brand shoes because this was something her parents could not afford to do. When Perez's summer vacation came, she continued working for Clean and Green, but this time it sent her to the office of Concerned Citizens of South Central Los Angeles (CCSCLA), where she was employed doing clerical work. When Perez arrived at CCSCLA, Cindy Migares, the office secretary, introduced her to Juanita Tate, the executive director of CCSCLA, and Tate gave her a big hug. Perez says, "I felt like if I was right at home." Perez met everyone in the office, including Melodie Dove, her supervisor at the time. Perez worked the entire summer at CCSCLA and learned a lot about how an office operates. She enjoyed working at CCSCLA so much that once the Clean and Green session had ended, she continued volunteering at CCSCLA on her own time.

In May 1998, Perez was hired by Melodie Dove in CCSCLA to work on a pilot program that would focus on environmental and workplace rights. The program was later named POWER (People Organized for Workplace and Environmental Rights). Working with the program really opened Perez's eyes. She had lived in South Central Los Angeles all her life and has never known all the injustice that her community faced. She had never known anything about lead poisoning and that the zip code she lived in, 90011, is the zip code with the highest number of children with lead poisoning in Los Angeles County. The issue that bothered Perez most was Jefferson New Middle School (JMS). "I never knew that LAUSD [Los Angeles Unified School District] would build a school on a contaminated site, a site that sits directly across the street from a federal Superfund site," Perez says. She wondered if this happened because students attending the school are all children of color. Maria subsequently stated, "I just could not understand this. I had attended schools in LAUSD for more than 9 years. Had I attended a school that was contaminated?"

Perez and the rest of the group went into action. They organized their first community meeting. The meeting was to inform the community about the issue surrounding JMS. They had people from different departments, each one of which had played a role in the building of JMS. The person Maria will always remember is Bill Piazza, who was the lead environmentalist on JMS at the time. The lies he told people from her community upset her and everyone else involved. Although Perez's parents have always taught her to respect her elders, that day, August 4, 1998, she forgot all the things her parents had taught her for a moment and began questioning Piazza because she felt that he was putting her community down.

After the community meeting, the youth continued organizing for JMS through outreach, door knocking, talking to people from the Southern California Regional Water Quality Control Board (SCRWQCB), the Air Quality Management District (AQMD), the Los Angeles Unified School District (LAUSD), the Department of Toxic and Substance Control (DTSC), and Senator Tom Hayden. "Meeting Senator Tom Hayden was a great experience. The best part of it was that he came to our office, to talk to us about JMS." After months of working on Jefferson Middle School, in March a reporter from the *Time* magazine came down and interviewed "the toxic crusaders," who included Maria Perez, Nevada Dove, and Fabiola Tostado, about the issue at JMS. Because of the *Time* article, they were also featured on CNN on April 18, 1999. "It felt weird appearing on TV, it never occurred to me that I would be doing this type of work at the age of fifteen. I am truly making a difference in my community. Yes, it is my community as long as I continue to live here." Perez was nervous about appearing on television, but knowing that other youth her age would see this and think that they could also make a difference where they live made her proud. After appearing in *Time* magazine and on CNN's *Newsstand Times*, Perez and the other crusaders were interviewed for *Seventeen* magazine and were featured in the September issue "20 under 20," which focused on young people who are working hard to bring about change in the community in which they live. Perez considers herself a youth environmental organizer. Her work is important not because she has received media attention but because she has helped to bring about positive changes in her community. Perez never planned to be an environmental organizer; in fact, she had never even heard of the term. However, she has learned that when one has the drive to make things different in one's community, there is no limit to what one can do. Perez's final words of advice to all are, "If you want to make a difference, nobody can stop you, no matter how young you are or what color your skin is. You can always make a change if you just try, and if you really want to."

—*Maria Perez*

Photo courtesy of Roger Tory Peterson Institute
of Natural History.

ROGER TORY PETERSON
(1908–1996)

The philosophy that I have worked under most of my life is that the
serious study of natural history is an activity which has far-reaching
effects in every aspect of a person's life. It ultimately makes people
protective of the environment in a very committed way. It is my opinion
that the study of natural history should be the primary avenue for creating environmentalists.

—Roger Tory Peterson

Since the death of Roger Tory Peterson on July 28, 1996, many people
have written and spoken about the relevance of Peterson's life in the context of the environmental movement of the twentieth century. The most
knowledgeable of these writers recognize the fact that Peterson ushered in
the education phase of the movement that is still going strong today. Since
1934, when Peterson published A *Field Guide to the Birds*, looking at the

outdoor world has never been the same. Before this book, ornithological works were written for students and scientists. Peterson gave the tools that once belonged only to the field biologist to the ordinary person, who could now easily learn about and understand the natural world. Citizen action evolved from this awareness. Today there are fifty-two volumes in the Peterson Field Guide Series, all either written or edited by Peterson. Without the foundation that Peterson laid, we would not have a countless number of birding associations or the tremendous number of wildlife refuges and perhaps not even the Endangered Species Act. Peterson also became well-known for his art, photography, and motion picture wildlife documentaries.

Peterson's beginnings shaped him into the man he became. He was born on August 28, 1908, in Jamestown, New York. Jamestown is in Chautauqua County, the westernmost county in the sprawling state of New York. Lake Erie borders the county on the north, and the Pennsylvania state line skirts the south. People from this region say that their roots are more midwestern than eastern. Jamestown is on the Allegheny Plateau and is surrounded by rich hardwood forests and fast-flowing streams.

Peterson was the son of the Swedish immigrant Charles Gustav Peterson and German immigrant Henrietta Bader. The influx of Swedes into Jamestown during the Industrial Revolution upset the status quo for a city comprised primarily of descendants from the landed English gentry. At the time, Jamestown was a city of mills for worsted woolens; the Swedes soon changed it into a city of furniture factories. There was much poverty in and discrimination against the Swedish community during this time. Peterson was often ridiculed by the other boys for his interest in birds and for being a Swede.

Charles Peterson came to Jamestown in 1873 at the age of two. The family's downward economic slide started just months later when his father died. Charles was forced to work in the woolen mills by the age of ten to support the family. With only a third-grade education, Charles became the breadwinner. As an adult, his expectation of his son Roger was to get a high-school education and then go to work in the mills. He was hard on Roger and found it difficult to understand the boy's curiosity about nature that occupied all of his time. As a youngster, Roger resented his father. It was not until he grew older that Roger fully appreciated "the odds that this man struggled against" (Devlin).

Peterson's mother, Henrietta (Nettie) Bader, was brought to America when she was four. A religious woman, she attended Trinity Lutheran Church with her children Roger and Margaret. When the new minister came to visit, Nettie remarked on how much Roger enjoyed birds and natural history. The minister said, "Well, that makes for unbelievers" (Devlin). Roger seriously questioned the church from then on. He grew up at 16 Bowen Street with his parents and sister. His paternal grandmother, an aunt, and six cousins also shared the Petersons' quarters. Some say that this was the reason Peterson was forever outside exploring the countryside.

Few boys had any special interest in nature, certainly not Peterson's intense, absorbing passion for everything from insects to birds. Most of Peterson's classmates were older because Roger had skipped two grades. In the cruel way of youngsters, some took to calling him "Professor Nuts Peterson." Peterson's mother, ever understanding, made him a butterfly net and even went to the druggist to explain why her son needed cyanide to preserve his specimens. His father was less understanding. But one need not think of Peterson as a lonely or unhappy child. He found great joy and happiness in all things wild and beautiful and still found time to play or scrap with the boys in the neighborhood.

At the age of eleven, birds "took over" his life. His seventh-grade teacher, Blanche Hornbeck, enrolled her students in the Junior Audubon Club, taught them about birds, and often walked them to a nearby forest where she used nature to teach writing, art, and science. It was during that year on a cold March morning that Peterson had an experience that was to shape the rest of his life. While he was hiking with a friend at nearby Swede Hill, the boys spotted a seemingly lifeless clump of brown feathers on a tree, very low to the ground. Although the northern flicker was merely sleeping, the boys thought that it was dead. "I poked it and it burst into color, with the red on the back of its head and the gold on its wing. It was the contrast, you see, between something I thought was dead and something so alive. Like a resurrection. I came to believe birds are the most vivid reflection of life. It made me aware of the world in which we live" (Devlin).

Peterson finished high school at sixteen. He received the highest marks in his class for art, history of art, and mechanical drawing. The description of Peterson printed next to his name in his class of 1925 yearbook proved prophetic: "Woods! Birds! Flowers! Here are the makings of a great naturalist."

During the summer of 1925, Peterson painted furniture at the Union National Furniture Company for eight dollars per week. He created decorative motifs of intricate Chinese subjects on exquisite lacquer-wood cabinets made there. Peterson's artistic talents were discovered that summer. The head of the Decorating Department, Willem Dieperink von Langereis, gave Peterson his first encouragement about being an artist and insisted that he go to art school. For the next two years, Peterson worked and saved his money. He left Jamestown for the Art Students League in New York City in 1927. In 1929, he advanced to the National Academy of Design.

While Peterson was working and saving money for art school, he very intently practiced art and photography, using birds as his subjects. Two of his earliest published photographs included northern cardinals in the 1925 Jamestown High School yearbook and black-capped chickadees in the 1926 yearbook. Both editions are highly sought after by collectors today. Roger was often at Jamestown's Prendergast Free Library researching his subjects and learning all he could about the natural world. He would usually read the *Auk* (the journal of the American Ornithologists' Union), *Wilson Bulletin* (the journal of the Wilson Ornithological Society), *Bird Lore* (the National Audubon Society magazine), and *National Geographic.*

In 1925 the *Auk* printed a notice about the next American Ornithologists' Union meeting to be held in New York City at the American Museum of Natural History. There was also to be a bird art show. Peterson submitted two paintings, and both were accepted. At the meeting, Peterson met Arthur A. Allen, founder of the Cornell Laboratory of Ornithology, Frank Chapman, curator of the Department of Ornithology at the American Museum of Natural History, and the renowned bird artist Louis Agassiz Fuertes. Within a year of this historic trip, Peterson had two more paintings displayed. In the Cooper Ornithological Club's first American bird art exhibition at its Los Angeles annual meeting, Peterson's oils of great-horned owls and screech owls were shown. He was seventeen and exhibiting with the great bird artists of the time: Allan Brooks, Bruce Horsfall, George M. Sutton, and Fuertes.

In the fall of 1931, Peterson joined the science department at Rivers Country Day School in Brookline, Massachusetts. Rivers School, a private day school, was the prep school for sons of "gentlemen" in the Boston suburb of Brookline, across the Charles River from Harvard. It was here that Elliot Richardson, later to become U.S. attorney general, got his first taste of natural history and painting. Richardson was one of Peterson's best students, winning both the art prize and the natural history prize.

In Boston, Peterson became a member of America's oldest ornithological organization, the Nuttall Club. It was here that he met fellow member Francis H. Allen. Allen, an editor at Houghton Mifflin Company, accepted Peterson's first book, *A Field Guide to the Birds.* On that morning in 1933, Allen did not need the binoculars he always kept on his office desk to realize at once that he was looking at a brand-new species of guidebook. A conservative contract for this "new find" was made because it had numerous drawings and four color plates, which were quite expensive to reproduce in those days. The publisher did not even pay royalties on the first 1,000 copies. Peterson was so thrilled that he did not care. The acceptance of the book was immediate. In the first week, the entire stock of 2,000 copies sold. Peterson received ten cents per copy of the second 1,000 copies. Today, over 7 million copies have been sold, and fifty-two field guides make up the Peterson Field Guide Series. "My identification system," explained Peterson, "is visual rather than phylogenetic; it uses shape, pattern, and field marks in a comparative way. The phylogenetic order, which is related to evolution, is not emphasized within families. Similar appearing species are placed together on plates and the critical distinctions are pointed out with little arrows" (Zinsser).

Former U.S. attorney general, and lifelong friend of Peterson Elliot Richardson once wrote:

When I first came to know Roger Tory Peterson some sixty-five years ago, he was already a brilliant field ornithologist, an inspiring teacher, and a gifted painter of birds. He was teaching art and nature studies at the Rivers School in Brookline, Massachusetts, and in the sixth and seventh grades I had the extraordinary good fortune not only to be introduced to birding by him but also to be taught drawing and painting. His great ambition, he told me once, was to excel Louis Agassiz Fuertes, in his eyes the finest bird painter who ever lived.

When in 1934 *A Field Guide to the Birds* first appeared it would have been impossible to foresee that it would occupy such a dominant role in Roger's life. Houghton Mifflin, his publisher, certainly did not anticipate the book's wide appeal: the first printing was only 2,000 copies. Since then some seven million have been sold, and that does not count the additional sales of his later field guides to wildflowers, mammals, insects, and the birds of other regions.

From 1934 on, thanks to Roger's genius for simplification and communication, any amateur could learn to distinguish one bird from another. Birdwatching soon became a hobby—even a sport—for vast numbers of people. From identifying birds emerged a love of birds, and from there it was only a short step to a more inclusive passion for nature. Given, moreover, the interdependence between the survival of species and the preservation of habitat there soon developed widespread awareness of the need to maintain a systemic ecological balance. To be sure, the conservation movement that began to gain headway after World War II owed much to other contributors—Rachel Carson's *Silent Spring*, for example. Still, Roger Tory Peterson, not only as artist and field-guide author, but also, as educational director, art director, and columnist for the National Audubon Society, was a towering influence. Indeed, Rachel Carson herself was indebted to his book, his early work on DDT, and his friendship.

The idea of creating an institute named for Roger Tory Peterson in Jamestown, New York, where Roger had grown up, occurred in 1976 to several of his friends. He had just been awarded the Linnaeus Gold Medal of the Swedish Academy of Sciences and had earlier received the Conservation Medal of the National Audubon Society. His classmate Lorimer Moe took the lead in enlisting broader support, and the Roger Tory Peterson Institute was formally incorporated in 1985. Conceived in part as a repository for his life work, it would also be committed to seeking innovative ways of linking art, nature and education. (*The Guide*, Summer 1996)

Nature artist Robert Bateman once said, "As the diversity of life on the planet is diminished, it's vital that we know and understand and, in fact, cherish [our] fellow living things . . . and Peterson has given us a way to do this." Robert McCracken Peck of the Academy of Natural Sciences of Philadelphia stated, "To want to protect wildlife and the habitats that support it, people first have to know what they are protecting. Beginning with birds and later expanding to everything from trees to tadpoles, Roger Tory Peterson has made natural history accessible" (The Roger Tory Peterson Institute).

Peterson often said, "My contribution has been as a teacher" (Zinsser).

His work inspires those who influence children—parents, teachers, and naturalists—to instill a love of nature in generations to come. The environmental movement of the twentieth century began with Theodore Roosevelt's U.S. Forest Service head Gifford Pinchot. Pinchot introduced the country to the word "conservation" and the reasons why natural resources can be exhausted by human greed. Roger Tory Peterson ushered in the education wave of the movement, which interprets the natural world to every person. We now have ready access to the "tools" to learn about nature. Rachel Carson inaugurated the activist wave where citizens use legislation, litigation, and lobbying to protect the environment. The three waves of the environmental movement were inevitable, considering the drastic rise in human population and the resulting degradation of the natural world. But it takes people who are ahead of their time, people with a passion for what they do, to change the status quo, to start a revolution. Roger Tory Peterson started a revolution by bringing the understanding of nature out of the halls of academia and into the homes of America.

BIBLIOGRAPHY

Devlin, J. C. (1977). *The World of Roger Tory Peterson: An Authorized Biography*. New York: Times Books.
The Guide. Newsletter of the Roger Tory Peterson Institute of Natural History. Jamestown, NY.
Peterson, R. T. (1934). *A Field Guide to the Birds*. Boston: Houghton Mifflin.
The Roger Tory Peterson Institute of Natural History. *http://www.rtpi org*
Zinsser, W. (1994). *Roger Tory Peterson: The Art and Photography of the World's Foremost Birder*. New York: Rizzoli.

—*Jim Berry*

GIFFORD PINCHOT
(1865–1946)

The forest is as beautiful as it is useful. The old fairy tales which spoke of it as a terrible place are wrong. No one can really know the forest without feeling the gentle influence of one of the kindliest and strongest parts of nature. From every point of view it is one of the most helpful friends of man. Perhaps no other natural agent has done so much for the human race and has been so recklessly used and so little understood.

—Gifford Pinchot

Gifford Pinchot was a chief forester and conservationist who has been described as a "catalyst" to the conservation movement and the father of scientific forestry in America. Pinchot was born at Simsbury, Connecticut,

Photo courtesy of USDA Forest Service, Grey
Towers National Historic Landmark, Milford,
PA.

on August 11, 1865. He grew up staying with relatives in Connecticut in
his early summers and living the rest of the time in New York City, where
his father had his business. The family was well-to-do and lived in an environment where the children spent the winter in New York City and lived
at the seacoast during the summer. Pinchot spent much of his youth at
Grey Towers near Milford, Pennsylvania. His father and mother endowed
the Yale School of Forestry and established at Milford the first forest experiment station in the nation to encourage the reforestation of denuded
lands.

Pinchot prepared for college at Phillips Exeter Academy and in the fall
of 1885 entered Yale University, from which he graduated in 1889. With
encouragement from his father, Pinchot decided to pursue professional forestry. However, he had to go abroad to study at the French National Forestry School in Nancy, France. At that time, professional training in forestry
was not available in the United States.

Pinchot returned to the United States after one year in France and took
a job surveying forest lands for the Phelps Dodge Company. He inspected
timber stands in several states, including Alabama, Arizona, California,
Pennsylvania, and Washington. Frederick Law Olmsted, the famous land-

scape architect and an old friend of Gifford's father, soon recommended him to George Vanderbilt. Vanderbilt hired the young forester to work on his Biltmore estate near Asheville, North Carolina. During his stay on this estate of more than 8,000 acres, Pinchot applied forest-management skills that called for cutting down trees without damage to the forest. In 1893, Pinchot published an account of his work at the Biltmore estate. The publication was the first to suggest a sustainability approach in forestry.

During the late 1800s, many people, including Pinchot, believed that if the rate of cutting down trees continued, the nation's timber supply would disappear in a scant sixty years. By 1885, much of the virgin forests of the eastern United States had been cut down to provide wood for construction and fuel or had been burned down to create farmlands. In those days, loggers would cut down all the marketable trees in the forests and leave the bare land exposed to water and soil erosion. There were no plans to save the limbs and smaller logs in the mills and no thought at all of reforestation.

Pinchot believed that there was a better way to manage forests and timber interests. In Europe, Pinchot observed forest-management practices—referred to as scientific forestry—in a number of countries. Loggers would cut down only certain trees that were marked in advance by professional foresters. The trees were cut down in a fashion that did little or no damage to other standing trees in the vicinity. No part of the tree was wasted at the mills. Limbs and small pieces were cut for fuel, and even the sawdust was used. The forest remained intact while producing forest products for the marketplace. When Pinchot began government service in 1898 as chief of the Division of Forestry (later renamed the U.S. Forest Service in the Department of Agriculture), he incorporated the European forest practices into his administration.

Under Pinchot, the U.S. Forest Service was in charge of all national forests. Its authority included not only trees but public lands, mining, agriculture, irrigation, and wildlife. Pinchot felt that there was a definite relationship between the forests and flood control, soil and erosion, minerals, and fish and game. His outstanding ability as an administrator generated strong loyalty from the small staff. Pinchot set out to prove that forestry could both produce timber for harvest and maintain the forest for future generations. The forest reserves, later renamed the national forests, were to be managed for the greatest good for the greatest number of people in the long run. Local questions were to be decided by local officials—a philosophy that has made the Forest Service, relatively speaking, one of the most decentralized agencies in the federal government. Pinchot used the press to get the word out about the nation's need for forestry, and his commentaries began to influence public opinion. Through his efforts and the backing of President Theodore Roosevelt, he was able to get all of the country's federal forest reserves (later renamed national forests) transferred to his own division, the U.S. Forest Service, from the General Land Office of the Department of the Interior by 1905.

Theodore Roosevelt and Gifford Pinchot were best friends. Their principles and philosophies helped mold Forest Service culture and values that have always aimed for conservation leadership, public service, responsiveness, integrity, a strong land ethic, and professionalism. Pinchot soon became a confidant and a member of the president's inner circle, advising him on all conservation questions and frequently writing his speeches and policy statements. Pinchot also served on a number of Roosevelt's commissions, including the Commission on the Organization of Government Scientific Work, the Commission on Public Lands, the Commission on Departmental Methods, the Inland Waterways Commission, and the Country Life Commission. Pinchot, under the leadership of Roosevelt, quickly began putting aside millions of acres as forest reserves. By the time Roosevelt left office, more than 254 million acres had been placed under the forest reserves.

Pinchot did have a few opponents. One was John Muir. Pinchot first met Muir in New York, and they traveled with the National Forestry Commission surveying the problems of the western forest reserves. At that time, they became good friends. The friendship was not to last, however. While Muir was returning from Alaska in 1897, he read a newspaper account reporting an opinion by Pinchot that sheep grazing in forest reserves did no harm. Muir had observed sheep firsthand as "hoofed locusts" literally destroying the flowery mountain meadows. When Muir found that Pinchot was staying at the same hotel in Seattle, he confronted him in the lobby. Surrounded by reporters and clutching the newspaper article in his hand, Muir asked Pinchot, "Are you correctly quoted here?" "Yes," Pinchot admitted. "Then," replied Muir, "I don't want anything more to do with you."

Muir went on to crusade for the preservation of wilderness, while Pinchot campaigned for utilitarian use of public lands. The deepening schism between Muir and Pinchot eventually grew into the great split between the preservation wing and the utilitarian wing of the conservation movement. Pinchot and Muir became major antagonists on the issue of damming Hetch Hetchy Valley in Yosemite National Park.

Richard Ballinger was another who had difficulties with Pinchot. President William Howard Taft, who took office in 1909, did not share Pinchot's philosophy. To Pinchot and others, Taft was not a conservationist. Taft appointed Ballinger secretary of the interior. Ballinger, who had a different opinion on what to do with some of the property in the public domain, proceeded to take some of the Alaskan public lands containing rich deposits of coal and turn them over to private investors. Pinchot was so angry about this situation that he went to Congress to denounce Taft and Ballinger's plans. His speech rallied the public, who pressured the president and Ballinger to drop the Alaskan plan, which eventually they did.

Clashes with Ballinger over conservation policy, hydroelectric power regulation, and Alaskan coal lands caused Pinchot's dismissal in 1910.

After Pinchot left the Forest Service, he retained his interest in conservation and forestry. He lobbied successfully for passage of the Weeks Act in 1911 and the Water Power Act of 1920. These laws allowed for expansion of forest reserves by purchase and began federal regulation of the power industry.

In 1921, Pinchot ran for governor of Pennsylvania and won. His term as governor was marked by vigorous enforcement of Prohibition and increased state regulation of electrical utilities. Grey Towers became his legal residence. His wife Cornelia transformed the 1880s mansion into a modern home more suited to their active lifestyle. This transformation enlarged the library, created a sitting room, and moved the dining room outdoors.

Pinchot was elected governor again in 1930. This was also the time of the Great Depression, and Pennsylvania was hit hard. With a flurry of innovation, he set up work camps throughout the state, which later became models for Franklin Roosevelt's Civilian Conservation Corps. Pinchot's favorite accomplishment during his second term was "taking the farmer out of the mud." The men in the relief camps built twenty thousand miles of paved country roads. Light but effective, they were adequate to enable farmers to get to their markets.

When Pinchot left office in 1935, he was seventy years old. During his last decade, he fought the transfer of the Forest Service from the Department of Agriculture to the Department of the Interior, an agency he insisted was still corrupt. During World War II, he developed for the navy a special fishing kit to help sailors adrift in lifeboats survive. The military commended him for saving countless lives.

Pinchot died in New York City in 1946. Shortly before his death, he completed a ten-year effort to write an autobiographical account of his work between 1889 and 1910 and his part in the development of forestry and conservation in the United States. *Breaking New Ground*, the title excerpted from a Roosevelt accolade, was published posthumously in 1947. Other writings that Pinchot authored included *The Fight for Conservation*, (1910) a dozen monographs on forestry subjects, a popular book on his journey to the South Seas, and approximately 150 published articles, reports, bulletins, lectures, and addresses.

Grey Towers was donated to the Forest Service of the U.S. Department of Agriculture in 1963 by Gifford Pinchot's son, Gifford Bryce Pinchot. Nestled among the Pocono Mountains of northeastern Pennsylvania, the estate overlooks the Delaware River Valley and the small town of Milford.

BIBLIOGRAPHY

Cutright, P. R. (1985). *Theodore Roosevelt: The Making of a Conservationist*. Urbana: University of Illinois Press.

Gifford Pinchot (1865–1946) Grey Towers National Historic Landmark. *http://www.pinchot.org/gt/gp.html*

Pinchot, G. (1947). *Breaking New Ground*. New York: Harcourt, Brace.

Pinchot, G. (1967). [1910]. *The Fight for Conservation*. Seattle: University of Washington Press.

Pinchot Institute for Conservation. *http://www.pinchot.org*

Sterling, K. B., Harmond, R. P., Cevasco, G. A., Hammond, L. F. (Eds.). (1997). *Biographical Dictionary of American and Canadian Naturalists and Environmentalists*. Westport, CT: Greenwood Press.

—John Mongillo

OCEAN ROBBINS
(1973–   )

> My dream is to see people of all backgrounds, races, religions and nations joining together for something we all have in common: planet Earth. I want to see each individual doing their own part to make the world a better place. I want to see the human race cherishing diversity, and working as a team to create the healthy future we deserve.
>
> —Ocean Robbins

In 1990, when Ocean Robbins was sixteen years old, he cofounded with Ryan Eliason, age eighteen, Youth for Environmental Sanity (YES!). The goal of YES! was to educate, inspire, and empower young people to take positive action for healthy people and a healthy planet. Since 1990, Robbins directed the organization and now serves as its co-president. YES! has been very successful since Robbins and Eliason started the program. It has reached 610,000 students in 1,200 schools in 43 states through full school assemblies. YES! has also organized and facilitated 66 weeklong summer camps in 7 countries, published 7 youth action guides, and led 150 daylong youth training workshops. The YES! Action Camp focus is comprehensive and cohesive, including community organizing, communication skills, healing of racism and sexism, conflict resolution, cultivation of peace in each of us and in our world, getting press coverage, fundraising, and public speaking. Robbins has spent seven years as the primary fundraiser and administrator for this nonprofit organization, which now has an annual budget of over $400,000. He has personally organized summer camps and workshops in Singapore, Costa Rica, Russia, Finland, and Canada and across the United States.

Robbins was born in 1973 in California. Even at an early age, he believed in being active in the community. When he was seven, he organized a peace rally in his elementary school. He ran his first marathon at the age of ten. His entrepreneurial talents began to emerge at the same time, when he was

Photo courtesy of Ocean Robbins.

the baker, salesman, and accountant for Ocean's Natural Bakery. When his door-to-door sales of natural baked goods began reaching hundreds of customers around the neighborhood, Robbins's work and life became the feature story on the front page of the *Santa Cruz Sentinel*. The headline read: "Boy Isn't Very Rich But He's Got Dough."

At ages fourteen and fifteen, Robbins planned and organized the environmental portion of two international youth summits in Moscow and another in Washington, D.C. He met with Raisa Gorbachev and numerous ambassadors and U.S. senators to discuss environmental concerns. His articles began to be published in major national magazines. At fifteen Robbins was cofounder of the Creating Our Future environmental speaking tour, on which he and three other participants spoke in person to more than 30,000 students, presented for 2,000 people at the United Nations, and opened for the Jerry Garcia band in San Francisco.

Robbins is coauthor of *Choices for Our Future: A Generation Rising for Life on Earth* (published in September 1994), which has sold more than 15,000 copies. He speaks widely, spreading a message of hope and inspiration to conferences, companies, and organizations. He has been interviewed on more than seventy local and national radio and television programs and in another fifty newspapers and magazines, including the

Washington Post, USA Today, and the *New York Times*. He has served on the board of directors for Friends of the Earth and EarthSave International. His ongoing efforts on behalf of a better world have been honored by many organizations, including the Giraffe Project (for people who stick their necks out for what they believe), the National Alliance for Animals (The Compassionate Youth Award), *E Magazine* (The Kid Heroes Hall of Fame), and EarthSave (Volunteer of the Year Award). He is an honored member of the U.S. registry of *Who's Who among Outstanding Americans*. Robbins lives in the mountains of Santa Cruz, California, with his wife, Michele Bissonnette, and his parents, Deo and John Robbins (author of the international best-seller *Diet for a New America* and founder of EarthSave).

BIBLIOGRAPHY

Robbins, O., & Solomon, S. (1994). *Choices for Our Future: A Generation Rising for Life on Earth*. Summertown, TN: Book Pub.

—*John Mongillo*

THEODORE ROOSEVELT
(1858–1919)

The object of government is the welfare of the people. Conservation means development as much as it does protection. I recognize the right and duty of this generation to develop and use the natural resources of our land; but I do not recognize the right to waste them, or to rob, by wasteful use, the generations that come after us.
—Theodore Roosevelt, "The New Nationalism" speech,
Osawatomie, Kansas, August 31, 1910

Theodore Roosevelt was a naturalist, writer, soldier, preservationist, and conservationist. He was also the governor of New York and the twenty-sixth president of the United States after the assassination of President William McKinley in September 1901.

Born on October 27, 1858, in New York City, Roosevelt was the second of four children. Even as a young boy, Roosevelt enjoyed the natural sciences. He collected insects, mammals, and birds, and at the age of thirteen, he took taxidermy lessons to became a taxidermist. He was an avid collector of birds and other specimens. He could walk through the local woods and identify birds even by their songs and Latin names. He could imitate their sounds as well. His father gave him his first gun in 1872, and hunting soon became another passion.

Roosevelt entered Harvard with the intention of becoming a professional field naturalist. However, he was not allowed to take electives his first year, so he took the required courses, such as Greek, Latin, and chemistry. He

Cartoon © 1999 J. N. "Ding" Darling Foundation.

became friends with Henry Davis Minot, whose interest in birds led them to birdwatching in the woods near Cambridge, Massachusetts. In college, he published two papers on birds. In his junior year, he took a zoology course and was first in his class.

In the 1880s, Roosevelt donated his boyhood collection to the Smithsonian Institution (622 bird skins) and the American Museum of Natural History (about 125 bird skins). He became a noted authority on North American mammals, about which he wrote extensively. He went to the Dakota Territory in September 1883 to hunt bison. Before returning home to New York, he became interested in the cattle business and established the Maltese Cross Ranch partnership. The next year, he returned to the Badlands and started a second open-range ranch, the Elkhorn. Roosevelt witnessed the decline in wildlife and saw the grasslands destroyed due to overgrazing.

In 1887, Roosevelt, George Grinnell, and some other sportsmen founded the Boone and Crockett Club, named after Daniel Boone and David Crockett, two of Roosevelt's heroes. Roosevelt had met previously with Grinnell to discuss the slaughter of large game animals in the west. Grinnell was the editor of the magazine *Forest and Stream* and had observed firsthand the overkilling of the animals. Grinnell and Roosevelt hoped that the organization would help preserve the animals from unsportsmanlike hunting. Roosevelt was named president of the Boone and Crockett Club. Its motto was "manly sport with the rifle," and its goal was the preservation and study of wildlife. The club was instrumental in the passage of legislation protecting Yellowstone Park and establishing the Forest Reserve System. Roosevelt and Grinnell edited and contributed to three books published by the Boone and Crockett Club. Roosevelt served as president of the club from 1888 to 1894.

When Roosevelt became president of the United States in 1901, the nation's natural resources had been depleted. More than one-half of the timber had been cut down. As president, Roosevelt's greatest contribution was his work for conservation. During his tenure in the White House from 1901 to 1909, he designated 150 national forests, the first 51 federal bird reservations, 5 national parks, the first 18 national monuments, the first 4 national game preserves, and the first 21 reclamation projects. Altogether, in the seven and one-half years he was in office, he provided federal protection for almost 230 million acres, a land area equivalent to that of all the East Coast states from Maine to Florida.

Roosevelt moved fast. In 1902, he established Crater Lake National Park. In the years ahead, he established Wind Cave National Park in South Dakota in 1903, Sullys Hill Park in North Dakota in 1904, Platt National Park in Oklahoma in 1906, and Mesa Verde National Park in Colorado in 1906. In 1902, Roosevelt also signed the Newlands Reclamation Act, leading to the first twenty-one federal irrigation projects, including Theodore Roosevelt Dam in Arizona. In 1903, Roosevelt proclaimed Pelican Island in Florida as the first federal bird reservation; a total of five bird reservations were established by the Roosevelt administration. In 1905, Wichita Forest in Oklahoma was made the first federal game preserve. Other federal game preserves established by Roosevelt included Fire Island, Alaska, in 1909, and National Bison Range, Montana, in 1909.

Under Roosevelt's leadership, the National Forest Service was established in 1905. In 1906, the Antiquities or National Monuments Act was signed, by which Roosevelt established the first eighteen national monuments, including Devils Tower (1906), Muir Woods (1908), the Grand Canyon (1908), and Mount Olympus (1909). Roosevelt appointed the Public Lands Commission in 1903, which led to new policies on use of the open range and federal lands; the Inland Waterways Commission in 1907, which studied river systems, waterpower, flood control, and reclamation; and the Country Life Commission in 1908 to study rural conditions. Roosevelt called the first national conference of governors, which met in 1908 to consider the problems of conservation. This led to the creation of thirty-eight state conservation commissions, the annual governors' conferences, and the appointment of the National Conservation Commission to prepare the first inventory of U.S. natural resources.

Roosevelt was surrounded by noted naturalists and conservationists of his time, including C. Hart Merriam, Henry Fairfield Osborn, John Burroughs, Gifford Pinchot, Frank M. Chapman, John Muir, and William Beebe. He was an enthusiastic supporter of Gifford Pinchot, head of the U.S. Forest Service, and a proponent of the Newlands Bill initiated by Pinchot. The bill expanded the nation's forest reserves, setting aside water-power sites and millions of acres of coal lands. Gifford Pinchot was a chief forester and conservationist who has been described as a "catalyst" to the conservation movement and the father of scientific forestry in America. Through his efforts and Roosevelt's backing, Pinchot was able to get all of the country's federal forest reserves (later renamed national forests) transferred to his own division, the U.S. Forest Service, from the General Land Office of the Department of the Interior.

Roosevelt and Pinchot were best friends. Their principles and philosophies helped mold Forest Service culture and values that have always aimed for conservation leadership, public service, responsiveness, integrity, a strong land ethic, and professionalism. The two men had much in common. They both grew up in New York City, came from wealthy, influential families, and had traveled extensively abroad. They could speak French and German and were naturalists who shared a love for the outdoors. By the time Roosevelt became president, their friendship had grown significantly. They rode, walked, thought, played, and swam together. Pinchot's unique relationship with the president gave him access to the Oval Office. He became the president's most trusted advisor on matters of conservation, which Roosevelt considered one of the great accomplishments of his administration. Later, in writing about his time at the White House, Roosevelt said, "Among the many, many public officials who under my administration rendered literally invaluable service to the people of the United States, Gifford Pinchot, on the whole, stood first" (Grey Towers).

In 1903, Roosevelt visited John Muir in Yosemite. There the two men had the opportunity to discuss a variety of environmental subjects that included Roosevelt's conservation programs. In 1915, Roosevelt recalled his warm relationship with Muir in an article he wrote for *Outlook* a year after Muir had died. Roosevelt said, "His was a dauntless soul. . . . Not only are his books delightful, not only is he the author to whom all men turn when they think of the Sierras and Northern glaciers, and the giant trees of the California slope, but he was also—what few nature-lovers are—a man able to influence contemporary thought and action on the subjects to which he had devoted his life. He was a great factor in influencing the thought of California and the thought of the entire country so as to secure the preservation of those great natural phenomena—wonderful canyons, giant trees, slopes of flower-spangled hillsides—which make California a veritable Garden of the Lord. . . . Our generation owes much to John Muir."

Roosevelt was an established author whose writings included books on outdoor life and nature such as *Hunting Trips of a Ranchman* (1885), *Ranch Life and the Hunting Trail* (1888), and *The Wilderness Hunter* (1893). He also wrote numerous introductions to books, articles, book reviews, and papers on outdoor life, nature, and conservation.

Roosevelt chose not to run for a third term. In 1909–1910, he led an expedition to Africa sponsored by the Smithsonian Institution, collecting some 5,013 mammals, 4,453 birds, and 2,322 reptiles and amphibians as well as thousands of fish, insects, plants, and other specimens. Many of the species and subspecies were new to science. Disturbed by the shifting political climate back home and the fear that his reforms were endangered, he returned to the United States, formed the Progressive party, and ran as its presidential candidate, but lost, coming in second in a three-party race.

In 1914, Roosevelt accepted an offer to speak and travel on a scientific expedition to South America. The American Museum of Natural History and the Brazilian government sponsored the trip. Plans were made to explore the Paraguay River, but when a Brazilian official asked them if they would like to explore the unexplored River of Doubt, Roosevelt and his party agreed. The 900-mile journey nearly killed him. In partnership with Brazil's leading explorer, Colonel Rondon, the party encountered snakes, piranhas, mosquitoes, waterfalls and rapids, malaria (which Roosevelt contracted), starvation, and murder, for one man killed another over food. During the trip, Roosevelt also broke his leg. His health worsened after his return to the United States. Colonel Rondon renamed the river the Rio Roosevelt, which is how it appears on maps today. Roosevelt later wrote, "The Brazilian wilderness stole ten years of my life." But, he said, "I am always willing to pay the piper when I have a good dance, and every now and then I like to drink the wine of life with brandy in it"(Grey Towers).

Roosevelt never fully recovered from his trip to Brazil. He returned to his Oyster Bay home at Sagamore Hill, where he fell ill and died quietly in his sleep on January 6, 1919, just five years after he returned from the River of Doubt. He was sixty years old. Charles Van Hise, conservationist, geologist, and president of the University of Wisconsin, wrote in 1911 that what Theodore Roosevelt had done as president to advance conservation and "to bring it into the foreground of the consciousness of the people will place him not only as one of the greatest statesmen of this nation but one of the greatest statesmen of any nation of any time" (Sterling et al.).

BIBLIOGRAPHY

Cutright, P. R. (1956). *Theodore Roosevelt, the Naturalist*. New York: Harcourt Brace.

Cutright, P. R. (1985). *Theodore Roosevelt: The Making of a Conservationist*. Urbana: University of Illinois Press.

Grey Towers National Historic Landmark Web site. *http://pinchot.org/gt/gp.html*

Miller, N. (1992). *Theodore Roosevelt: A Life*. New York: Quill William Morrow.

Pinchot, G. (1947). *Breaking New Ground*. New York: Harcourt, Brace.

Roosevelt, T. (1996). [1885]. *Hunting Trips of a Ranchman*. New York: Modern Library.

Roosevelt, T. (1999). [1893]. *The Wilderness Hunter*. Birmingham, AL: Palladium Press.

Roosevelt, T. (2000). [1888]. *Ranch Life and the Hunting Trail*. Fairfield, CA: James Stevenson Publisher.

Sterling, K. B., Harmond, R. P., Cevasco, G. A., & Hammond, L. F. (Eds.). (1997). *Biographical Dictionary of American and Canadian Naturalists and Environmentalists*. Westport, CT: Greenwood Press.

Theodore Roosevelt Organization. *http://www.theodoreroosevelt.org*

Whitehouse Web site. *http://www.whitehouse.gov/WH/glimpse/presidents/html/tr26.html*

—*John Mongillo*

KIRKPATRICK SALE
(1937–   )

Kirkpatrick Sale has been a writer for most of his life on a variety of subjects in large part unified by a philosophy of decentralism and an underlying belief that the health of an ecosystemic world, and of the human species within it, depends on small, self-sufficient, cooperative, interacting units.

Born on June 27, 1937 in Ithaca, New York, the son of academic parents, Sale attended Swarthmore College and Cornell University and at the latter was the editor of the student newspaper and a leader of the first campus "student power" demonstration, in 1958, a forerunner of similar protests in the 1960s. He worked at several newspapers and wrote for various periodicals, then became associate editor of the *New Leader*, a weekly political magazine in New York City.

In 1961–1962, sensing that the "Africa Decade" of the United Nations would produce interesting new political byproducts on that continent, he traveled throughout the sub-Saharan region as a freelance correspondent for several newspapers and a contributor to magazines in both Africa and the United States. He returned to America to write a book on the newly independent Ghana, *The Land and People of Ghana* (1963), and then went back to that country to teach European history at the University of Ghana from 1963 to 1965.

Returning to New York City, he worked for three years as an editor on the *New York Times Magazine*, attempting to alert a mainstream audience to the immensely creative ferment then surfacing on college campuses

Photo courtesy of Shirley Branchini.

among students and young faculty. Largely frustrated in moving the establishment bastion, he left to become a freelance writer and to produce a book on the most dynamic of the sixties' organizations, the campus-based Students for a Democratic Society. According to his analysis, published as *SDS* in 1973, the organization began promisingly, discarded the outworn socialism of its parent organization, the League of Industrialized Democracy, and tried to forge an authentically American vision of a decentralized, grassroots polity guided by "participatory democracy"; failing to win much headway with that vision except among a minority on campuses, and faced with an escalating war in Vietnam, it soon turned more toward resistance and revolution and eventually collapsed in factionalism, leaving only a few remnant groups to celebrate the end of the war when it came in 1973.

Sale's next canvas was wider, but it depended upon an analysis that had been forged in SDS in its better days, arguing that the country was increasingly dividing into two camps, one the traditional "yankees" of the Northeast and the other the upstart "cowboys" of the South and West, with different styles and different approaches to politics, economics, and ethics. Emerging in 1975 as *Power Shift*, Sale's book expanded and buttressed this analysis, showing how the new cowboy—or "Southern Rim," "Sunbelt"—establishment was becoming dominant, drawing population and business, and gaining control of Congress and the White House (Nixon, Carter, and Reagan, Bush, and Clinton to follow). The attempt was to get a wider public to understand the nature of American life and its fundamental divisions into widely different geographic regions (and subregions), and a number of people responded to this point, largely by forming regional think tanks and governors' groups to try to assert separate rights and shares of the national pie.

This led to the work that some people have seen as Sale's most important analytically, *Human Scale*, called by the British critic John Papworth "the bible of decentralism" (book's jacket). It is a long and detailed description of the failures of gigantic institutions—social, political, and economic—and a careful examination, with historical, anthropological, psychological, and experiential buttresses, of the success of small states, institutions, societies, schools, businesses, and communities "built to a human scale." Lewis Mumford, a Sale mentor in the decentralist tradition, said that "if anything can arrest the total disintegration of world civilization today it will come through a miracle: the recovery of 'the human scale,' described" in this "encyclopedic book" (book's jacket).

Although there was an important ecological dimension to *Human Scale*, it tended to concentrate on human societies, and it was not until Sale's next work, *Dwellers in the Land: The Bioregional Vision* (1985), that he branched out to encompass a whole-scale ecological perspective. This book was an attempt to provide a brief framework or philosophy for the then-young bioregional movement and thus a grounding for the more far-seeing parts of the wider environmental movement of the 1980s. In an interesting preface he added to the paperback edition of this book in 1991, Sale wrote:

I have been led to this consideration of the shape of the bioregional vision, inevitably as it were, by the trajectory of my previous work: on American radicalism, on American regionalism, on the abject failure of American giantism. It expresses for me not merely the newest and most comprehensive form of the ideals of decentralism, participation, liberation, mutualism, and community that I have expounded in all my work—but, as it stems from the most elemental perception of the crises of the planet, the ideals of ecological sanity, regional consciousness, speciate humility, and global survival. It is for me, therefore, not merely a new way of envisioning and enacting a very old Americal ideal, but also a crucial, and perhaps virtually the only possible, means of arresting the impending ecological apocalypse.

As Sale later told several interviewers, it was in the course of being an activist in the bioregional and environmental movements, including work with a green organization in New York and on the national level, that he decided that the source of the industrial world's ecological madness went back far into the past of Western civilization and could not be cured or ameliorated until a wider understanding of this source was held at large. Hence, and in anticipation of the upcoming celebration of the 500th year since the voyage of Columbus and "the discovery of America," he set out to do a full-scale revisionary history of that event and its role in shaping Western consciousness and implanting Western attitudes to nature and the people of nature in the New World.

Conquest of Paradise: Christopher Columbus and the Columbian Legacy (1990) was treated with general respect by reviewers and readers, since it was a careful reconstruction of the four Columbian voyages, newly translating sources and documents and using full range of Columbus's writings. But the radically new picture of the famous explorer that it documented, showing him to be a gold-driven autocrat, ruthlessly murderous to the Indian population (and none too kind to fellow Spaniards), acting out the European perception of nature as simply a treasurehouse for plundering by the strongest, upset many traditional groups (schoolteachers, for example) and many conservative historians. It was, however, welcomed by most Native American groups and spokespeople, who saw it as a valuable tool in their efforts to make the quincentennial events an educative and transformative experience instead of a crass celebration of Western dominance. In the event, the book, along with a number of other less thorough books and articles, served to take the bloom off the quincentennial affairs almost everywhere in the United States and in many quarters forced a painful reexamination of the historical record and the Columbian legacy.

Something of the same sort of historical reexamination was the inspiration for Sale's next work, *Rebels against the Future: The Luddites and Their War on the Industrial Revolution: Lessons for the Computer Age*, in 1995. Again a painstaking historical reconstruction, in this case of events far more obscure, it told the story of the cloth workers of middle England who rose up under the banner of a mythical King Ludd to oppose the machines and factories being imposed upon them by the Industrial Revolution in the years 1811–1813, and how they were eventually crushed by massive military force from the British government. Its later chapters suggested that a second Industrial Revolution was being wrought by the computer chip and detailed how a neo-Luddite resistance was responding to it in the United States and throughout the world, in part learning the lessons of the Luddites.

At the heart of this work, and of Sale's articles and speeches for the next few years, was an argument that, in effect, the modern world's submission to technology has created a vast and powerful technosphere that threatens to obscure humans' connections to the biosphere and hence make its manipulation, exploitation, and ultimate destruction easier and swifter. This ecological dimension of the technological crisis was underscored by Sale's repeated reminder of Herbert Read's sage advice: "Only a people serving an apprenticeship to nature can be trusted with machines. Only such people will so contrive and control those machines that their products are an enhancement of biological needs, and not a denial of them."

Although Sale has had a home in Cold Spring, outside a little village in upstate New York, for several decades, he made it his full-time residence only in 1995. He has thus been removed from some of the activist pursuits of earlier years, although he was a board member of the adult political education center the Learning Alliance from its inception in 1980 to its demise in 1997 and has remained a member of the bioregional movement. He has also been a member of the board of the E. F. Schumacher Society in Great Barrington, Massachusetts, since its founding in 1980, has been a frequent speaker at its forums, and has twice given the annual Schumacher Lectures, which appear in *People, Land, and Community* (1997).

This geographical detachment may have contributed to his concentration in recent years on aspects of what he calls "ecoism," or nature-based spirituality, a view of the world based on the idea of a living, sacred earth and the tenets of deep ecology that hold all species as equally precious and protectable. He has written a novel, *As a Devouring Fire* (forthcoming), about a small community of ecoists devoted to the living earth and her works and facing the final ecocatastrophe, and has published several articles dealing with ecoist and "ecocentric" themes. Sale is widowed and has two grown children and a grandchild. He has appeared in *Who's Who in America* since the 1990 edition.

BIBLIOGRAPHY

Hannum, H., ed. (1997). *People, Land, and Community*. New Haven, CT: Yale University Press.

Read, H. E. (1955). *Grass Roots of Art*. London: Faber and Faber.

Sale, K. (1963). *Land and People of Ghana*. Philadelphia: Lippincott.

Sale, K. (1973). *SDS*. New York: Random House.

Sale, K. (1975). *Power Shift*. New York: Random House.

Sale, K. (1980). *Human Scale*. New York: Random House.

Sale, K. (1990). *Conquest of Paradise: Christopher Columbus and the Columbian Legacy*. New York: Alfred A. Knopf.

Sale, K. (1991). [1985]. *Dwellers in the Land: The Bioregional Vision*. Philadelphia: New Society Press.

Sale, K. (1993). *The Green Revolution: The American Environmental Movement, 1962–1992*. Ed. Eric Foner. New York: Hill & Wang.

Sale, K. (1995). *Rebels against the Future: The Luddites and Their War on the Industrial Revolution: Lessons for the Computer Age*. Reading, MA: Addison-Wesley.

—*Kirkpatrick Sale*

Photo courtesy of George B. Schaller.

GEORGE SCHALLER
(1933–)

To be a conservationist means being a politician, diplomat, educator, sociologist, fundraiser, and, almost incidentally, a biologist. No university class teaches this combination; it requires experience.

—George Schaller

George Schaller is a field biologist and author of many articles and books. He is the director of science at the Wildlife Conservation Society in New York. One of his major accomplishments was persuading the Chinese government to set aside large areas of Tibet as nature preserves.

In late 1999, he was on his way to the Kamchatka Peninsula in the Russian Far East to study snow sheep, a species found only there. Uncontrolled hunting had reduced members greatly. His task was to evaluate the situation. He and his Russian coworkers flew along the rugged coastline, the Bering Sea on one side and snow-capped volcanoes on the other. He realized that any conservation initiative in this sparsely populated and wild region would be a difficult challenge. But he also knew that no matter how great the problems, they had to try. Schaller believes that in an age when a rapidly growing human population consumes and wastes ever more natural resources, pollutes and destroys the environment, and exterminates species, we desperately need to protect the remnants. This is a task to which everyone must contribute. Some can do so by becoming involved in the environment professionally as a scientist, lawyer, journalist, or teacher, to name a few occupations. Others can join local conservation groups, donate funds, and, above all, become aware of problems and responsibilities.

Schaller has given himself the task of trying to help protect wildlife and its habitat in various parts of the world. For nearly fifty years, he has sought little-known creatures in the ice-bound Himalaya, the Amazon rainforest, and the East African savannahs. By studying such animals as mountain gorillas, snow leopards, and giant pandas, he has tried to collect the kind of information that would contribute to their survival. Basic research is essential before realistic conservation efforts can be made. How many animals are there? How far do they travel? How fast do they reproduce, and what causes death? What are the attitudes of local people to animals? To obtain answers to these and many other questions may take several years of living in the wild.

Studying exotic wildlife in remote places sounds romantic to those who read accounts of Schaller's work or watch television documentaries: distance lends enchantment. However, life afield is often difficult for a naturalist. Schaller describes that one lives under simple conditions in tents or huts, isolated in an alien culture, usually unable to speak the language well. Local people may be suspicious of outsiders and reluctant to help. The weather may turn the work into misery, whether through bitter cold or torrential rain. A researcher needs tenacity, patience, and a capacity for enduring discomfort.

Schaller describes how exhilarating it is to be in the wilderness—to come face to face with a mountain gorilla, one of our closest kin, or watch a tiger striding through the forest, elegant, powerful, beautiful. There is in many of us a spirit of adventure, a longing for empty spaces, something that beckons us to search for the unknown beyond the distant horizon. Schaller, for example, experienced such feelings when he was in quest of wildlife at 16,000 feet on the Tibetan Plateau. He traveled into unknown terrain on the backs of Bactrian camels much as explorers did there a century ago.

Schaller's parents bear little responsibility for his being a naturalist. He was born in 1933 in Germany, and there was no inspiration in his family background—his father was a diplomat—to prod him toward natural history. Like many a child who ultimately selects this profession, he liked to roam forests, observe birds, and investigate his natural surroundings. After he came to America at the age of fourteen, where he lived in Missouri, he kept many pets, from black snakes to opossums. His cousin took him camping in Canada's wilderness. But he had no idea how these boyhood interests could be translated into a profession; in fact, he had no idea that such professions existed. Guidance counselors in high school, where he was a mediocre student, bored by most subjects, recommended that he become an interior decorator or forget about college for a life as a mechanic.

Fortunately, his cousin suggested that he attend the University of Alaska, where he had gone for a while, because it was surrounded by wilderness: black bears and moose visited the campus. At that time, in 1951, Alaska was still a territory rather than a state. The university was small and intimate, with only 350 students. Several professors nurtured his interest in the natural sciences. Graduate students in wildlife management were studying the habits of a variety of animals, from martens to walrus, and he learned from them. Biologists from elsewhere visited the campus and made him aware of conservation issues. He took all possible courses in zoology, botany, anthropology, and related topics such as mathematics and writing. However, textbooks and lectures cannot really teach natural history. One must learn in the field, experiencing, practicing, delighting in the joy of discovery while probing the intricate ecological relationships between animals and their habitat. During the long Alaskan summers, he gained such experience by following caribou on their migrations and studying birds on the Arctic tundra. The university gave his life a focus.

Next he had to enter a graduate program, one that would lead to a Ph.D. degree, as is needed in any scientific field. For that, one must find a professor, a mentor, who guides one's research for a thesis and selflessly promotes one's interests and aspirations. Again he had great luck. John Emlen, a well-known ornithologist at the University of Wisconsin, accepted him as a graduate student. His first project under Emlen's direction was to study bird behavior in the laboratory as well as outdoors. Even then he was uncertain whether his future was as an academic in a university, a biologist in a state game department, a curator in a natural history museum, or whatever. Soon he also had someone else to share the future. He had met Kay at the University of Alaska. After their marriage, Kay liked to camp, travel, and share in his work, and she has contributed immensely to all his projects ever since. Beyond that, she has provided stability, encouragement, and love in an unsettled life as they have moved around the world.

Schaller became a naturalist partly by accident and partly by desire, intrigued by the lives of other creatures. As a member of an expedition to northern Alaska with Olaus Murie, then president of the Wilderness Society, and through reading such books as Aldo Leopold's *Sand County Almanac*, he became aware that conservation must also be part of every study. Schaller believes that to describe the life history of an animal or plant is not enough: there is a responsibility to protect the species and help it survive. He noted, too, that the best naturalists are also fine writers who not only publish solid scientific treatises, but also express facts and ideas in clear and evocative prose. Science deals mainly in facts and figures; conservation is, in the final analysis, a moral issue of the rights of all species to exist and of other values. Schaller wanted to emulate those who combined science and conservation—a field that today is termed conservation biology—and who communicated their findings to both peers and the public.

The opportunity to study a rare animal, one that combined research and conservation, came with the suggestion that a project on mountain gorillas in the Congo was timely. John Emlen obtained grant funds and came to Africa to help Schaller initiate the work in 1959. Schaller and his wife stayed on. Daily they climbed through forests to look for gorilla groups whose members they learned to recognize individually and in whose lives they participated almost as if they were part of their extended family. In 1960, the Congo, which had until then been a Belgian colony, became independent. Civil war followed. The Schallers had to terminate their study, but others have since continued the work intermittently in spite of continual unrest to the present. Schaller wrote a scientific monograph as well as a popular book, *The Year of the Gorilla*. The study emphasized that research is easy, but conservation is difficult. Countries in political turmoil have priorities other than conservation. Schaller commented that farmers want to turn forests into much-needed fields. How does one convince them of the value of gorillas and forests, of the fact that their future depends on a healthy environment? That is a basic challenge. Local people have to be involved in conservation if species and habitats are to endure. Conservation is an endless task. We can never turn our backs on gorillas and consider them saved. Continuity in effort on behalf of these magnificent animals is crucial. To be a conservationist means being a politician, diplomat, educator, sociologist, fundraiser, and, almost incidentally, a biologist. No university class teaches this combination; it requires experience.

The gorilla project made it clear to Schaller that it is easier to get public support to help save a spectacular species than an obscure one. But by protecting the one species we protect all the others, the thousands of plants and animals, large and small, that make up the intricate web of life in an area. Since then, his studies have concentrated on large and rare animals, especially carnivores, such as tigers and pandas, because they fascinate him and because he wants to chronicle the lives of these persecuted animals while they still exist.

After the gorillas, the Schallers observed wildlife in India for two years, this time accompanied by their two small sons. The title of his book *The Deer and the Tiger* conveys what animals he studied there. Next came a three-year period on the great Serengeti plains of Tanzania where he focused on lions and the impact of their predation on the wildlife. The Schallers had an idyllic life among the vast herds of wildebeest and zebra, and lions roared around their house at night. Kay taught school to the boys. After Schaller published his research results in a monograph, *The Serengeti Lion*, and also a popular book, the Schallers moved to Pakistan. There, and in nearby Nepal, Schaller roamed the high ranges—Himalaya, Hindu Kush, Karakoram—to study snow leopards, as well as various wild sheep and goats, and to stimulate governments to establish reserves for the protection of the flora and fauna. He wanted to prevent the mountains from becoming lifeless, from turning into *Stones of Silence*, as he titled one of his books. This was followed by a project in the great swamps of the Mato Grosso in Brazil where Schaller delved into the lives of jaguars, alligators, capybaras (at one hundred pounds the world's largest rodent), and other species. To his delight, his adolescent sons were now old enough to assist him as coworkers and companions.

Ever since the gorilla study, his work has been sponsored by the Wildlife Conservation Society (formerly the New York Zoological Society). The society has been the perfect professional home for Schaller. Although it runs New York's zoos, it has with extraordinary vision also been deeply involved in wildlife research and conservation since 1895, long before these were issues of general concern. At present, the society has projects in about fifty countries. The society works with governments and local biologists to gather knowledge, establish and manage reserves, educate local people about the value of protecting the environment, and train conservation staff. Schaller has the rare and unusual freedom to select projects anywhere in the world as long as they promote conservation—and the necessary funds can be obtained from foundations and other sources.

Rare, conspicuously colored in black and white, at home in remotest China among dense bamboo thickets, the giant panda remained a mystery. Schaller was intrigued about its obscure life, but with China closed to Americans, a study seemed beyond reach. Then, quite suddenly, China opened its door to the outside world. In 1980, the Schallers traveled to panda country to begin a four-year collaboration with a team of Chinese researchers. They were isolated in a mountain camp, and it was at first a difficult cultural experience both for the Chinese and for them. There were misunderstandings based on language and different perceptions about the project and what it intended to achieve. Some of their coworkers had never seen foreigners, much less shared their lives with them. But they overcame obstacles, as described in Schaller's book *The Last Panda*. In this project, as in all others, they were emotionally involved, dedicated to the animals, the people, and the area. As always, it was not a job but a mission. When one completes a project, one also wants to leave something behind, to make some contribution to the country beyond an offer of facts. The greatest satisfaction comes from stimulating someone to carry on the work with the same spirit of dedication and commitment. One of Schaller's Chinese coworkers, a professor at Peking University, continued the panda project and trained students who now, years later, are themselves training students. The impact of Schaller's initial collaboration will, to his delight, continue to influence conservation long after he has been forgotten.

Schaller has continued working in China for the past decade and a half. There at high altitudes on the Tibetan Plateau is a unique fauna of wild yaks, wild asses, Tibetan antelopes, wolves, bears, and other animals. *Tibet's Hidden Wilderness*, as he called a recent book, pictures the beauty of these unbounded uplands where only a few nomadic pastoralists live. Other projects in other lands have held his attention too, such as the rare wild camels in the Gobi Desert of Mongolia and little-known and new species of deer in the highlands of Laos. He need not enumerate them all. He has become a nomad, wanting to go where he has not been before, always searching for remote fragments of nature that need an advocate for conservation.

Schaller is now in his late sixties. His sons are university professors, one in molecular biology and the other in social psychology. People ask him when he plans to retire. His wife teases him that since he has never had a real nine-to-five job, there is nothing from which to retire. Indeed, his work is both his hobby and his profession. He wants to keep going. Schaller believes that to be successful, one must love his or her work. He has been honored—the Cosmos Prize in Japan, the Tyler Environmental Prize in the United States.

What motivates Schaller? There is no simple answer because motivations are complex—a desire to escape daily constraints, a solitary nature, a compulsion to do well, a sense of wonder at the marvelous complexity and beauty of the natural world. Perhaps above all, he wants to contribute to conservation, a gentle science that helps society by protecting some of the splendor of this earth for future generations. As Schaller notes, the snow sheep of Kamchatka are unaware of their precarious future. Their fate depends on us.

BIBLIOGRAPHY

Nichols, M., & Schaller, G. (1989). *Gorilla: The Struggle for Survival in the Virungas*. New York: Aperture Foundation.

Schaller, G. (1964). *The Year of the Gorilla*. Chicago: University of Chicago Press.

Schaller, G. (1967). *The Deer and the Tiger: A Study of Wildlife in India*. Chicago: University of Chicago Press.

Schaller G. B. (1972). *The Serengeti Lion*. Chicago: University of Chicago Press.

Schaller G. B. (1973). *Golden Shadows, Flying Hooves*. New York: Knopf.

Schaller, G. (1980). *Stones of Silence: Journeys in the Himalaya*. New York: Viking Press.

Schaller, G. B. (1993). *The Last Panda*. Chicago: University of Chicago Press.

Schaller, G. B. (1997). *Tibet's Hidden Wilderness: Wildlife and Nomads of the Chang Tang Reserve*. New York: Harry N. Abrams.

Schaller, G. B. (1998). *Wildlife of Tibetan Steppes*. Chicago: University of Chicago Press.

—*George Schaller*

Photo courtesy of Peggy M. Shepard.

PEGGY SHEPARD
(1946–   )

Peggy Shepard is a cofounder of the West Harlem Environmental Action (WE ACT), New York City's first African-American and Latino environmental justice organization, established in 1988. The other cofounders include Vernice Miller and Chuck Sutton. Shepard was born on September 16, 1946 in Washington, D.C. and grew up in Trenton, New Jersey. She is the oldest of six children, with three sisters and twin brothers. As a youngster, she was interested in starting and organizing clubs. She enjoyed writing poetry and reading and wanted to be a writer when she grew up.

Shepard learned volunteering from her mother, who was very active in the community. Her mother belonged to a number of clubs and volunteering organizations that gave scholarships or provided certain services to communities. Shepard's father was a doctor and obstetrician, but he volunteered at a planned-parenthood organization and helped to create a free health center in Trenton, New Jersey. Shepard describes him as being very committed to his work and very dedicated, a workaholic. One of her brothers is involved in an organization that provides educational scholarships and tennis scholarships for young Afro-American youth in Washington, D.C.

After her high school years, Shepard attended Howard University in Washington, D.C. After being the first black newspaper reporter at the *Indianapolis News*, she came to Time-Life Books in New York City in 1971 to pursue a publishing career as a text and photo researcher. Later she was an editor for magazines such as *Redbook, Essence*, and Earl Graves Publishing Company.

Shepard decided to leave magazines and became a speech writer at the New York State Division of Housing and Community Renewal, where she then began to get more involved politically. She also moved to Harlem in the early 1980s and got involved in the 1985 Jesse Jackson campaign. She was the public relations coordinator for the Manhattan campaign under David Dinkins and Bill Lynch. Dinkins ultimately became mayor, and Lynch became deputy mayor. It was a turning point in her life when she began to work in politics. She became the female Democratic district leader of the West Harlem part of the Harlem 70th Assembly District in 1985.

During that year, the community came to Shepard and asked her what she was going to do about the North River Sewage Treatment plant, which was ready to open officially in 1986. It was then that she really began to focus on the plant and realized that there was a problem. With Chuck Sutton, the male Democratic district leader, she began to convene monthly meetings with the city and state demanding government accountability and to discover the extent of the problem facing the community. Shepard states that the new treatment plant in her neighborhood was a prime symbol of environmental racism, in which communities of people of color are targeted for waste facilities and other polluting industries. In the 1970s, the plant was initially sited to be built in another area, on West 70th street on the Hudson River, but the developers influenced the city planning commission to move it from the original site because they felt that the property was too valuable. As a result, the plant was to be sited in West Harlem, and the local residents were not able to overcome the city planning commission's decision. Shepard states that of the city's fourteen plants, this is the one closest to people's homes. That is really when she got involved in environmentalism.

After the $1.1-billion plant went on line, it was designed to treat 170 million gallons of waste daily. In a short time, foul odors and emissions became a problem for the community. In 1988, the high levels of emissions were exacerbating respiratory diseases in the community. It was during the same year that Shepard, Chuck Sutton, and Vernice Miller started West Harlem Environmental Action. For this ad hoc community group to challenge the city on a project that cost a billion dollars and impacts sewage on the whole west side of Manhattan was not a small task. Because the plant was mandated by the Clean Water Act, the founders of WE ACT realized that their strategy could not be to shut the plant down but to ensure that it was maintained well. They conducted several protests, including civil disobedience on the highway, in front of the plant and drew media attention to their cause. In 1992, WE ACT sued the city over the plant's sloppy operation. In 1993, the lawsuit was settled. That settlement was considered a huge victory for the community because the city agreed to pay $1.1 million into an environmental benefit fund to address the health concerns of the community. The settlement also called for the plaintiffs, including WE ACT, the Natural Resources Defense Council (NRDC), the Hamilton Grange Day Care Center, and seven individuals, to play a role in monitoring and policing the emissions from the plant.

WE ACT organized a group called the Earth Crew. For five years, it has been a group of Latino and African-American boys and girls thirteen to seventeen years old. WE ACT has taught them environmental leadership skills through seminars on environmental issues like lead and air pollution. They have taken on their own projects, such as greening spaces into small vest-pocket parks. They have engaged in an air-monitoring study that WE ACT executed cooperatively with the Columbia School of Public Health. They initiated and participated in a study that was published in the *American Journal of Public Health*. This study looked at the impact of diesel emissions on junior-high-school students because there are many bus depots in their community. West Harlem has over one-third of the city's bus fleet of 4,200 and six out of the eight bus depots in Manhattan. Most of the depots are adjacent to public schools and residential housing. So WE ACT and Columbia studied fifty to one hundred young people, taking urine tests and testing pulmonary functions to see if they found biomarkers for diesel emissions. They in fact did find that 75 percent of the students had biomarkers for diesel emissions in their urine and decreased lung function.

WE ACT has gained a national reputation and recognition around environmental health and community-based environmental health research. WE ACT has developed a "model" community academic partnership with the Columbia School of Public Health. These partnership projects are to a large degree funded by the National Institute of Environment Health Sciences (NIEHS). WE ACT does much public speaking promoting model partnerships of community and university, and NIEHS considers WE ACT and Columbia a good model partnership. For instance, eight new children's environmental health centers have been funded by NIEHS and the U.S.

Environmental Protection Agency (EPA) around the country, two of them are in northern Manhattan, one at Mt. Sinai and the other at Columbia. Shepard is the codirector of the Community Outreach and Education Program for Columbia's Children's Environmental Health Center. Shepard believes that these kinds of activities give WE ACT local and national visibility and ensure that it really works with a diverse group of people and stays close to the community. WE ACT has seven full-time and two part-time staff members.

Shepard believes that there is a difference between the environmental justice movement and the environmental movement. The environmental movement looks at impacts on natural resources, while the environmental justice movement looks primarily at impacts on public health and human beings, not only in terms of physical health but also with regard to mental health, stress, and aesthetics benefits in the community. Therefore, there is a need to ensure that all communities have a livable quality of life with clean air and clean water. Those things that environmental justice advocates are concerned about include transportation, sanitation, and housing issues. As an example, in an urban center, housing is a key environmental issue because of indoor air pollution and indoor allergens. Lead and asbestos are key issues in housing.

Today, WE ACT is an integral part of the Northeast Environmental Justice Network of which Shepard is co-chair with Pen Loh of ACE in Roxbury, Massachusetts. Besides its focus on the sewage treatment plant and public health issues, it is also involved in sustainable economic development and historical preservation issues in the West Harlem neighborhood as well as public policy issues related to the environment on the state and national levels. WE ACT has also expanded its catchment area to the northern Manhattan communities of East, West, Central Harlem, and Washington Heights, a population numbering about 500,000 people.

Shepard has recently been elected Vice Chair of the National Environmental Justice Advisory Council to the EPA. She will become Chair in 2001.

BIBLIOGRAPHY

WE ACT Web site. *http://www.weact.org*
West Harlem Environmental Action Web site. *http://www.crp.cornell.edu/projects/westharlem/intro.html*

—John Mongillo

KAREN SILKWOOD
(1946–1974)

I'm not sure she saw herself as a hero. Karen Silkwood was a person
who had courage, and was an activist in the truest sense of the word.
—Meryl Streep (E-Wire)

Karen Silkwood was a laboratory analyst at the Kerr-McGee nuclear fuels
production plant in Crescent, Oklahoma, and a member of the Oil, Chemical, and Atomic Workers' Union. Silkwood was an activist who was critical of allegedly inadequate safety provisions at the Kerr-McGee facility. In
November 1974, at the age of twenty-eight, she died in a fatal one-car
collision as she traveled to meet David Burnham, a *New York Times* in-
vestigative reporter, to talk about radioactive leaks at the facility. Specu-
lations over foul play in her death were never substantiated, but did lead
to a federal investigation into security and safety at the Kerr-McGee plant
and a National Public Radio report about the plant's large quantity of
misplaced plutonium. Since her death, Silkwood's story has achieved worldwide fame as the subject of many books, magazine and newspaper articles,
and a major motion picture.

Silkwood was born in Longview, Texas, in 1946, the daughter of William and Merle Silkwood. She was raised in Nederland, Texas, and studied
medical technology at Lamar State College in Beaumont on a scholarship
from the Business and Professional Women's Club. In 1965, she married
William Meadows, with whom she had three children. She left her husband
in 1972 and went to Oklahoma City, where she was employed briefly as
a clerk in a hospital before being hired as a metallography laboratory technician at the Cimarron River plutonium plant of the Kerr-McGee Corporation. Silkwood joined the Oil, Chemical, and Atomic Workers' Union
and in 1974 became the first female member of the union bargaining committee in Kerr-McGee history.

On her first union assignment to investigate health and safety issues at
the plant, she discovered evidence of spills, leaks, and missing plutonium.
As the public's concern for the environment increased in the 1970s, Kerr-
McGee faced litigation involving worker safety and environmental contamination. According to the Oil, Chemical, and Atomic Workers' Union, the
Kerr-McGee plant had manufactured faulty fuel rods, had falsified product
inspection records, and had risked employee safety. Silkwood testified before the U.S. Atomic Energy Commission that she had suffered radiation
exposure in a series of unexplained incidents. In September 1974, Silkwood
flew to Washington, D.C., to speak with union officials about the problems
she had uncovered.

During the next two months, Silkwood gathered additional information about the plant. A union official eventually arranged for Silkwood to discuss her findings with a reporter. During the two weeks prior to her death, Silkwood reportedly had been gathering supporting evidence for her union's claims that Kerr-McGee was negligent in maintaining plant safety. During the same period, she was the victim of a number of unexplained exposures to the radioactive element plutonium 239, approximately 300 micrograms of which were ultimately found on surfaces and food in her apartment. The amount of plutonium discovered at Silkwood's home raised concerns. Kerr-McGee arranged for Silkwood, her roommate, and Silkwood's boyfriend to be tested at the U.S. Department of Energy's Los Alamos National Laboratory. They underwent whole-body and lung radiation counts, total urine and feces collection, and monitoring of external alpha activity. They were informed by the chief of the Laboratory Health Division that all of their test results had fallen well within safe limits.

On the evening after her return from New Mexico, Silkwood attended a union meeting in Crescent, Oklahoma. When the early-evening meeting ended at about 7 P.M., she left alone in her car, headed for Oklahoma City. Her car was found by the highway patrol smashed into a concrete culvert about seven miles south of Crescent, with Silkwood dead at the scene of multiple injuries. Though investigators from the highway patrol concluded that Silkwood's accident had occurred as a result of her falling asleep at the wheel, a Dallas accident-reconstruction firm found that her car had been bumped from behind and forced off the road. Allegedly, she had had in her possession documents that would prove that Kerr-McGee had falsified quality-control records concerning the nuclear fuel rods it manufactured at the Crescent plant. However, those records were never found.

After Silkwood's death, an autopsy was performed on her at the University Hospital in Oklahoma City. Blood tests performed as part of the autopsy showed that at the time of her death, Silkwood's blood contained 0.35 milligrams of methaqualone (Quaalude) per 100 milliliters, almost twice the concentration that would induce drowsiness. An additional 50 milligrams of undissolved methaqualone remained in her stomach. Organs from her body were examined by a Los Alamos Tissue Analysis Program team at the request of the Atomic Energy Commission (AEC) and the Oklahoma City medical examiner. The highest concentrations of radioactive material were found to be in the contents of her gastrointestinal tract, which demonstrated that she had ingested plutonium prior to her death. Results of lung analyses also confirmed that Silkwood had received very recent exposure to plutonium. Her household goods reportedly were buried in a nuclear waste dump because of the high levels of radiation found in her apartment. It took three months to decontaminate the apartment.

Silkwood's story emphasized the hazards of nuclear energy and raised questions about corporate accountability and responsibility. The Kerr-McGee nuclear fuels plant closed in 1975. On behalf of Silkwood's three children, her estate filed a civil suit against Kerr-McGee for the allegedly inadequate plant health and safety measures that permitted her exposure to radiation. The first trial ended in 1979 with the jury awarding Silkwood's estate $10.5 million in personal-injury and punitive damages. At the time, Eleanor Smeal, president of the National Organization for Women, proclaimed, "This victory is in honor of Karen Silkwood, who broke the stereotype that women on the job will tolerate exploitation" (The Feminist Majority Foundation). The 1979 finding was later reversed by the federal court of appeals in Denver, Colorado, which instead awarded a mere $5,000 for the personal property destroyed during the decontamination of Silkwood's apartment. In 1986, twelve years after Silkwood's death, the suit was again headed for retrial when it was finally settled out of court for $1.3 million. Under the 1986 agreement, the Kerr-McGee Company admitted no guilt for the automobile accident that killed Silkwood. That same year, *Life* magazine's editors included Silkwood in their list of "*Life* Magazine's 50 Most Influential Baby Boomers."

In December 1999, actress Meryl Streep, who played the title role in the 1983 movie *Silkwood*, sang "Amazing Grace," a song from the film, during a twenty-fifth-anniversary fundraising tribute to Silkwood and a commemoration of her death. "Those of us who knew her through her union work or from the portrayal of her struggle in the film 'Silkwood' will always be moved by her courage," Streep wrote in a preevent letter to participants. (Associated Press 1999b). During the tribute, Streep added, "I'm not sure she saw herself as a hero. Karen Silkwood was a person who had courage, and was an activist in the truest sense of the word. I think the best way to honor her is to support the people who are fighting now to do what is right" (E-Wire). Proceeds from the event, which also featured appearances by Silkwood's father, son, and former nuclear plant coworkers, will benefit the Labor party's "Just Health Care" Campaign, a union-sponsored program to reform health care. Jim Key, a union health and safety official who was also recognized at the event, said, "Everything Karen Silkwood was saying 25 years ago about the lies told to workers and communities throughout the nuclear industry is being proven true today" (E-Wire). Another participant, Anthony Mazzocchi, who had worked with Silkwood and her local union, chimed in, "Karen Silkwood is the personification of thousands of people who work every day in these difficult jobs in high-risk industries, trying to improve conditions for their co-workers" (E-Wire).

BIBLIOGRAPHY

Associated Press. (1999a, March 7). Silkwood's children break their silence about her life, death. *Dallas Morning News*, p. 49A.

Associated Press. (1999b, December 12). Streep to sing tribute to Silkwood. Available: *http://www.elibrary.com*

E-Wire. (1999, December 22). Karen Silkwood memorial event helps spotlight current workplace issues. *http://ens.lycos.com/e-wire/Dec99/22Dec 9901.html*

The Feminist Majority Foundation/Women's Web World online. The Feminist Chronicles, May 18, 1979. *http://www.feminist.org/research/chronicles/ fc1979.html*

Friedman, R. (Ed.). (1996, June). The baby boom turns 50: The 50 most influential boomers. *Life.* 19(7A). Available: *http://www.pathfinder.com/Life/boomers/ 50boomers04.html#40*

The General Libraries at the University of Texas at Austin/Texas State Historical Association. The Handbook of Texas online. *http://www.tsha.utexas.edu/ handbook/online*

Nuclear Abolition Network. Let the Facts Speak online. *http://www.prop1.org/ 2000/accident/facts3.htm*

Public Broadcasting Service, *Frontline.* (1995). The Karen Silkwood story. Available: *http://www.pbs.org/wgbh/pages/frontline/shows/reaction/interact/ silkwood.html*

Rashke, R. (1981). *The Killing of Karen Silkwood: The Story behind the Kerr-McGee Plutonium Case.* Boston: Houghton Mifflin.

Streep to sing at Silkwood event. (1999, December 13). *Minneapolis Star Tribune,* p. 5B.

United Auto Workers International Union online. Karen Gay Silkwood and the Kerr-McGee Nuclear Corp. *http://www.uaw.org/special/women/women6. html*

—*Bibi Booth*

CATHRINE SNEED
(1954–   )

Perhaps we cannot raise the winds. But each of us can put up the sails so that when the wind comes we can catch it.

—E. F. Schumacher

Cathrine Sneed is the founder (1992) and director of The Garden Project in San Francisco. The United States Department of Agriculture called the Project "one of the most innovative and successful community-based crime prevention programs in the country. But its value goes far beyond crime prevention, providing job training, employment, environmental appreciation and beautification—often for the first time—for people and places in need."

Originally from a suburb outside of Newark, New Jersey, Sneed came to California in the early 1970s with her two young children. She enrolled

Photo courtesy of Cathrine Sneed.

in law school and took classes with a professor named Michael Hennessey, who would soon become Sheriff of San Francisco and ask Sneed to join his staff.

Sneed decided to leave law school and take a job as a counselor at the San Francisco County Jail. She had originally dreamed of being a lawyer, but while working on capital cases during law school, she realized she wanted to do something to prevent people from going to jail or returning. Her work as a counselor was difficult and yielded little positive results. No matter how much she talked to the prisoners or how many suits she gave them for job interviews when they left jail, they continued to return.

Taking her inspiration from John Steinbeck's *The Grapes of Wrath*, Sneed sought to give the prisoners faith in their own ability for personal growth by trying to instill in them a sense of purpose and connectedness. With little to start with, Sneed began the program with what she did have— land, work to be done, and people who wanted to do it. Sneed believes in the words of her friend, the author and poet Wendell Berry, who writes "Out of a history so much ruled by the motto Think Big, we have come to a place and a need that requires us to think little."

Prisoners began to work growing vegetables and herbs that were to be donated to local soup kitchens and family shelters. Still, the prisoners continued to return to jail. With criminal records, and with no job experience or skills, most had no alternative but to return to crime to support themselves. Those that did try to carry the lessons they learned in the garden into their new lives found it difficult without the support of others making a similar transition. Once, a prisoner who was about to be released asked Sneed if he could stay at the jail. He told her he wanted to stay at the farm because it was the best thing that had ever happened to him. On the outside, he had no one and nothing.

Recognizing the need for a post-release program, Sneed began The Garden Project. Sponsored by the San Francisco Sheriff's Department, The Garden Project offers structure and support to former offenders through on-the-job training in gardening and tree care, counseling, and assistance in continuing education. While participants earn a living wage, they also learn valuable skills and give back to their communities—continuing to donate food and planting street trees. Meanwhile, former prisoners work to get their lives back: They work to meet the conditions of their parole or probation, attend school, work with counselors to address personal issues, and re-establish contact with their families and children. Sneed's work has shown that bringing people out of the jails and off the streets into the garden can be transformative in both the human and natural realms.

Today, food grown by the Project feeds hundreds of families and seniors each week. San Francisco enjoys the more than 10,000 street trees planted by Garden Project participants. Best of all, people without hope are given a second chance and proving they can change: 75 percent of Garden Project participants do not return to jail, compared to 55 percent of non–Garden Project participants.

People often ask Sneed how she inspires the people she works with to change. She says that it isn't her—it's working with the plants. One of the most common observations from the seniors who visit The Garden Project's garden, in a poor and working class area of San Francisco called Bay View Hunter's Point, is that they used to fear the young people who work there. Sneed tells the seniors that those young people were afraid too. Sneed believes that during those days spent quietly working the earth, seeing a previously unknown cycle of growth and renewal, nurturing and persisting, these young people experience release and change. The Garden Project grows not only plants but people, too. And those people grow other people: the participants bring their families and friends to the garden; they bring the food they grow home to their families. They are proud to donate the food to seniors and families in need and to explain the different kinds of lettuce they grow.

Sneed says that she did not know the effects beginning the garden would have, she only knew that she had to begin somewhere. Beginning somewhere is enough: The economist E. F. Schumacher once wrote, "Perhaps we cannot raise the winds. But each of us can put up the sails so that when the wind comes we can catch it."

Sneed and her work have been featured in *The New York Times, The Economist, The Chicago Tribune, The Los Angeles Times, US News and World Report*, and *The San Francisco Chronicle*, among other publications, and the A&E Channel's "Uncommon Americans" and the Lifetime Channel. Cathrine has spoken across the country and has been honored with such awards as the National Caring Award, the Hero for the Earth Award, and the National Foundation for the Improvement of Justice Award.

—Makeda Best

WALLACE EARLE STEGNER
(1909–1993)

Wallace Earle Stegner was a novelist, teacher, and conservationist. He was born in Lake Mills, Iowa, on February 18, 1909, and spent his early years in North Dakota, Washington, Montana, Wyoming, and Saskatchewan before his family settled in Salt Lake City, Utah, in 1921. He married Mary Stuart Page in 1934 and had one son, Stuart Page, in 1935.

Stegner entered the University of Utah in 1927 and received a B.A. degree in 1930. He later received an M.A. in English literature from the University of Iowa in 1932 and a Ph.D. in English in 1935. Stegner began teaching in 1934 at Augustana College in Rock Island, Illinois. He continued his academic career during the 1930s as an instructor at the University of Utah and at the University of Wisconsin.

From 1939 to 1945, Stegner was Briggs-Copeland Instructor of Writing at Harvard University. In 1945, he became professor of English at Stanford University. Stegner worked as West Coast editor for Houghton Mifflin Company from 1945 to 1953 and served as editor-in-chief of the magazine *American West* from 1966 to 1968. Stegner's first novel, *Remembering Laughter* (1937), won the Little Brown Prize and brought its author to national attention. At Harvard, Stegner wrote *On a Darkling Plain* (1940), *Fire and Ice* (1941), *Mormon Country* (1942), *The Big Rock Candy Mountain* (1943), which was heavily influenced by his family experiences, and *One Nation* (1945), dealing with racial minorities, for which he received the Houghton Mifflin Life-in-America Award and shared the Anisfield-Wolfe Award.

Following Stegner's founding of Stanford's Creative Writing Program in 1946, the program produced such writers as Larry McMurtry, Robert Stone, Wendell Berry, and Tom McGuane, among others. In Stegner's words, "I had a sense . . . that I was seeing American literature before it was in print" (Sterling et al.). Throughout life, he continued to address old and new themes, such as the complicated aspects of human nature, past and present; death; human isolation and inequality; skimming of the land; and reverence for nature.

Stegner has been referred to as "the dean of western writers." Through his writings and lectures, he brought attention to the fragility of the environment and the need for its protection. He became a champion of the U.S. conservation movement. Stegner wrote more than two dozen works of fiction, history, and biography, much of which deal with the West. As a result, he contributed to changed perceptions of the West and raised the nation's consciousness about the environment.

Stegner lobbied for regulations to prohibit construction of dams in the national parks and campaigned to support the Wilderness Act, which was eventually signed into law by President Lyndon Johnson in 1964. The editor of *Wilderness* magazine called Stegner "the modern equivalent of Thoreau" for his environmental advocacy through literature. In the 1960s, Stegner became special assistant to Secretary of the Interior Steward Udall and served as a member of the National Parks Advisory Board. His numerous awards, grants, and fellowships included three O. Henry Memorial Short Story Awards, the Rockefeller Fellowship for writers in the Far East (1950–1951), Guggenheim Fellowships (1950 and 1959), and a Wenner-Gren Foundation Grant (1953). He participated as a fellow of the Center for Advanced Studies in the Behavioral Sciences (1955–1956) and as writer-in-residence at the American Academy in Rome (1960). He received a Senior Fellowship from the National Endowment for the Humanities in 1972, the Robert Kirsh Award for Life Achievement in 1980, and the Life Time Achievement Award from the PEN USA Center West and a Senior Fellowship from the National Endowment for the Arts in 1990.

A member of the National Institute of Arts and Letters, the American Academy of Arts and Sciences, Stegner also served on the Board of Directors of the Sierra Club and the Governing Council of the Wilderness Society. His many works included *The Preacher and the Slave* (1950); *The Women on the Wall* (1950); *Beyond the Hundredth Meridian* (1954); *This Is Dinosaur* (1955); *A Shooting Star* (1961); *All the Little Live Things* (1967); *The Sound of Mountain Water* (1969); *The Uneasy Chair* (1974); and *Crossing to Safety* (1987). With his son Page and Eliot Porter, Stegner published the essay collection *American Places* (1981). His next collection of essays was *Where the Bluebird Sings to the Lemonade Springs: Living and Writing in the West* (1992). For his novel *Angle of Repose* (1971), he received the Pulitzer Prize for fiction, and for *The Spectator Bird* (1976), the National Book Award for Fiction. Stegner died in Santa Fe, New Mexico, on April 13, 1993.

BIBLIOGRAPHY

Porter, E., Stegner, W., & Stegner, P. (1981). *American Places*. New York: Dutton.

Stegner, W. E. (1940). *On a Darkling Plain*. New York: Harcourt, Brace and Company.

Stegner, W. E. (1941). *Fire and Ice*. New York: Duell, Sloane and Pearce.

Stegner, W. E. (1942). *Mormon Country*. New York: Duell, Sloane and Pearce.

Stegner, W. E. (1943). *The Big Rock Candy Mountain*. New York: Duell, Sloane and Pearce.

Stegner, W. E. (1945). *One Nation*. Boston: Houghton Mifflin.

Stegner, W. E. (1950a). *The Preacher and the Slave*. Boston: Houghton Mifflin.

Stegner, W. E. (1950b). *The Women on the Wall*. Boston: Houghton Mifflin.

Stegner, W. E. (1954). *Beyond the Hundredth Meridian. John Wesley Powell and the Second Opening of the West*. Boston: Houghton Mifflin.

Stegner, W. E. (1955). *This Is Dinosaur: Echo Park Country and Its Magic Rivers*. New York: Knopf.

Stegner, W. E. (1961). *A Shooting Star*. London: Heinemann.

Stegner, W. E. (1967). *All the Little Live Things*. New York: Viking Press.

Stegner, W. E. (1969). *The Sound of Mountain Water*. Garden City, NY: Doubleday.

Stegner, W. E. (1971). *Angle of Repose*. Garden City, NY: Doubleday.

Stegner, W. E. (1974). *The Uneasy Chair. A Biography of Bernard DeVito*. Garden City, NY: Doubleday.

Stegner, W. E. (1976). *The Spectator Bird*. Garden City, NY: Doubleday.

Stegner, W. E. (1987). *Crossing to Safety*. Franklin Center, PA: Franklin Library.

Stegner, W. E. (1992). *Where the Bluebird Sings to the Lemonade Springs: Living and Writing in the West*. New York: Random House.

Stegner, W. E. (1996). [1937]. *Remembering Laughter*. New York: Penguin Books.

Sterling, K. B., Harmond, R. P., Cevasco, G. A., & Hammond, L. F. (Eds.). (1997). *Biographical Dictionary of American and Canadian Naturalists and Enviromentalists*. Westport, CT: Greenwood Press.

—*John Mongillo*

EDWIN WAY TEALE
(1899–1980)

> It is the conservationist who is concerned with the welfare of all the land and the life of the country, who, in the end, will do the most to maintain the world as a fit place for human existence.
>
> —Edwin Way Teale

Edwin Way Teale, one of the best-loved naturalists and nature writers of his generation, captured the imaginations and hearts of Americans in his many books about nature. He has been ranked with John Muir, Henry David Thoreau, and John Burroughs as one of the best and most influential nature writers in American history. He was also an accomplished photographer who pioneered new techniques for creating close-up images of insects and other wildlife.

Teale was born in Joliet, Illinois, in 1899, the only child of a railroad mechanic. His lifelong interest in natural history grew out of childhood summers on his grandparents' northern Indiana farm, a period recollected in his 1943 memoir *Dune Boy*. At the farm, he had many experiences that honed his observation skills and developed his keen sensitivity to all aspects of the natural world. In his 1942 book *Near Horizons*, Teale described a typical childhood summer adventure that took place in a field of rye growing between his grandfather's farm and a dune field. Teale crawled into the dense forest of rye, hollowed out a cave, and lay very still. All day, he watched the activities of the small creatures around him—ants, beetles, flies, toads, and snakes—using his imagination to picture life as it must have appeared to them. He felt that he had discovered a world that went unnoticed by most people. "Even the commonest of insects, once we enter the Alice-in-Wonderland realm they inhabit, become engrossingly interesting. Their ways, their surroundings, their food, their abilities are so foreign to our own that imagining ourselves in their places becomes an exciting adventure of the mind" (Nature Writing).

Teale graduated from Earlham College in Richmond, Indiana, in 1922 and received his master's degree in English literature from Columbia University in 1926. He worked in New York as a magazine staff writer and editor for *Popular Science* from 1928 to 1941. It was not until about 1930 that Teale seriously incorporated photography into his investigations in natural history. His photographic studies of insects resulted in the publication in 1937 of his first nature book, *Grassroot Jungles*.

Teale and his wife, Nellie, leased land near their Baldwin, Long Island, home to serve as a garden where Teale could study and photograph insects. He developed new techniques for close-up photography so he could show his readers the worlds he discovered. During this early period of nature writing, the Teales' only son, David, was killed in battle in World War II. In *North with the Spring*, Teale says that he and his wife coped with their grief during those days of war and loss by keeping a dream alive: they planned a series of books based on the progression of the seasons through America. Teale left his New York career in October 1941 to fulfill that dream. The first of the books, *North with the Spring*, was published in 1951. In it, Teale recounts his and Nellie's departure from their New York home on February 14 to drive their black Buick to the Everglades in Florida. Once there, they turned around and followed the northward advance of spring on a 17,000-mile journey through the eastern United States, from southern Florida to the Canadian border. They traveled through twenty-three states over 130 days. Teale said at the time, "My wife and I dreamed of knowing something of all phases [of the movement of spring], of seeing, first hand, the long northward flow of the season" (University of Connecticut Library).

Over the next twenty years, the Teales followed the other three seasons in a similar manner, traveling to almost every U.S. state. These travels resulted in a quartet of books describing all four American seasons: *North with the Spring* (1951), *Autumn across America* (1956), *Journey into Summer* (1960), and *Wandering through Winter* (1965). The books provide a fascinating chronicle of America, treating not only its changing seasons, but also its little-known people and places during a pivotal decade and a half in a rapidly changing America. The inscription in each of the four books reads, "Dedicated to David Who Traveled with Us in Our Hearts."

In 1959, Teale and his wife escaped the increasing suburbanization of their Long Island home for a 130-acre overgrown farm near Hampton, Connecticut, which they named Trailwood. Teale had always wanted to devote time to exploring the natural world of one place; the farm gave him that opportunity. "A miracle," Teale wrote in his diary that first year, "almost exactly what we had envisioned. Here are opportunities for everything" (University of Connecticut Library). Over the years, the plants and animals of the farm unfolded their secret worlds to the Teales. They came to know and to be a part of the land and the community of living things at Trailwood. Teale's experiences became the subject of one of his most popular books, *A Naturalist Buys an Old Farm* (1974), which also touches on his voluminous correspondence with fans, writers, and naturalists such as Rachel Carson and Roger Tory Peterson. Teale and his wife lived at Trailwood for the rest of their lives.

Teale published more than thirty books over his lifetime, all illustrated with his own photographs; ten of them were completed at Trailwood. *Wandering through Winter* (1965), the final volume in his natural history of the four seasons in America, won the 1966 Pulitzer Prize for nonfiction. Teale's concern for conservation underlies his work, yet his writing never appears political or argumentative. He said, "It is the conservationist who is concerned with the welfare of all the land and the life of the country, who, in the end, will do the most to maintain the world as a fit place for human existence. . . . Those who wish to pet and baby wildlife love them, but those who respect their natures and wish to let them live their natural lives, love them more" (Connecticut Audubon Society).

From 1947 to 1948, Teale was president of the American Nature Study Society, America's oldest environmental education organization. He was the recipient of numerous awards, including the John Burroughs Medal in 1943 and the Ecology Award of the Massachusetts Horticultural Society in 1975. He was an elected fellow of the American Association for the Advancement of Science and the New York Academy of Sciences and an associate of the Royal Photographic Society.

Teale kept an unusually detailed record of his life and work, carefully preserving his diaries, field notes, correspondence, and rough drafts. Shortly before his death in 1980, he donated his literary manuscripts to the University of Connecticut Libraries. Nellie Teale later donated additional papers and a large part of the Teales' personal library. These materials are an important primary source for understanding America's growing interest in natural history and the environment during a period of rapid urbanization.

After Nellie Teale's death in 1993, Trailwood passed to the Connecticut Audubon Society, which maintains it as a public sanctuary and a memorial to the Teales and their appreciation for the natural world. Visitors can follow trails over the sanctuary's pond, wetland, meadow, and woodland habitats. There the Audubon Society also maintains a small natural history museum that includes Teale memorabilia.

BIBLIOGRAPHY

American Nature Study Society online. American Nature Study Society: Leading the way to the future through environmental education. *http://hometown. aol.com/anssonline/index.htm*

Biography Online. Teale, Edwin Way. *http://www.biography.com*

Connecticut Audubon Society online. Trail Wood. *http://www.ctaudubon.org/centers/trailwood/trailwoo.htm*

Natural Resources Defense Council online. The story of *Silent Spring*. *http://www.nrdc.org/nrdc/health/pesticides/hcarson.asp*

Nature Writing online. Favorite writers: Edwin Way Teale. *http://www.nature writing.com/frfavwrit.htm*

Teale, E. W. (1937). *Grassroot Jungles*. Boston: Little, Brown and Co.

Teale, E. W. (1942). Favorite writers: Edwin Way Teale. *Near Horizons: The Story of an Insect Garden*. New York: Dodd, Mead and Co.

Teale, E.W. (1951). *North with the Spring*. New York: Dodd, Mead and Co.

Teale, E. W. (1960). *Journey into Summer: A Naturalist's Record of a 19,000-Mile Journey Through the North American Summer*. New York: Dodd, Mead and Co.

Teale, E. W. (1986). *Dune Boy: The Early Years of a Naturalist*. Bloomington: Indiana University Press.

Teale, E. W. (1990a). [1956]. *Autumn Across America: A Naturalist's Record of a 20,000-Mile Journey Through the North American Autumn*. New York: St. Martin's Press.

Teale, E. W. (1990b). [1965]. *Wandering Through Winter: A Naturalist's Record of a 20,000-Mile Journey Through the North American Winter*. New York: St. Martin's Press.

Teale, E. W. (1998). [1974]. *A Naturalist Buys an Old Farm*. Storrs, CT: Bibliopola Press.

University of Connecticut Bookstore online. *A Naturalist Buys an Old Farm:* Edwin Way Teale. *http://www.bookstore.uconn.edu/teale.html*

University of Connecticut Library online. Edwin Way Teale: About the naturalist. *http://www.lib.uconn.edu/Exhibits/carroll/ewteale*

—Bibi Booth

MARIA TELKES
(1900–1995)

> The heating system [for the first solar-heated house] was developed by
> Dr. Maria Telkes. It was cheap and effective and promptly ignored.
> —Women of Achievement

Maria Telkes was a physical chemist and biophysicist who designed the
first residential solar heating system, developed a solar-powered distillation
system to convert seawater to fresh water, and invented many other patented solar-powered devices. Known by some authors as the "Sun Queen,"
Telkes was the world's most famous female inventor in the field of solar
energy. Many of her original designs still serve as the basis for solar devices
under development today.

Telkes was born in Hungary in 1900 and first became interested in the
potential of solar power as a high-school student there. Photovoltaics, solar
cooking, solar space and water heating, solar distillation, and thermoelec-
tricity were some of her areas of curiosity. These interests ultimately led to
a fifty-year career of extensive research and development of solar energy
devices. After she earned a doctorate in physical chemistry from the University of Budapest, Telkes came to the United States.

From 1939 on, Telkes held positions in solar energy research and development at the Massachusetts Institute of Technology (MIT) and the New
York University College of Engineering, among other academic institutions.
At MIT, she researched phase-change materials for their potential application as energy storehouses. In 1948, she used her findings about the latent
heat of fusion of salt hydrates to design a heating system for an experimental, solar-heated house in Dover, Massachusetts. Built with funds provided by wealthy Boston sculptor Amelia Peabody and designed by
architect Eleanor Raymond, the house is still in use today. In the mid-
1950s, the International Solar Energy Society held its first meeting in Phoenix, Arizona, to bring together solar energy pioneers from around the
world. Telkes took advantage of the opportunity to share news about advances in the development of solar energy applications and ideas for its
promotion as an alternative energy resource.

During the oil crisis of the early 1970s, solar energy gained worldwide
attention as a possible alternative to fossil fuels, and some of Telkes's early
designs began to receive serious consideration. This led to the 1980 construction of the Carlisle Solar House in Carlisle, Massachusetts, a joint
effort of MIT and the U.S. Department of Energy. The Ford Foundation
also issued a $45,000 grant to Telkes to continue the development of her

solar oven. This grant allowed her to give her invention diverse capabilities: her solar ovens are designed to be adaptable to any tribal or ethnic cuisine, can be used safely by children, will not scorch foods, and liberate the cook from the need for constant stirring of food. These features have particularly benefited rural women in underdeveloped nations, who are typically responsible for the gathering of often-scarce fuel wood and the labor-intensive preparation of meals. Telkes's solar energy inventions even led to the development of a faster way to dry crops. Her additional solar-powered appliances include a system for distilling salt water for use on life rafts. She also designed, built, and tested solar thermoelectric generators for terrestrial and space uses.

In 1952, the Society of Women Engineers, which encourages women to achieve their full potential in careers as engineers and leaders, presented Telkes its Achievement Award in recognition of her contributions to the utilization of solar energy and her advocacy of solar energy over nuclear energy. In 1977, Telkes received the Charles Greeley Abbot Award from the American Solar Energy Society for her significant contributions to the field of solar energy. The Maria Telkes Papers, which are housed at the Arizona State University Architecture and Environmental Design Library, comprise a diverse collection of correspondence, reports, research notes, and files about the late scientist and her solar-powered inventions.

BIBLIOGRAPHY

American Solar Energy Society online. The Charles Greeley Abbot Award. *http://www.ases.org/about/abbot.html*

Arizona State University Libraries, Special Collections, online. Telkes, Maria, 1900–1905. *http://www.asu.edu/caed/AEDlibrary/libarchives/solar/telkes.html*

Inventors Assistance League, Women in Science online. Maria Telkes, Chemist and solar engineering. *http://www.inventions.org/culture/science/women/telkes.html*

Society of Women Engineers online. 1952: Maria Telkes. *http://swe.org/SWE/Awards/achieve3.htm#y1952*

University of Florida, Solar Energy and Energy Conversion Laboratory, online. The solar touch. *http://www.me.ufl.edu/SOLAR/pg5.htm*

Virginia Tech Virtual Museum of Women in Electrical Power Engineering. Dr. Maria Telkes (1900–1995). *http://www.ee.vt.edu/~museum/women/maria/a_maria._html*

Women of Achievement online. Maria Telkes. *http://dpsinfo.com/women/history/cal4.html*

—Bibi Booth

HENRY DAVID THOREAU
(1817–1862)

We need the tonic of wildness . . . We can never have enough of nature.
. . . We need to witness our own limits transgressed, and some life pas-
turing freely where we never wander.
     —Henry David Thoreau (*Walden and Other Writings*)

Henry David Thoreau was an author, naturalist, and prophet for ecological
thought and the value of wilderness. He articulated the idea that humans
are part of nature, and that we function best, as individuals and societies,
when we are conscious of that relationship. Thoreau also emphasized the
dynamic ecology of the natural world. Though largely overlooked during
his lifetime, Thoreau earned his place in history in 1845, when he moved
to Walden Pond "to live deliberately." A committed individualist, he championed the human spirit against materialism and social conformity. His
time at Walden, slightly over two years, demonstrated the natural harmony
that was possible when a thinking human lived simply.

Over the past century and a half, millions have read and been inspired
by *Walden; or, Life in the Woods*, his thoughts on his sojourn there. Many
consider Thoreau to be the father of the American conservation and environmental movements: he argued passionately against the careless destruction of wilderness. "In wildness is the preservation of the world," he wrote,
and with such statements he helped shape the thinking of modern environmentalists (University System of Maryland). The environmental movement
of the past thirty years has embraced Thoreau as a guiding spirit, and he
is valued for his early understanding of the idea that nature is made up of
interrelated parts. Nature is Thoreau's first great subject; the question of
how humans should live is his second.

Henry David Thoreau was born in 1817 in the village of Concord, Massachusetts. He was John and Cynthia Dunbar Thoreau's third child of four.
Thoreau's mother loved the outdoors and was socially conscious and
assertive, helping to form the Concord women's antislavery society. Thoreau's father, who was quiet and industrious, managed several stores
around Massachusetts, which caused the family to move three times during
Thoreau's childhood; they returned to Concord in 1823 and never moved
again. When the stores eventually failed, Thoreau's father started a pencil-
making business.

Thoreau's childhood was spent exploring the woods and fields around
his home. In elementary school, he was curious and intelligent, and his
serious disposition earned him the nickname "Judge." When he was eleven,

Henry and his older brother, John, Jr., enrolled in the Concord Academy. In 1833, Thoreau entered Harvard College, whose curriculum at the time focused on theology. He studied four languages and spent much time in the library, amassing over 5,000 pages of notes, which he often referred to in later years as source material for his writings. In December 1835, Thoreau took advantage of a policy at Harvard that allowed students a thirteen-week leave of absence to teach school to help pay tuition; he worked in Canton, Massachusetts, during this time. Though he was never pleased with the teaching methods at Harvard, he did make good use of the lifelong borrowing privileges at Harvard College Library that were conferred with his degree. During his Harvard years, he was also exposed to the writings of the transcendental philosopher, essayist, and poet Ralph Waldo Emerson, who later became his chief mentor and friend.

After receiving his bachelor's degree in 1837, Thoreau returned to Concord and was hired to teach at the Center School. After two weeks of teaching, he angrily resigned after having been reprimanded for refusing to whip students as a disciplinary measure. Thoreau then helped his father at the pencil factory until he established his own school, which had an enrollment limited to twenty-five and emphasized experiential learning. His brother, John, taught there with him. The school operated for two years, closing in 1841 because of John's declining health.

Thoreau was a skilled, self-taught surveyor, and when he was in need of additional income, he would hire himself out to landowners to assist with property assessments and settlement of boundary disputes. In 1838, he began lecturing on the Lyceum circuit from Maine to New Jersey, which he continued throughout his life. Though these lectures were not lucrative, they were important to Thoreau's process of composition; most of the ideas and themes in his essays and books were first presented to the public in lectures.

In 1841, he was invited to join Ralph Waldo Emerson's household, where he lived intermittently until 1843. He served as handyman and assistant to Emerson, helping to edit and contributing poetry and prose to the transcendentalist magazine, the *Dial*. In 1842, he traveled to Staten Island, New York, to serve as a tutor to the children of Emerson's brother, while also cultivating the New York literary market. His literary activities in New York were not successful, however. Confirmed in his distaste for city life and disappointed by his failure, he returned home to Concord in late 1843.

In January 1842, Thoreau's brother died of lockjaw. Devastated, Thoreau decided to write a book commemorating a two-week canoe trip he had taken with his brother two years earlier. Seeking a quiet place to write, Thoreau took a friend's suggestion and built a small, crude cabin on a piece of land owned by Emerson on the north shore of sixty-one-acre Walden Pond. He started work on his cabin in March 1845, and on July 4, he moved in. During his time there, Thoreau wrote his tribute to his brother, *A Week on the Concord and Merrimac Rivers*, and otherwise filled his time by observing nature, working in his garden, talking with visitors, reading, and writing in his diary. Aiming to lead a life free of materialistic pursuits, he supported himself by growing vegetables and by surveying and doing odd jobs in nearby Concord. But most of all, he walked and thought. In his two years living in his little cabin in the woods, he brought himself to a state of conscious living, where thought and action were harmoniously combined. Although many believe that Thoreau was a recluse during his Walden period, he was actually no stranger to society during this time. He had frequent dinners with family and friends and also had friends and an occasional neighbor visit him at his cabin. He had only three chairs there: "one for solitude, two friendship, and three for society" (University System of Maryland).

In late July 1846, a little more than one year into his Walden Pond stay, Thoreau walked into Concord to have a shoe repaired. He encountered the town constable, who asked Thoreau to pay his overdue poll tax. At this point, Thoreau was intentionally several years behind in his tax payments because he objected to the use of such revenues to finance America's war with Mexico and support the enforcement of slavery laws. Thoreau again refused to pay his taxes and spent that night in jail. He said later, "It costs me less in every sense to incur the penalty of disobedience to the State than it would to obey" (Ecolink). Later that evening, a woman, probably Thoreau's aunt, paid the delinquent bill. When Thoreau found out that his taxes had been paid, he was outraged and demanded to remain in jail. He was nevertheless released the next morning. The evening he spent in jail prompted Thoreau to write what became one of his most famous essays and one of the most important political essays in history, "On the Duty of Civil Disobedience," which was a call for passive resistance to unjust laws. First published in 1849 under the title "Resistance to Civil Government," the essay asks all people to question both their own actions and the actions of the state. The essay has since had far-reaching influence on social movements and their leaders, including Mohandas Gandhi and Martin Luther King, Jr.

Thoreau left his cabin at Walden Pond in September 1847. "Perhaps it seemed to me that I had several more lives to live, and could not spare any more time for that one," he said in *Walden*'s last chapter. After seven drafts, *Walden* was finally published in 1854; it was received well, but still took five years to sell the 2,000 copies in its first print run.

In the summer of 1850, Thoreau was again living in his family home, considering what to do next. "I feel ripe for something, yet do nothing," he wrote in his journal (Associated Press 1999b). During this time, Thoreau was becoming less of a transcendentalist and more of an activist and above all, a dedicated abolitionist. In his journal, Thoreau wrote about a number of times when he assisted fugitive slaves on their way north. He hid them, drove them to the train station, bought their tickets, and sometimes even accompanied them to the next station. He lectured and wrote against slavery, with "Slavery in Massachusetts," a lecture delivered in 1854, serving as his harshest criticism. Having met abolitionist John Brown in 1857, he was the first person in America to write in defense of Brown's ill-fated raid on the federal arsenal in Harpers Ferry, West Virginia. His "Plea for Captain John Brown," written in 1860, is among his most passionate works.

As Thoreau grew older, his attentions turned more toward the systematic observation and recording of natural history in the area around Concord. He kept thorough journals of his observations, and the citizens of Concord began to regard him as the town naturalist. Thoreau sought to remove himself from the center of his observations and let the natural objects he studied speak for him, intending to be less romantic and more scientific. He spent the last decade of his life compiling data for a comprehensive natural history of Concord.

In December 1860, Thoreau caught a cold while counting tree rings in a snowstorm near Concord and eventually developed bronchitis. His lungs had long been tubercular, and he was housebound for many weeks. In the summer of 1861, his doctor advised a change of climate, so Thoreau traveled to Minnesota, only to become homesick and return to Concord two months later. His health steadily worsened, and he died at home in 1862 at the age of forty-five.

Shortly before his death, Thoreau had wrapped up the unfinished manuscript of his Concord nature observations, a portion of which was entitled *Wild Fruits*, and placed it in a wooden chest, where he kept thousands of pages from other projects. Much jumbling and shuffling occurred as the papers passed from owner to owner, finally ending up in storage at the New York Public Library. That confusion, complicated by Thoreau's notoriously poor handwriting, kept *Wild Fruits* unavailable until only recently. *Wild Fruits: Thoreau's Rediscovered Last Manuscript* was published in October 1999, nearly 140 years after Thoreau's death. His other posthumously published works include *Excursions*, published in 1863, *The Maine Woods* (1864), *Cape Cod* (1865), and *A Yankee in Canada* (1866).

Established in 1941, the Thoreau Society is the oldest and largest organization devoted to an American author. Its mission is to honor Henry David Thoreau and to stimulate interest in and foster education about his life, works, and philosophy and his place in his world and ours. In 1965, Walden Pond was designated a National Historical and Literary Landmark; about 750,000 people visit it annually. Walden Pond is the centerpiece of the approximately 425-acre Walden State Reservation, administered by the Commonwealth of Massachusetts, and is the crown jewel in the much larger Walden Woods ecosystem, now being incrementally secured and protected by the Walden Woods Project.

Though Thoreau's cabin was long ago dismantled, it lingers in the American consciousness as an image of perfection. The spot by the north shore on which Thoreau's cabin once stood is marked with a cairn and nine stone posts marking the locations of the cabin walls. Travelers to the site, including poet Walt Whitman, have paid homage to Thoreau by laying stones on the cairn.

BIBLIOGRAPHY

Associated Press. (1999a, September 7). Thoreau gets back to nature yet again: Best of new work on par with "Walden." *Washington Times*, p. A8.

Associated Press. (1999b, September 7). Book by Thoreau, naturally: His deciphered "Wild Fruits" to be published. *Newsday*, p. A8.

Biography online. Thoreau, Henry (David). *http://www.biography.com/cgi-bin/biomain.cgi*

The Columbia Encyclopedia (5th ed.). (1993). Thoreau, Henry David. New York: Columbia University Press.

Compton's Encyclopedia online. Thoreau, Henry David. *http://www.comptons.com*

Ecolink online. Henry David Thoreau: A biography. *http://www.envirolink.org/enviroethics/thoreaubio.html*

Ecology Hall of Fame online. Ecology Hall of Fame: Henry David Thoreau 1817–1862. *http://www.ecotopia.org/ehof/thoreau/index.html*

Encylopaedia Britannica online. Thoreau, Henry David. *http://www.britannica.com*

Gray, P. (1999, November 29). Unregarded berries: Wild Fruits, Thoreau's "new" work, is as fresh as when he wrote it more than 130 years ago. *Time*, p. 81.

Infoplease online. Thoreau, Henry David. *http://www.infoplease.com*

Schnur, S. (1999, August 11). Thoreau's life on Walden's simple shores. *Christian Science Monitor*, p. 18.

Stamberg, S. (1998, June 6). Interview with Jane Langdon, Thoreau Institute. National Public Radio Weekend Saturday. Washington, D.C. Available: *http://www.elibrary.com*

Thoreau, H. D. (1970). [1866]. *A Yankee in Canada, with Anti-Slavery & Reform Papers*. Westport, CT: Greenwood Publishing Group.

Thoreau, H. D. (1984). [1865]. *Cape Cod*. Orleans, MA: Parnassus Imprints.

Thoreau, H. D. (1992a). [1863]. *Excursions*. Temecula, CA: Reprint Services Corporation.

Thoreau, H. D. (1992b). *Walden and Other Writings*. New York: Barnes & Noble Books.

Thoreau, H. D. (1995). [1864]. *The Maine Woods*. New York: Viking Penguin.

Thoreau, H. D. (1999). *Wild Fruits: Thoreau's Rediscovered Last Manuscript*. New York: W. W. Norton.

University System of Maryland online. Cybersaunter—Henry David Thoreau. *http://usmb12.usmd.edu/thoreau*

The Walden Society online. Thoreau, the man. *http://www.walden.org/thoreau*

—*Bibi Booth*

Photo courtesy of John Todd.   Photo courtesy of Nancy Jack Todd.

JOHN TODD AND NANCY JACK TODD
(1939– and 1939–)

In Steven Lerner's book *Eco-Pioneers*, the author describes John Todd as a Canadian-born marine biologist who has pioneered a new way of treating sewage and other wastes. The body of his work encompasses the whole field of applied ecology. Todd is the president of Ocean Arks International and a research professor and distinguished lecturer at the University of Vermont. Presently, he teaches ecological design and living machines and oversees an ecological design studio. He is also the cofounder and a director of Living Technologies, Inc., the cofounder and a director of IASIS Corporation, and a partner in John Todd Research and Design, Inc. He is working on developing an eco-industrial park for the city of Burlington, Vermont. In recent years, his work has expanded toward comprehensive ecological design and includes the design of resorts and new towns.

Todd's strongest early influences were his parents, who gave him a deep love for nature and landscape. As a young teenager, he was profoundly influenced by Louis Bromfield, the novelist, writer, farmer, and earth steward. Bromfield's Malabar Farm series of books shaped his life by showing him that the healing of degraded and despoiled landscapes was practical and realistic.

Todd was trained in agriculture. He received his bachelor's degree from McGill University. He did graduate training in parasitology and tropical medicine, earning his M.S. at McGill University. He received his doctorate in fisheries at the University of Michigan. His early research, including his research as an assistant scientist at the Woods Hole Oceanographic Institution, involved the behavioral ecology of fishes. This work was summarized in his *Scientific American* article "The Chemical Languages of Fishes" (May 1971).

In 1969, Todd cofounded the New Alchemy Institute to create a science and engineering based upon ecological precepts. In 1980, he founded Ocean Arks International. Throughout the 1980s, he developed an ecological design field that included energy, architecture, waste, food, and ocean-transport components. This work was published in a series of books: *The Village as Solar Ecology* (1980), *Tomorrow Is Our Permanent Address* (1980), *Reinhabiting Cities and Towns: Designing for Sustainability* (1981), and *Bioshelters, Ocean Arks, City Farming: Ecology as the Basis of Design* (1984). In 1984, he began developing technologies for treating wastes and purifying water. In 1987, he was recognized by the United Nations Environmental Programme for his "contributions toward protecting the Earth's environment."

Todd is the inventor of the concept of a Living Machine based upon ideas derived from the ecologist H. T. Odum. A Living Machine is a contained ecosystem that is designed to accomplish a specific task or series of tasks. Living Machines have been designed by Todd to generate fuels, transform wastes, grow foods, heal damaged aquatic environments, and regulate climate and air quality in buildings. As an example, in waste-treatment Living Machines, wastewater is guided through a series of cylinders, each one with a different ecosystem, each with a different combination of microbes, plants, and animals. As the water passes through the various systems, different contaminants from bacteria to heavy metals to industrial waste are removed. As liquid waste moves along in the process, toward the midsection of the tank, higher plants, snails, and fish are introduced into the systems, and the diversity of life increases. At the end of the series, the water is purified to high standards through ecological processes. The algae and plankton used to clean the water also provide feeds and nutrients for beneficial byproducts, including trees, flowers, and fishes. Todd's early work at the New Alchemy Institute pioneered ecosystems-based aquaculture with tilapia, a mild-tasting fish originating in Africa. This work culminated in the integration of aquaculture and agriculture in solar-based structures called bioshelters.

Sewage-treatment Living Machines have been tested in a wide variety of climates and wastes. For example, the sewage-treatment plant constructed on the north coast of Scotland at Findhorn treats the sewage from the Findhorn Foundation community. The challenge of this project was to treat wastewater from the residential community of 300 pe (person equivalents) while providing capacity for large seasonal fluctuations in flow resulting from over 10,000 visitors per year. Wastewater is pumped into buried anaerobic reactors where microbial communities begin digesting the waste.

After this initial digestion process, wastewater flows to a closed aerobic reactor and then into open tanks in a greenhouse, where the rest of the treatment occurs. The visual effect is stunningly beautiful. The whole system looks like a tropical garden in a conservatory. The open aerobic tanks contain aerobic ecologies and are planted with diverse biology. Clarifiers settle out the biosolids, which are returned to the anaerobic reactors. Patented fixed-film reactors called "Ecological Fluidized Beds" polish organic material and suspended solids from the effluent. Final effluent is suitable for reuse after ultraviolet disinfection.

Todd's Living Machines for waste treatment have been built or are under construction in nine countries, including Scotland, England, Czechoslovakia, Hungary, India, Brazil, Australia, the United States, and Canada. On April 22, 1996, he was recognized with the Environmental Merit Award from the U.S. Environmental Protection Agency for his work on the development of Living Machines. He was also awarded the EPA's first Chico Mendes Memorial Award for his work with Living Machines.

Todd is the author of over 200 technical and popular articles on biology and planetary stewardship. He sits on a number of environmental and technical boards, including the Massachusetts Foundation for Excellence in Marine and Polymer Sciences, and has cofounded three environmental companies: Living Technologies, Inc., Burlington, Vermont; Ecological Engineering Associates, Inc., Marion, Massachusetts; and John Todd Research and Design, Inc., Falmouth, Massachusetts.

Nancy Jack Todd was born in South Africa of Canadian parents. She attended schools in South Africa, Canada, the United States, and Argentina and graduated in arts from the University of Western Ontario. As a child of World War II, Nancy Jack Todd developed a deep abhorrence of war and violence, and she became active in the peace movement during the war in Vietnam. She subsequently became involved in environmental issues as well. A lifelong love of literature and of the natural world led her to environmental writing, and she is currently the editor and publisher of *Annals of Earth*, the ecological journal published by Ocean Arks International. She also serves as the vice president of Ocean Arks. She was a cofounder of the New Alchemy Institute, where she edited and published the *Journal of the New Alchemists*. She is the author of several books and has contributed articles to many environmental journals and anthologies.

In 1998, Nancy Jack Todd received the Charles A. Anne Morrow Lindbergh Award for balance between technological innovation and environmental preservation with John Todd. In the same year, she was awarded the Bioneers Award in Ecological Design with John Todd. She is the coauthor of *From Eco-Cities to Living Machines: Precepts of Ecological Design* (1994), *Bioshelters, Ocean Arks, City Farming: Ecology as the Basis of Design* (1984), and *Tomorrow Is Our Permanent Address* (1980). Her next book is on the history and application of ecological design and is entitled

Annals of New Alchemy. Nancy Jack Todd is also deeply committed to women's and children's issues and to the entwined concerns of women and ecology. She has lectured and taught as a guest lecturer at a number of universities and colleges in the United States and Europe.

BIBLIOGRAPHY

Lerner, S. (1997). *Eco-Pioneers: Practical Visionaries Solving Today's Environmental Problems.* Cambridge, MA: MIT Press.
Todd, J. (1971, May). The chemical languages of fishes. *Scientific American.*
Todd, J., & Todd, N. J. (1980). *Tomorrow Is Our Permanent Address: The Search for an Ecological Science of Design as Embodied in the Bioshelter.* New York: Harper & Row.
Todd, N. J., & Todd, J. (1984). *Bioshelters, Ocean Arks, City Farming: Ecology as the Basis of Design.* San Francisco: Sierra Club Books.
Todd, N. J., & Todd, J. (1994). *From Eco-Cities to Living Machines: Principles of Ecological Design.* Berkeley, CA: North Atlantic Books.
                                                —*John Todd and Nancy Jack Todd*

FABIOLA TOSTADO
(1983–   )

Fabiola Tostado was born on August 26, 1983, in Los Angeles, California, the daughter of Refugio Tostado and Patricia Dominguez, but was raised by her aunt and uncle, Natividad and Gonzalo Jimenez. Her childhood was spent moving constantly between Seattle, Washington, and Los Angeles. As a child, she always dreamed of the day when she would see a change in her neighborhood as a whole. She wanted to take part in making sure that every human being received the quality of life that all humans deserve. She finally settled down when she moved into her father's home in 1996.

Struggling in school because of the constant moving, Tostado finally showed her potential when she entered the eighth grade. While she was attending George Washington Carver Middle School, she joined a group called Clean and Green, a program that teaches youth the importance of beautification in one's community. At Clean and Green, she was offered the opportunity to work at a community-based organization called Concerned Citizens of South Central Los Angeles (CCSCLA). When the Clean and Green position was over, she decided to continue at CCSCLA as a volunteer. During the time of volunteering, she became aware of the contamination at Jefferson Middle School. This lit a spark in her and made her angry and upset. The fact that inner-city youth were made to choose between an education or an early grave set Tostado into action. During her volunteering, she also became aware of lead poisoning and its effects. Her heart opened up to all those children who might be exposed to lead poi-

Photo courtesy of James A. Brown.

soning. She believes that no one takes into consideration that these children may be lead poisoned; instead, they are placed in special education classes, many are kicked out of school because of their behavior problems, and some are sent to jail.

With all the time she spent at CCSCLA and the things she was learning, her passion grew, and CCSCLA became her second home. At CCSCLA, she became part of a new family where everyone backed each other up, assisted each other, and listened to one another. After a period of volunteering, she was given a position as a youth intern for the People Organizing for Worker and Environmental Rights (POWER) project. This project made fighting for her community not only her passion, but also her job. She worked long, hard hours organizing around the Jefferson Middle School issue and came face to face with Los Angeles Unified School District (LAUSD) officials. Even though she felt belittled and talked down to, she always respectfully expressed her concerns and asked questions until she felt satisfied with the response.

Overcoming obstacles was hard at first because of the quiet person she was. However, at a community meeting where she felt that a LAUSD representative was continuously lying to the community, she exploded. She asked him why their emergency plan said that they would transport students, in case of an emergency, to a hospital that had not been in existence for over twenty years. After nailing him to the wall, the next time she talked to him, she learned that he no longer worked on the JMS, and this motivated her to continue to fight. Working for CCSCLA, she has made presentations on prevention of lead poisoning, recycling of used motors, Jefferson Middle School, and teen worker rights. During this time, she has also met Senator Tom Hayden and has been personally invited by him to attend different meetings.

Tostado soon got over the frustration of older community members not listening to her because she was young when she realized that getting through to at least one person could make a difference. She figured that some may listen and others will not, but she would continue to work as hard as she could to reach those who would. She believes that being young is the best time in a person's life, and part of that time should be used in fighting for what one believes is right. The advice she gives to youth is to never give up on what they believe in no matter who puts them down. From past experience, she knows not to listen to negative people. In life, you will come across those who will think that you are wasting your time and would advise young people to spend their time doing something "fun." Tostado says, "Don't let that get you down. Others might not understand how you feel about something, you can try to explain but don't spend too much time on trying to convince them. Someday they will see a change and acknowledge what you have accomplished."

—*Fabiola Tostado*

PHIL WILLIAMS
(1945–   )

Phil Williams is founder of the International Rivers Network (IRN) in Berkeley, California. Although Williams now lives in the United States, he was born and raised in England. His father was a junior officer in the Royal Navy, so for all of his childhood, he lived in seaports. Williams had always been fascinated by moving water. One of his earliest memories was playing in the brook that ran through his grandfather's farm, building a mud dam and then demolishing it to watch the surge of water reclaim the stream bed below. When he was twelve his father gave his brother and him a small wooden sailing dinghy that was moored in a tidal creek and used to explore the estuaries and marshes around Portsmouth Harbor, where his father was stationed. In those days, Portsmouth was Britain's main navy port and was full of all kinds of warships from different countries. Williams recalls that one time an American fleet came to visit, and all the English school kids would hang around the dockyard gates because the American sailors would give them bubble gum, something they had not seen before.

When Williams was a teenager, his parents paid extra fees for him to go

Photo courtesy of Philip B. Williams and International Rivers Network (IRN).

to Portsmouth Grammar School, a private boys' school. He hated it there because it was very strict and very boring. In those days, the only things that seemed to count were whether one was good at sports, had upper-class parents, or was absolutely brilliant. He was none of these and consequently was a bit of a loner. But a few good things did come out of his high-school education: He was forced to do well enough in his exams to go to university, and he developed an intense dislike of the bullying and hypocrasies he saw around him. By the time he was sixteen, he was very much an outsider and had adopted decidedly left-wing views, contrary to almost everyone else in the school. When he was seventeen, he volunteered to work in the Labour party's general election campaign. For a middle-class boy, canvassing the poor working-class neighborhoods of Gosport, his hometown across the harbor from Portsmouth, was a real education.

The British educational system is different from the American one because it is very intense and very specialized. Williams's favorite subjects were history, geography and English literature, but because he was poor at languages, he had to specialize in science. When it was time to go to university, he chose to study civil engineering almost as an act of rebellion because everyone considered engineers to be low-class, but also because he wanted to literally do something constructive for society.

At Sheffield University, Williams felt free at last. The year was 1963, and in the first few months he was there, the Beatles and Bob Dylan came to town. Sheffield is an industrial city in northern England where the people are very different from the southerners in Williams's hometown. They were friendly and not class conscious. He found engineering pretty boring, but he worked hard enough to get a reasonable degree. Only in the last year did he find a couple of courses he really liked, hydrology and geology, because they dealt with the natural world.

By the time Williams graduated in 1966, just as in the United States, tremendous cultural changes were happening, and he wanted to be part of the action. He decided to avoid getting a job by becoming an engineering graduate student in London. He had very little money, but there was always a party to go to. He was there when the Rolling Stones played Hyde Park, and when Bill Clinton demonstrated against the Vietnam War outside the U.S. embassy. Eventually, in 1970, he received a Ph.D. in hydraulics engineering and had to figure out what to do next. His research was rather conceptual and abstract and did not seem much use to anyone. In those days, he thought that to earn the title of "doctor," one should be brilliant, and he did not feel brilliant, just reasonably intelligent and capable of hard work when it was needed. He knew that he did not want to be a professor; and the British engineering firms were very conservative, just like high school. So he took a plane to Salt Lake City, Utah, and then bought a bus ticket and ended up in San Francisco.

Deciding to emigrate and leave everything behind at the age of twenty-four was a bit easier than it sounded. His father had been in the navy, but his mother was equally as adventurous. When she was twenty-two, she had emigrated to Canada and had worked as a governess in a remote town in the Rocky Mountains. Williams was brought up with stories of grizzly bears and Native Americans. The first thing he wanted to do when he had saved enough money was to see North America, and he was able to do this when he was nineteen, traveling all across Canada and America in a bus. He loved the open space, the wilderness, the straightforward, friendly people, and the feeling of freedom. But for some reason, one place attracted him more than any other. He decided to go back to San Francisco.

He managed to get his first real job working for Bechtel, a big engineering firm in San Francisco that had gotten started by building Hoover Dam in the Great Depression. He was put to work on a new technology, designing pipelines to pump ground-up coal from strip mines to power plants. The company sent him to work on a new mine being opened at Black Mesa on the Navaho-Hopi Indian reservation in northern Arizona. Although the work was technically challenging, he found himself alienated from the remote arid landscape and from his fellow engineers, who were mainly from Texas and did not appreciate the peace symbol on his hard hat. Then interesting things started to happen. On the big sign leading to the camp, someone scrawled "Honky go home!" across the words "Peabody Coal." The firm had emergency shutdowns of the pipeline because someone was shooting holes in it. One day the engineers were told to stay inside the camp because trouble was expected. The "trouble" was a group of students and Hopi Indians in traditional dress who came to perform a ceremony to stop the our strip mine desecrating their ancestral land.

This demonstration had a profound effect on Williams. For the first time, he came face to face with people who really cared about their land. He started to see that the social issues he cared so much about also had an environmental dimension. Protecting the environment was as important to people's well-being as decent working conditions or the rule of law. He grew increasingly disenchanted with what he was doing. It was not good enough for him to do engineering well; he felt that he had to be responsible for its consequences. In 1972, he resigned his position as senior engineer at Bechtel.

He did not know what to do next. It seemed that everywhere he looked, engineers were assisting destructive forces, building nuclear power plants no one needed, freeways that gutted cities, or big dams that destroyed the environment. He tried different things and worked at a public radio station for a while, but had pretty much given up on his area of expertise, hydraulics engineering, until one day he got a piece of junk mail. In 1972, hardly anyone sent junk mail, so he read it. The mail was from the Environmental Defense Fund (EDF) a new group set up as a "coalition of scientists and lawyers to defend the environment." As its office was in nearby Berkeley, he dropped by to see if it needed help. Immediately he was sucked into his first dam fight, the effort to stop the Auburn Dam on the American River above Sacramento. His role was to critique the technical planning of the dam, and for the first time, he could see how his technical training could be put to good use defending the environment. This campaign was his introduction to U.S. water politics, the sad story of how corrupt politicians and spineless bureaucrats have let special interests steal the water and destroy the rivers under the pretense of "progress" and "jobs," a story grippingly told in Marc Reisner's *Cadillac Desert*.

In the 1970s there were plenty of dams and water projects to fight. Most of the battles were lost, but gradually people started to realize that dams not only destroyed the environment but were a huge waste of taxpayers' money and did not stop flood damage or improve local economies. It became clearer and clearer that big dams were what Ralph Nader said they were, "a careless technology," drying up and polluting rivers, destroying fisheries and wetland ecosystems, and adversely altering one of the most important elements in our life-support system: fresh, free-flowing water.

The campaign that had the greatest influence on Williams was the fight to stop the New Melones Dam on the Stanislaus River, flowing out of the Sierra through a spectacular limestone canyon. This time he was helping a new group called Friends of the River. Its members all worked together closely and would often take two- or three-day raft trips down the canyon. Through these trips, being there, touching and seeing the flowing water,

Williams grew to love this river and then all rivers. This changed the way he saw his work. Instead of it just being an altruistic fight against injustice, his work became also a fight for "my" rivers in the same way the Hopi fought to protect their land. Over nine years, the Stanislaus campaign came close to winning three times, but big-money political contributions won out, and by 1983, the dam was built and the canyon was flooded. This was a bitter defeat for the group, but Williams swore never to forget the Stanislaus, and its spirit continues to be one of the driving forces in his efforts to bring the dam-building lobby to account. This experience also led him to seek ways that engineers could develop constructive ways of managing rivers for both humans and the environment. This in turn led him to form a hydrology consulting firm in 1976, dedicated to advancing these ideas.

While Williams was helping EDF, Friends of the River, and other environmental groups as a technical expert in their campaigns, he could not help noticing glowing accounts in the engineering press of more and more big dam projects being promoted in other countries, using the same lies, sometimes by the same people, that had justified destroying America's rivers. He became increasingly aware that the fight to save rivers was a global fight, and just as he and others were starting to stop new dams in the United States, similar corrupt politicians and special interests were pushing for new dams in Third World countries, often encouraged and subsidized by aid bureaucracies like the World Bank.

In 1983, Williams decided to travel to Tasmania to meet Bob Brown, the leader of a recent successful fight to stop the Franklin Dam being built in this beautiful wilderness area of Australia. He was struck by how similar the Stanislaus and Franklin campaigns had been and asked Brown if he thought that these experiences would be useful in helping other fights around the world. With Brown's encouragement, he took the next step and talked to his friend Brent Blackwelder, now president of Friends of the Earth and then the leading antidam lobbyist in Washington, D.C. In those days, Blackwelder organized an annual dam fighters' conference in Washington, and in 1983 he asked Williams to put on a special workshop on "Damming the World's Rivers." He was amazed at the response: the room was packed. He thought that people had only just started to think about global environmental issues, and from the stories that were told, it was clear that global destruction of rivers was not just an ecological issue, but a major social and human rights issue as well. Millions of people were being flooded out by huge reservoirs or forced off their land to make way for irrigation projects that poisoned the soil after a few years. He had heard stories of massacres when tribal peoples resisted, of people starving when their farmlands were no longer regularly flooded, of diseases spread by stagnant reservoirs, of fishermen jobless when the fish disappeared, and of massive dam failures in China that had killed hundreds of thousands of people but had been successfully hushed up. Williams notes that all these stories are documented in Patrick McCully's book *Silenced Rivers*.

Williams started making contact with more dam fighters in other countries. He found experts like Vijay Paranjpe, who was opposing massive dams in India, and activists and students in Austria who, in the freezing winter of 1983, had camped out in the Hainburg Forest to stop the Danube being dammed again. He met Edward Goldsmith and Nicholas Hildyard in England, who were editing the first book about big dams. By 1984, he was confident that it was time to develop a network to support each other in their activities and launch joint campaigns against the international interests promoting big dams. His friend Mark Dubois, who had been one of the leaders of the Stanislaus campaign, volunteered to help out. After unsuccessfully trying to persuade established environmental groups to take on this campaign, Williams decided to go it alone in 1985. He remembers that the founding meetings of the International Rivers Network, the Rainforest Action Network, and David Brower's Earth Island Institute all occurred within a few weeks of each other in an empty warehouse building in San Francisco, and many of the activists all attended each other's meetings.

Williams and his fellow activists had no money; they were only a group of about ten volunteers with an idea. They decided that the first step was to put out a newsletter that later became *World Rivers Review*. They would meet every Tuesday evening on the Berkeley campus and walk around until they found an empty classroom where they could work. Gradually things started to improve. They were able to raise a little money to hire an editor, but they could do little to respond to the increasing number of letters they got from all around the world asking for help or information. Then they were handed a golden opportunity. The International Commission on Large Dams, lobby for the big dams, decided to hold its international meeting in San Francisco in 1988. With the support of local foundations, Williams, Dubois, and others organized a counterconference focused on the true story of big dams. Their conference was much more interesting and attracted good coverage in the media. The result was that not only did they scare the dam builders, but for the first time they were able to invite dam fighters from about twenty different countries. When they got together, they agreed on a joint position and international campaign strategy.

IRN was off and running. Owen Lammers joined the organization as executive director, and within a year IRN had organized demonstrations outside the World Bank headquarters in Washington, D.C. In the last ten years, IRN has grown to an operation of twenty persons supporting campaigns against dams in India, Canada, Slovakia, Laos, China, Russia, Brazil, Spain, Chile, Honduras, Japan, Malaysia, Argentina, Bulgaria, Vietnam,

Belize, Korea, Nepal, Turkey, and Pakistan. The IRN often works behind the scenes to help environmental and social action groups work together against dams.

Williams notes that there are only a finite number of rivers in the world, and year by year the amount of free-flowing rivers diminishes as more dams are built, but he thinks that it is fair to say that IRN has slowed things down, and if they keep fighting, as they are, to stop projects like the mammoth Three Gorges Dam on the Yangtze, they will turn the trend around. Williams believes that with the Internet, it is becoming easier to share information and coordinate action with other activists. IRN is promoting the value of living rivers and demanding restoration of destroyed ecosystems and reparations for people whose lives have been devastated by the destruction of their river valley.

In July 1998, the IRN decided that the best defense is a good offense and initiated its "River Revival" program to support groups who are now campaigning to demolish dams such as the Snake River dams in Idaho. In 1998, Williams was in Moab, Utah, for a meeting of activists who were organizing to tear down Glen Canyon Dam and restore the Colorado River. These activists are people who could be characters out of Edward Abbey's famous novel *The Monkey Wrench Gang*. Late that evening, in the bunkhouse, Williams started telling them his story of the Hopi Indians at Black Mesa long ago. The man next to him, about his age, suddenly interrupted, "I know that demonstration, I was there, I organized it!" Williams said, "It sure makes you think: when we act for something we believe in, we can change each other's lives."

BIBLIOGRAPHY

Abbey, E. (1997). *Monkey Wrench Gang*. New York: Avon Books.
Fradklin P. (1996). *A River No More*. Berkeley: University of California Press.
International Rivers Network Web site. *www.irn.org*
McCully, P. (1996). *Silenced Rivers*. London: Zed Press.
Pearce, F. (1992). *The Dammed*. London: The Bodley Head.
Reisner, M. (1993). *Cadillac Desert*. New York: Penguin Books.
World Rivers Review. Bimonthly publication of the International Rivers Network.
—*Phil Williams*

EDWARD O. WILSON, JR.
(1929–   )

Species are going extinct in growing numbers; the biosphere is imperiled; humanity is depleting the ancient storehouses of biological diversity. . . . The worst thing that can happen, *will* happen . . . is the loss of genetic and species diversity to the destruction of natural habitats.
—Edward O. Wilson, Jr.

Edward O. Wilson, Jr., is an entomologist who is the founder of the field of sociobiology and the world's leading expert on the biology of ants. Wilson is internationally regarded as the preeminent biological theorist of the late twentieth century and one of the greatest naturalists in American history. He was named one of America's twenty-five most influential people by *Time* in 1996, which noted that Wilson's most enduring influence may be as an "ecological Paul Revere." Alarmed by the stunning loss of species around the world in the twentieth century, he has become an eloquent and powerful activist for species preservation. "The loss of biodiversity," he has said, "is the folly our descendants are least likely to forgive us" (University of Portland).

Born in Birmingham, Alabama, to Inez Freeman and Edward Osborne Wilson, Sr., Wilson grew up in the South, a nature-loving only child. He collected insects and kept snakes, spending much of his time alone in the woods and swamps along Mobile Bay and in northern Florida along the Gulf of Mexico. When he was seven, his parents divorced, and he spent periods living with each of them. Edward, Sr., an accountant for government agencies, frequently moved from one southern town to another. Edward, Jr., attended sixteen different schools, but discovered a constant in the woods and on the beaches. "A child comes to the edge of deep water with a mind prepared for wonder," he later wrote in his autobiography, *Naturalist*. "Hands-on experience at the critical time, not systematic knowledge, is what counts in the making of a naturalist" (11–12).

Wilson's choice of a career as an entomologist was influenced by a freak fishing accident that occurred when he was seven. He had hooked a pinfish and pulled it ashore too sharply; one of its spines punctured his right eye, effectively blinding it. He subsequently also lost most of his hearing in the upper registers, probably because of a hereditary defect. Wilson partly attributes his career as an entomologist to these physical losses. "I had to have one kind of animal if not another, because the fire had been lit, and I took what I could get," he says. "I would thereafter celebrate the little things of the world, the animals that can be picked up between thumb and forefinger" (Seagren).

It was while Wilson was exploring the National Zoo and nearby Rock
Creek Park in Washington, D.C., at the age of ten that he dedicated himself
to the world of insects. He says, "To put the matter as simply as possible:
most children have a bug period, and I never grew out of mine" (Wilson
1994, 34). He joined the Boy Scouts of America when he was twelve and
advanced to Eagle Scout with palm clusters, the highest rank, in three years.
At thirteen, he made his first major entomological discovery in Mobile,
Alabama: a species of fire ant that subsequently spread throughout the
South. Based on the challenge and opportunity presented by flies, he initially chose to study them, but World War II interrupted his supply of
Czechoslovakian insect pins. Since ants could be preserved in alcohol in-
stead, Wilson switched his entomological focus.

Wilson earned his bachelor's and master's degrees in biology at the University of Alabama at Tuscaloosa. In 1951, his writings on the little-known
habits of certain ants prompted a fellow entomologist to urge Wilson to
transfer to Harvard University, whose museum housed the world's largest
ant collection. In 1954, while earning his Harvard doctorate in biology,
Wilson won a fellowship to travel to Cuba, Mexico, and the South Pacific,
where he collected more than one hundred new species, several of which
were later named for him. Six weeks after his return in 1955, he married
Irene ("Renee") Kelley, with whom he had one daughter, Catherine, in
1963.

After his appointment to the Harvard biology faculty in 1956, Wilson
made a series of important discoveries, including the determination that
ants communicate primarily through the transmission of chemical substances known as pheromones. He also discovered that these insects have
a social caste system. In the course of revising the classification of ants from
the South Pacific, he formulated the theory that speciation and species dis-
persal are linked to the varying habitats that organisms encounter as their
populations expand. As he accumulated ant data, Wilson began to see patterns. For example, he found that the number of ant species on an island
tended to correlate closely with island size. In 1961, Wilson took a sab-
batical from Harvard and traveled to South America. He found the rain
forests of Suriname to be full of previously unstudied ants. He also visited
the islands of Trinidad and Tobago, just offshore from Venezuela. On the
larger Trinidad, he found more species of ants than on the smaller Tobago,
a finding that was consistent with his correlation of island size with number
of ant species.

Contemplating the qualitative types of work he had been doing, Wilson
began to seek something more. "It just didn't seem enough to continue
enlarging the natural history and biogeography of ants," he recounted in
his short memoir, *In the Queendom of the Ants*, which was published with
little fanfare in 1985 (Quammen, 68). "The challenges were not commen-
surate to the forces then moving and shaking the biological sciences." His

new acquaintance with the noted ecologist Robert MacArthur and other mathematical ecologists had persuaded Wilson that "much of the whole range of population biology was ripe for . . . rapid advance in experimental research; but this could only be accomplished with the aid of . . . mathematical models." Returning to Harvard, he enrolled in a mathematics class, an associate professor among a crowd of undergraduates. Eventually, after further remedial classes, he mastered enough math to feel comfortable.

Wilson expressed his idea to MacArthur that he thought that biogeography could be made into a rigorous, analytical science. Wilson had found that there were striking regularities in the existing data that no one had explained. For instance, with each tenfold decrease in land area came a twofold decrease in species diversity. Among the ant species of Asia and the Pacific Islands, Wilson had noticed a second pattern. Newly evolved species seemed to originate on the large landmasses of Asia and Australia and to disperse from there out to the remote islands. As these dispersing species colonized these small places, they seemed to supplant the older, native species that had arrived there before them. New species were continually arriving, old species were continually going extinct, and the net effect was no gain or loss in the number of ant species. It looked to Wilson like some sort of natural balance.

During 1961 and 1962, Wilson and MacArthur brainstormed over this notion of a biogeographical balance. They scrutinized Wilson's ant data from the South Pacific and looked at patterns of distribution among bird species in the Philippines, Indonesia, and New Guinea. They became convinced that the number of species lost from an island, during a given time period and under ordinary circumstances, is roughly equal in number to the species gained by that island over the same period: the rates of losses and gains tended to cancel each other out, resulting in a dynamic stability. In addition, they found that remote islands harbored fewer species than nearer islands, because remote islands received fewer immigrants while suffering just as many extinctions. They called this phenomenon the distance effect. Area and distance thus combined their effects to regulate the balance between immigration and extinction.

Wilson and MacArthur also created an intricate mathematical model that could foretell the particulars of species equilibration on a given island. It also, in a broader sense, offered ecologists a radically new way of understanding ecosystems all over the planet, particularly as human activities began to fragment them into islandlike pieces. In 1967, they published their findings as *The Theory of Island Biogeography*. The book provided the scientific foundation for all subsequent discussions about the decline of ecosystems.

In 1971, Wilson published *The Insect Societies*, his definitive work on ants and other social insects. The book provided a comprehensive picture, addressing the ecology and population dynamics of innumerable species in addition to their societal behavior patterns. Wilson's research for the book led him to the belief that human behavior might also result partly from genetic evolution, rather than from cultural forces or learning alone. This belief led to his founding of the field of sociobiology, the study of the genetic basis of the social behavior of animals. In 1975, Wilson published *Sociobiology: The New Synthesis*, which propounded his controversial theory. Wilson was accused by some critics of inventing a "pseudoscientific" justification for the racist and sexist status quo, and he was denounced in print as a fascist and a proponent of biological determinism. At a 1978 meeting of the American Association for the Advancement of Science, several protesters stormed the stage and dumped a pitcher of water over Wilson's head. "I thought my views were self-evident, but they weren't acceptable in the '70s," says Wilson (Reed). In his next book, *On Human Nature*, he argued that most domains of human behavior—from child care to sexual bonding—are the result of deep biological predispositions consistent with genetic evolution. The book won a Pulitzer Prize in 1979.

Despite initial resistance, the discipline of sociobiology has since been absorbed into the scientific mainstream. Wilson and others have defended and refined sociobiology over the years to such a point that new generations of evolutionary scientists accept it as a given. When the *National Review* included *Sociobiology* on its list of "The 100 Best Non-Fiction Books of the Century" more than two decades after its publication, panelist Michael Lind stated, "Darwin put humanity in its proper place in the animal kingdom. Wilson put human society there, too" (*National Review*).

For a 1980 article entitled "Resolutions for the 80s," the editors of *Harvard Magazine* asked seven Harvard professors to identify what they considered to be the most important problems facing the world in the coming decade. Wilson wrote: "Species are going extinct in growing numbers; the biosphere is imperiled; humanity is depleting the ancient storehouses of biological diversity. . . . The worst thing that can happen, *will* happen . . . is the loss of genetic and species diversity by the destruction of natural habitats" (Wilson 1980a). Wilson felt that this article marked his debut as an environmental activist. He was, he believes, unforgivably late in arriving.

Wilson was awarded the National Medal of Science in 1976 for his work on the organization of insect societies and the evolution of social behavior in insects and other animals. Wilson is the only person to have received both the National Medal of Science and the Pulitzer Prize in literature. *The Ants*, the definitive book on the subject, won him a second Pulitzer Prize in 1991. Among the other honors Wilson has received are Japan's International Prize for Biology, the Crafoord Prize of the Royal Swedish Academy of Sciences, the French Prix du Institut de la Vie, Germany's Terrestrial Ecology Prize, and Saudi Arabia's King Faisal International Prize in Science. His 1994 book *Journey to the Ants* won Germany's Science Book of the Year award, and his 1994 autobiography, *Naturalist*, was cited as one of the best books of the year by the *New York Times Book Review*. He has also received many other prizes, medals, and awards from around the world, and more than twenty honorary degrees.

Wilson's latest work, 1998's *Consilience: The Unity of Knowledge*, proposes that knowledge is ultimately a whole, that unity lies in the interlocking of different disciplines of learning, and that it must be scientists who judge what is true and what is not. Wilson's goal in *Consilience* is to convince readers of the need to unify all knowledge through the joining together of the great branches of learning.

Today, Wilson lives with his wife in Lexington, Massachusetts. In part because of his public battles, Wilson guards his privacy fiercely; only the urgency of species loss, says Wilson, keeps him in the public arena. He warns that "humanity is entering a bottleneck of overpopulation and environmental degradation unique in history." "We need to carry every species through the bottleneck," he says. "Along with culture itself, they will be the most precious gift we can give future generations" (Reed).

BIBLIOGRAPHY

Alabama Academy of Honor online. Edward Osborne Wilson. *http://www.archives. state.al.us./famous/academy/e_wilson.html*

Bennett, P. (1998, April 13). From Ants to Einstein: "Wired" interview with Edward O. Wilson. Available: *http://www.wired.com/news/news/story/11564. html*

Biography online. Wilson, Edward O. (Osborne). *http://www.biopraghy.com/ cgi-bin/biomain.cgi*

Blume, H. (1995). Interview with Edward O. Wilson: *Aliens. Boston Book Review*. Available: *http://www.bookwire.com/bbr/interviews/edward-wilson.html*

Encyclopaedia Britannica online. Wilson, Edward O. *http://www.britannica.com*

E. O. Wilson is on top of the world: Harvard University professor of biology looks at ants and unity of all knowledge [interview]. (1998). *Psychology Today, 31* pp. 50–57.

Gergen, D. (1998, June 30). Uniting knowledge: "The NewsHour with Jim Lehrer" interview with Edward O. Wilson. Available: *http://www.pbs.org/newshour/ gergen/june98/wilson_6-30.html*

Gibson, J. W. (1998, May 24). Sociobiologist goes for the sword in the stone. *Dallas Morning News*, p. 8J.

Holldobler, B., & Wilson, E. O. (1990). *The Ants*. Cambridge, MA: Belknap Press.

Holldobler, B., & Wilson, E. O. (1994). *Journey to the Ants*. Cambridge, MA: Harvard University Press.

Lester, T. (1998, March 18). A conversation with Edward O. Wilson. Atlantic Unbound, Books and Authors. Available: *http://www.theatlantic.com/ unbound/bookauth/ba980318.htm*

MacArthur, R. H., & Wilson, E. O. (1967). *The Theory of Island Biogeography*. Princeton, NJ: Princeton University Press.

Monmaney, T. (1998, July 9). Divisive ideas on "Unification": An interview with Edward O. Wilson. *Los Angeles Times*. Metro Section, p. 5.

National Review. (1999, May 3). The 100 best non-fiction books of the century. Available: *http://www.elibrary.com*

Quammen, D. (1996). Life in equilibrium. *Discover, 17* (3), pp. 66–76.

Reed, S. (1993, October 11). E. O. Wilson: The famed biologist says saving other species is the only way to save ourselves. *People*, pp. 123–124.

Roorbach, B. (1994, November 13). King of the anthill. *Newsday*, p. 38.

Seagren, L. (1995, March 17). Author's bugs will charm you. *Capital Times* (Madison, WI), p. 11A.

Sheppard, R. Z. (1990, September 3). Splendor in the grass: A new book by noted entomologists looks to the ant for behavior's roots and discovers the iron laws of the superorganism. *Time*, p. 78.

University of Portland Web. Commencement 1997. *http://www.up.edu/academics/commencement97/honorary-wilson.html*

Wilson, E. O. (1971). *The Insect Societies*. Cambridge, MA: Harvard University Press.

Wilson, E. O. (1978). *On Human Nature*. Cambridge, MA: Harvard University Press.

Wilson, E. O. (1980a, January–February). Resolutions for the 80s. *Harvard Magazine*, pp. 20–25, 70.

Wilson, E. O. (1980b). [1975]. *Sociobiology: The New Synthesis*. Cambridge, MA: Belknap Press.

Wilson, E. O. (1994). *Naturalist*. Washington, DC: Warner Books/Island Press.

Wilson, E. O. (1998). *Consilience: The Unity of Knowledge*. New York: Alfred A. Knopf.

—*Bibi Booth*

SUSAN WITT
(1946–   )

Susan Witt has served as executive director of the E. F. Schumacher Society since its founding in 1980. The Schumacher Society is a national, non-profit organization that promotes and supports grassroots initiatives to develop more self-reliant regional economies. It maintains a 7,000 volume computer-indexed library that may be searched over the Web (*www.schumachersociety.org*). The Society's annual lecture program hosts such speakers as Wendell Berry, Jane Jacobs, and Ivan Illich. The lecturers are collected in the book *People, Land, and Community*, edited by Hildegarde Hannum.

Ernest F. Schumacher is best known for reshaping our assumptions about economics. His collection of essays entitled *Small Is Beautiful: Economics as If People Mattered* argues that an economic system in which goods are produced locally for local consumption is inherently more accountable to the unique social and ecological conditions of a region. When these improved human and environmental conditions are factored into overall re-

Photo courtesy of Susan
Witt. © Clemens Kalischer.
Used by permission.

turn, then vital local economies also prove to be the most sustainable
economies.

Schumacher's reasoning challenged prevailing economic theory that big-
ger was better, especially relating to the manufacturing and distribution of
goods. The publication of *Small Is Beautiful* in the United States in 1974
made Schumacher a popular hero for a new generation of activists. He died
prematurely in September 1977 and was mourned by thousands.

Susan Witt has no formal background in economics; she was a literature
teacher at a small private Waldorf high school in New Hampshire and
loved the world of great books that surrounded her. She came to believe,
however, that some of the world's most pressing social and environmental
problems could be solved only by changes in the economic system, and she
wanted to help make that change. In 1977, at age thirty-one, she received
a small inheritance from her grandfather that enabled her to volunteer for
organizations working to bring a renewed moral dimension to economic
activity. When she turned on the radio and happened to hear Robert Swann
speaking about E. F. Schumacher from the Cambridge Forum, her course
was set.

Swann's organization, the Institute for Community Economics, had
sponsored Schumacher's historic 1974 United States tour. Swann had spent
World War II in federal prison as a conscientious objector. There he had
taken a correspondence course with Arthur Morgan on the "small com-

munity." His studies convinced Swann that centralized control of land and money was the basis of conflict between nation-states; as a result he became committed to strong, democratically managed, regional economies. He earned his living as a carpenter and contractor, building affordable housing, but he devoted much time to the civil rights and antinuclear movements. Troubled by the difficulty African-American farmers were having in gaining affordable access to farmland, Swann developed the concept of the community land trust (CLT) and, with Slater King, implemented a CLT in Albany, Georgia, the first in the country.

The Institute's co-founder, Ralph Borsodi, was a published economist and a friend of Vinoba Bhave, a close associate of Gandhi. Borsodi enlisted Swann in a plan to create a loan fund for farmers in small rural villages throughout India. In the process, he taught Swann about commodity-backed currencies, intentional communities, and homesteading. In 1972, at the age of ninety, Borsodi decided that instead of writing another book about commodity-backed currency, he would issue one. With Swann's help, he worked with a bank in his hometown of Exeter, New Hampshire, to issue a *Constant*. The *Constant* circulated widely in Exeter for a year before Borsodi's doctor recommended that he give up the project. Nevertheless, the experiment drew much media attention to the subject of local currencies.

Borsodi died the same month as Schumacher, but his legacy was still strong at the Institute for Community Economics when Witt joined as a volunteer in September of 1977. Discussion at the Institute's offices often centered around the history of national currencies and the need to create alternative structures. In 1978 the Institute founded the Community Investment Fund to invest in worker-owned businesses, cooperatives, community land trusts, and other socially responsible projects; Witt was its staff person.

In 1980, Susan Witt moved with Robert Swann to the Berkshire region of Massachusetts to help an emerging community land trust develop a proposal to gather and manage woodlots for sustainable timber harvest. Swann believed that cordwood would be a suitable backing for a regional currency. Forestland held by a democratically based regional land trust might well be the reserve needed to launch a cordwood note. The move to the Berkshires was meant to be temporary, but before long Witt and Swann were renovating a garage into a home on land owned by the newly formed Community Land Trust in the Southern Berkshires. The land is on the north slope of Jug End Mountain, close to the Appalachian Trail, and overlooks the Housatonic River Valley. The house is surrounded by an old apple orchard. Witt and Swann pruned the overgrown trees to make them productive again, planted grapes, pears, cherries, and raspberries, and put in a big garden. They were establishing roots.

The same year a group of Schumacher's friends who were connected with the British Schumacher Society asked Swann and Witt to establish an E. F. Schumacher Society in North America. They were ready for the task. Grounded in the Berkshire community with its rich history of independence and cultural inventiveness, the new organization was able to pursue both the conceptual development and practical implementation of tools for a sustainable economy. Witt and Swann started a small study group in their home to discuss community economics and the role of people, land, and money in building local economies.

Witt's work at the Institute for Community Economics had challenged her to think about socially responsible investment vehicles. Now in a rural area, she saw first-hand the problems small business owners had in accessing credit. How could socially concerned consumers wishing to support a local economy be of help to small business in the region? Witt considered starting a community development credit union, but the history of credit unions showed a pattern of consumer loans for new cars, trips, appliances, and home improvements, whereas Witt wanted to devote her energies to facilitating the productive loans necessary to foster new locally-owned businesses.

Therefore, in late 1980, she developed a proposal for collateralizing loans to small businesses. The plan was to separate two aspects of banking—management of funds and risk taking. Consumers would deposit funds in special savings accounts at a local bank. By advance agreement, the depositor would permit a nonprofit corporation to use the money in the savings accounts to collataralize loans to businesses that otherwise would not qualify for bank loans. The bank would handle all paperwork and tax reporting—a function performed efficiently and well by banks.

After carefully examining Witt's micro-credit proposal, the community economics study group formed the Self Help Association for a Regional Economy (SHARE) and negotiated with a locally-owned and -managed Berkshire bank to implement the program. With the help of the organization's board members, Witt drafted the by-laws and incorporation documents, developed the agreement papers, designed pamphlets, and launched a local publicity campaign. SHARE made over twenty small loans in its first three years of operation, without a default. Many were to women without credit histories who were starting new businesses. Loans supported the purchase of stainless steel equipment used in cheese processing at a goat farm, construction of a barn for a draft horse team trained to haul logs from the forests, farm equipment for a local organic farm, a piano for a piano teacher, a knitting machine for a sweater designer, and similar productive purposes.

From SHARE emerged *SHAREcroppers*, a local newsletter connecting Berkshire consumers and farmers. The twenty-five families in the local Pumpkin Hollow Food Buying Club purchased quantities of organic root crops and squash each year, delivered from farms in California. Witt knew that Berkshire farmers were fully capable of growing all of these crops if given the opportunity. She placed an ad in the 1983 Spring issue of *SHAREcroppers* offering to purchase a ton of organic carrots, 1,500 pounds of onions, and other crops to be delivered in the fall. Local farmers stepped forward. The downpayment from the twenty-five families helped the participating farmers and guaranteed a future local supply of organic vegetables for the buying club.

Jan Vander Tuin read about SHARE and *SHAREcroppers* in a Rodale publication. He visited the Schumacher Society in 1984 to discuss an approach to farming he had seen in Switzerland whereby consumers contract to pay a yearly income to farmers in exchange for a share of the harvest. Witt introduced Vander Tuin to Robyn Van En who owned nearby Indian Line Farm and was looking for a partner on her farm. Out of that partnership grew the first Community Supported Agriculture (CSA) farm in this country. Vander Tuin leased the orchard abutting the Schumacher Society's office on Community Land Trust land. A champion biker, he developed a pedal-powered press to turn the apples into cider. The first distribution at Indian Line Farm was a share of the 1985 cider production. There are now over one thousand CSA farms around the country, thanks to Robyn's perseverance and commitment to the concept.

The board of directors of the Community Land Trust had become concerned about the high cost of housing in the Berkshires. Bob Swann designed a quadruplex building in the belief that multiple construction would produce more affordable home ownership possibilities. At a 1985 Fourth of July party for Community Builders, a cooperative Swann had founded, Witt learned that the neighboring land was for sale. Located on the Great Barrington water and sewer line, it was just the kind of site suited for a multiple-unit building. So began a four-year, $1.5 million building project, resulting in seventeen units of affordable housing for local year-round families. As administrator of the Community Land Trust, Witt oversaw the land purchase, land-use planning, permitting process, construction financing, contracting, lease development, and sale of units at Forest Row. Never having done this kind of work before, she depended on the cooperation of local bankers, lawyers, builders, land-use planners, and town officials to work with the nonprofit to accomplish its goals. In the process, strong community partnerships were built. The project was completed without government subsidies—a true local citizen initiative.

The future homeowners at Forest Row were involved in every stage of the process. Many of the families worked on their own units: putting up sheetrock, building cabinets, painting, laying tile. They worked with Witt and other Schumacher Society staff to create a second-mortgage fund to reduce monthly costs on their new homes. Called the Fund for Affordable Housing and capitalized by loans from the second-home community, the fund enabled five families who otherwise would not have had sufficient down payments to qualify for bank financing to purchase units. Publicity for the Fund raised consciousness in the area about the high cost of housing. Again, out of project necessity rather than formal training, Witt learned to write loan documents, create payment schedules, and oversee the collection and distribution of the second-mortgage funds.

The Deli is a popular hangout in Great Barrington for students from Simon's Rock College, for construction crews breaking for lunch, for the second-home community looking for a great sandwich, and for the lawyers, bankers, shopkeepers, and others working on Main Street. In September of 1988, Frank Tortoriello, the Deli's owner, went to Witt to inquire about a loan from SHARE. His lease at his Main Street storefront was terminating. He had a new site under contract, but it would require renovation, and he needed a loan to do the work. The banks rejected his application; maybe SHARE would help him. Because the Deli had an established group of devoted customers, Witt suggested he turn to them for a loan, rather than to SHARE members, by issuing his own currency. "Deli Dollars" went on sale on October 23rd. Schumacher Society member Martha Shaw designed the note, a Great Barrington shop donated the printing, and Witt trained Frank and his crew in handling the scrip. Deli Dollars were dated over a year to prevent a flood of redemptions during the first week of operation. The community was enthusiastic to help. Frank raised $5,000 in thirty days.

Deli Dollars started a movement. In the next three years Witt worked with two farm stands to issue Berkshire Farm Preserve Notes, with a general store to issue Monterey General Store Notes, and with a Japanese restaurant to issue Kintaro Notes. In 1991 a *Washington Post* reporter interviewed Witt and the scrip issuers for a front-page story. ABC, NBC, CNN, CBS, and Tokyo TV called in quick succession, excited to see consumers taking responsibility to help keep local businesses operational. Paul Glover saw the TV coverage and contacted the Schumacher Society about issuing scrip in his hometown of Ithaca, New York. Glover spent a week at the Schumacher Library developing a plan for Ithaca Hours. There are now over $60,000 worth of Ithaca Hours in circulation, stimulating the growth of small home-based business in the region. The Ithaca Hours program has stimulated the growth of local currencies around the country. Witt and Glover edit *Local Currency News*, and the Schumacher Library continues to be a vital resource for community groups on the history, theory, and practice of decentralized issuing of scrip.

While CSAs provide a way for consumers to support the yearly operating costs of the farmer, who pays for the initial high cost of farmland? In 1997 Robyn Van En died tragically of asthma, and her Indian Line Farm came up for sale for $155,000. If it were to be affordable for farmers, the community would have to help. Witt worked with the Berkshire Highland Program of The Nature Conservancy, the Community Land Trust in the

Southern Berkshires, and two young farmers to structure an innovative partnership to purchase the farm. With donations from area residents, the Community Land Trust purchased the land for $50,000, the farmers obtained a bank loan of $55,000 to purchase the buildings, and The Nature Conservancy paid $50,000 in exchange for conservation restrictions on the land. The farmers lease the land for organic vegetable production. Should they decide to move, the lease requires them to resell the buildings to the Community Land Trust at their replacement cost adjusted for deterioration. In this way the ecological integrity of the land is maintained and the buildings remain affordable for future farmers. The Indian Line Farm legal documents are a model for citizen involvement in ensuring a future for local food production.

It is these stories of her Berkshire neighbors working together to shape the future of their local economy that Witt describes in articles and talks around the country. She finds their striving and their struggles compelling and hopeful in an age so influenced by large-scale, impersonal corporate activity.

In 1994 Schumacher's widow donated his personal library of books and papers to the E. F. Schumacher Society. Witt expected a collection of books on economics. Instead she found Goethe, Martin Buber, Selma Lagerlof, Gandhi, Simone Weil, Plato, Walt Whitman, Dante, Kropotkin, Solovyev, and books on all the world's great religious traditions. It was through pondering works of a moral and spiritual nature that Ernest Schumacher arrived at a more humane and ecologically responsible economic theory. His library reflected Witt's own choice of reading and helped affirm the path that had led her to a passionate interest in renewing local economies in general and her own Berkshire economy in particular. In economic life we apply labor to transform natural resources into goods for exchange with our neighbors. That transformation and exchange can be life renewing or life destroying. Susan Witt is committed to helping shape local institutions that make economic processes more visible, place-based, and accountable to community. She believes that a system of vibrant and diverse local economies best meets the challenge to conduct our affairs on earth in a manner responsible to the natural world and to other sentient beings.

BIBLIOGRAPHY

Benello, C. G., Swann, R., & Turnbull, S. (1997). *Building Sustainable Communities: Tools and Concepts for Self-Reliant Economic Change*. Croton-on-Hudson, NY: Bootstrap Press.

Ithaca Hours online. Hometown Money Starter Kit. *http://lightlink.com/ithacahours*

E. F. Schumacher Society Web site. *http://www.schumachersociety.org*

Schumacher, E. F. (1999). [1974]. *Small Is Beautiful: Economics as If People Mattered*. Point Roberts, WA, and Vancouver, BC: Hartley and Marks.

Solomon, L. (1996). *Rethinking Our Centralized Monetary System: The Case for a System of Local Currencies.* Westport, CT: Praeger.

Steiner, R. (1993). *Economics. The World as One Economy.* Bristol, England: New Economy Publications.

Witt, S. (2000). Printing money, making change: The promise of local currencies. In Anderson, C., & Runciman, L. (Eds.), *A Forest of Voices: Conservations in Ecology.* Mountain View, CA: Mayfield.

Witt, S., & Swann, R. (1996). Land challenge and opportunity. In Vitek, W., & Jackson, W. (Eds.), *Rooted in the Land.* New Haven, CT: Yale University Press.

Witt, S., & Swann, R. (1997). Local currencies: Catalysts for sustainable regional economies. In Hannum, H. (Ed.), *People, Land, and Community: Collected E. F. Schumacher Society Lectures.* New Haven, CT: Yale University Press.

—*Susan Witt*

HAZEL WOLF
(1898–2000)

I think the best an organizer can do is get people together, inspire them to want to stay together, and share with them any particular knowledge you might have to help them do that. And then that's the end of it, as far as I can see. If you want to *run* the thing, then you're not an organizer—you're just on an ego trip.

—Hazel Wolf

Hazel Wolf was an environmental activist, a strong advocate for human rights, and a longtime crusader for the Seattle chapter of the National Audubon Society. In January 2000, Wolf passed away at the age of 101.

Hazel Wolf was born in March 1898 in Victoria, British Columbia, Canada. Her mother was an American, and her father was a British seaman who passed away when Wolf was ten years old. She grew up in Canada. At the age of sixteen, just out of the eighth grade in Canada, she had her first job in a law office. She recalled in one of her speeches, "In those days one didn't have to have a Ph.D. to work in a law office. One needed only to be able to type and be an expert with shorthand. Shorthand, as you know, is a now a lost art. I received $25 a month, $15 of which I paid my mother for board, $5.00 a month on a bicycle. The rest went for candy no doubt." Later on, she married and soon separated from her husband. In 1923, Wolf moved to Seattle, Washington, with her young daughter. She did a variety of jobs during that time and lived for a time on welfare in a boardinghouse in the city. She was a union organizer for the Works Progress Administration (WPA).

From 1949, when she became an American citizen, until she retired at

Photo by Greg Kinney.

age sixty-seven, Wolf worked for Seattle civil rights attorney John Caughlan. In 1962, she joined the National Audubon Society and from then on played a prominent role in environmental efforts in local, national, and international arenas. In addition to cofounding the Seattle Audubon Society, where she worked as secretary for twenty-six years, she set up more than twenty other local chapters, such as the Black Hills Audubon Society. She also established a chapter in British Columbia.

Wolf worked on a wide range of issues—feminism, human rights, labor, and environmental protection. One of Wolf's proudest achievements was organizing a conference of environmentalists and Native Americans called the Indian Conservationist Conference in 1979. She visited twenty-six tribes throughout the state of Washington, listening to their concerns. She then urged them to attend a conference she was planning. The conference was a great success and helped found several alliances, including a legal fight against a proposed northern-tier pipeline.

Wolf was also a member of the Sierra Club, Greenpeace, and the Earth Island Institute. Her endeavors to improve environmental safety in low-income inner-city housing were done through the Community Coalition for Environmental Justice, which she also cofounded. In her late eighties and nineties, she still found time to go hiking and made overnight trips to local mountain areas. She was kayaking at one hundred and traveled to many places throughout the country to speak and lobby for tougher environmental laws.

In honor of her hundredth birthday, the Audubon Society created a Hazel Wolf "Kids for the Environment" fund to foster environmental appreciation among young people. The program for students ages twelve through fifteen concentrates on such activities as tree planting or cleaning up a cluttered stream to bring back the fish.

Wolf was the recipient of a number of other conservation awards, including the Audubon Medal for Excellence in Environmental Achievement in 1977. She also received the Washington State Department of Game's award for services in protection of wildlife in 1978 as well as the State of Washington Environmental Excellence Award, the People's Daily World's Newsmaker Award, and the Washington State Legislature Award for environmental work. The Women in Communications group bestowed on her its top honor, the Matrix Award for Women of Achievement.

In 1996, on her ninety-eighth birthday, Washington State governor Mike Lowery declared March 10 "Hazel Wolf Day." In June of that year, Seattle University conferred upon her a doctorate in humanities honoris causa.

In 1995, in an interview with the author Studs Terkel, Hazel Wolf expressed her desire to live in three centuries and said, "Then, I am going." She achieved her goal—she was born in the nineteenth century and died in the third millennium. At the time of her death, the president of the National Audubon Society said, "Our grief is tempered only by the fact that Hazel's life was so full of joy and accomplishment. She was an inspiration to all of us who knew her. She challenged us to be better conservationists and better human beings. She will be greatly missed."

BIBLIOGRAPHY

Biography Online Database. Hazel Wolf Web site. *http://www.tripod.com/~hazel wolf/*

National Audubon Society. (1997). Hazel Wolf to Receive National Audubon Society Medal for Excellence. *http://www.audubon.org/local/cn/97spring/wolf. html*

Wolf, H. (1997, October). Minnesota Justice Foundation [speech]. Available: *http:// www.tripod.com/~hazelwolf/*

Wolf, H. (1998, March). The Big Bash Scottish Rite Temple [speech]. Available: *http://www.tripod.com/~hazelwolf/*

Wolf, H. (1999, September). Great Women In My Life, Hazel Wolf High School [speech]. Available: *http://www.tripod.com/~hazelwolf/*

—*John Mongillo*

Photo © Don Ingle.

JOAN LUEDDERS WOLFE
(1929–   )

Joan Luedders Wolfe is a renowned advocate for environmental issues and has been recognized internationally for her contributions to environmental protection. Wolfe was born in 1929 and raised in an urban environment, Highland Park, a city inside Detroit, Michigan. However, as she says, she was fortunate as a child to be exposed to open spaces during camping experiences. She loved the outdoors but knew little about the environment in which she lived. However, as a young adult, she was impressed with a book on pesticides that rocked the world at the time: *Silent Spring*, written by Rachel Carson. Carson's lessons came close to the conditions near her married home in Dearborn. Carson had reported that Dearborn was one of the places that had been sprayed with pesticides without the population being aware of the dangers. Wolfe's children had a pet toad that they fed with worms from their yard. One day the toad died mysteriously of a nervous disorder that made it twitch involuntarily and flip across its terrarium. When she read Carson's book she inferred that the toad had died of pesticide poisoning. Since Wolfe was also a passionate birdwatcher, Carson's book sensitized her to environment issues. She subsequently became an ardent environmentalist.

The first environmental organization Wolfe belonged to was the Audubon Club. She and her husband also joined other environmental organizations. However, the Wolfes realized that although each organization was working hard, they were not coordinating their efforts. Each group was trying unsuccessfully to get something different done. Wolfe's first big effort in getting something done occurred in 1966 when she decided to get a group of organizations to sponsor an all-day "teach-in" at a large local church. That work led Wolfe to start the West Michigan Environmental Action Council (WMEAC) in 1968 to work with all kinds of organizations, including church and social clubs, PTA groups, labor unions, and student groups. It may have been the first large, broad-based environmental organization of its kind in the United States. With Wolfe acting as its first president, the WMEAC led the passage of the landmark Michigan Environmental Protection Act of 1970, which was a model for legislation in nine other states. The bill allowed a citizen to go to court to stop environmentally harmful projects. But it was not an easy bill to pass. Wolfe remembers that all of the big corporations were against it, and only two state legislators would even sign on to the bill. However, through a big effort and cooperation by many citizens and organizations, the WMEAC got the support to pass the bill, a bill many had said could not be passed.

Under Wolfe's leadership, the WMEAC worked on a case that standardized Michigan's enforcement policy on air pollution. The organization also joined a major and successful national safety case regarding the use of plutonium in breeder reactors. Wolfe led the passage of another model law, Michigan's Inland Lakes and Streams Act. Acting as a lobbyist for citizen concern gave her a lot of insight into the honesty, courage or treachery of the individual legislators she dealt with. In 1973, Wolfe was appointed to the Michigan Natural Resources Commission and was the first woman in the United States to serve on a similarly powerful NRC.

Wolfe believes that young people and others can make a difference. The best way to learn is to volunteer in an organization that is working on environmental problems. She stressed that learning includes making appointments and having the willingness to talk to and listen to people who are involved with environmental issues. It helps to hear all sides of an issue as well, but to also consider the source in evaluating what you hear.

Wolfe received her B.A. degree in economics from the University of Michigan and is a consultant to groups on how to organize and be more effective. She has received international recognition and many state and local awards for her work. She was awarded an honorary doctorate in public service from Western Michigan University and is in the Michigan

Women's Hall of Fame. She is the author of *Making Things Happen: How to Be an Effective Volunteer*.

BIBLIOGRAPHY

Wolfe, J. L. (1991). [1981]. *Making Things Happen: How to Be an Effective Volunteer*. Washington, DC: Island Press.

—*Joan Luedders Wolfe*

HOWIE WOLKE
(1952–   )

Howie Wolke is a wildland conservationist and writer. He is also a wilderness guide and avid outdoorsman who loves to explore wild country. Backpacking, river running, canoeing, backcountry skiing, hunting, fishing, and cloud gazing are among his favorite activities. He also enjoys singing and strumming guitar tunes and listening to Emmylou Harris sing like a bird. His writing, public speaking, and conservation work are based upon the philosophy that all life and the wild habitat that sustains it have intrinsic value. Since humans are merely one member of the biotic community, they must learn to eliminate behaviors resulting in the wholesale extinction of species and wild habitats. In other words, Wolke is firmly rooted in efforts to save and restore as much wilderness as possible in the United States and elsewhere on earth.

Wolke has authored two books on the vanishing American wilderness, has written numerous articles and editorials for various publications, has been a columnist for the *Missoula Independent* and various conservation journals, and is an experienced public speaker. He is a former Wyoming representative for Friends of the Earth and is a cofounder of Earth First!, which he left ten years after its inception.

Wolke has no rational explanation for his fanatical love of wild things and places other than the "Neanderthal gene," a metaphorical recessive trait that crops up in some individuals, including those, like himself, with little recent family interest in wildlands. Wolke was born in Brooklyn in 1952; his family moved to the outskirts of Nashville, Tennessee, six years later. During these years, Wolke roamed and explored the green hardwood forests of central Tennessee and the Cumberland Plateau, and as early as he can remember, the sight of a bulldozer clearing the woods upset him. When he was old enough to consider adulthood, Wolke decided that he would be a forest ranger, protecting and exploring the forest.

In Tennessee, and later when his family moved to Pittsburgh, Wolke's father was a traveling hosiery salesman who occasionally took little Howie on summertime pantyhose runs through the South and the Midwest. Along

Photo by Marilyn Olsen.

the highway, Wolke would gaze into the woods, up toward a ridgetop, or down a river with a sense of awe, mystery, and youthful wonder. "What's it like on that hillside, in those woods, along that river, over that ridge?" In 1966, Wolke's sister Laurie was born, his only sibling.

The next career move of Wolke's father in 1968 took the family to central New Jersey, where Wolke spent his last two high-school years playing football and hooky and doing just enough schoolwork to placate his parents and teachers. He joined the Sierra Club and The Wilderness Society and read extensively about environmental issues. He spoke out for nature in school, where most fellow students thought him a bit weird, nicknaming him "Smokey." He longed for a greener, wilder place and, after graduating from high school in 1970, entered the University of New Hampshire's forestry school.

It was a shock to learn that the forestry curriculum was primarily concerned with training young foresters to produce sawlogs. It was a shock to hear professors rail against the idea of protecting some areas from logging as designated Wilderness. It was a shock to learn that many professors and forestry students were hostile toward the very environmental organizations that Wolke so admired. He quickly learned that the U.S. Forest Service functioned primarily as an arm of the timber industry. Halfway through his junior year, Wolke gave up on forestry and finished college majoring in environmental conservation with an option in wildlife ecology. He graduated in December 1974 and shortly thereafter loaded his compact car and headed for Wyoming.

In the mid-1970s, Wolke did volunteer work for the Sierra Club and The Wilderness Society and in late 1976 became the Wyoming representative for Friends of the Earth. He lived in Jackson, Wyoming, and augmented his stipend of sixty dollars a month with occasional work as a ranch hand, oilfield roughneck, construction laborer, and bar bouncer. During these years, Wolke learned the ropes of conservation activism: reviewing Environmental Impact Statements (EISs) and related documents, testifying at public hearings, filing administrative appeals, organizing grassroots support for protecting wilderness, and writing lots of articles, editorials, and letters to editors. He quickly discovered that without strong efforts to educate and organize at the local and regional levels, lobbying and legal actions were doomed. In other words, he concluded that the battle for wilderness would be won or lost primarily at the grassroots level, not in the smoke-filled backrooms of Congress or the Forest Service. He discovered that wilderness advocates make enemies who occasionally threaten and attempt to intimidate; yet he also found that in general, the public supported and respected wildland advocates who held their ground and utilized both passion and technical knowledge. He also concluded that the most effective wildland activists were those who spent their free time exploring the wilds.

Wolke explored the wilds of the American West with zeal. He walked across and through the Absarokas, the Beartooths, the Wind Rivers, Yellowstone's backcountry, the central Idaho wilds, Utah's sandstone canyons, Arizona's cactus deserts, and more. He read works of Aldo Leopold, Robert Marshall, and John Muir. Paul Ehrlich's *Population Bomb* was a major early influence, and he read, digested, and reveled in the descriptive prose of Edward Abbey, bard of the wilderness movement. In 1979, Wolke hung out his shingle as a wilderness backpacking guide, which remains today his primary livelihood.

In April 1980, Wolke and some friends headed for the remote deserts of northern Mexico and, after lots of camping, hiking, and drinking south of the border, founded Earth First! with Dave Foreman and Mike Roselle while driving across southwestern New Mexico. Perhaps the major impetus for Earth First! was the Forest Service's Roadless Area Review and Evaluation (RARE II) of 1978–1979, in which the agency studied most of its unprotected roadless areas and decided to recommend just a tiny realm (15 million out of about 80 million acres) of this priceless "resource" to be protected as designated Wilderness. The vast majority of these wildlands would be open to logging, mining, road building, all-terrain vehicles, and other forms of industrial activity. Wolke was dismayed not only by the Forest Service's unyielding dedication to industrial development, but by the conservation movement's refusal to seriously challenge RARE II, at least partially because various former employees of environmental groups had been appointed to positions in the Carter administration's Interior and Agriculture departments. What was needed, Wolke and Foreman believed, was a new group that would be the cutting edge, the radical arm of wildland conservation.

Earth First! was originally designed to strengthen the movement, to make it easier for groups like the Sierra Club to take stronger, more effective positions, and to advocate far more wilderness than most conservationists were willing to publicly support. In these early days, EF! was focused on wildlands and was very much part of the conservation movement. Its original platform was primarily a list of big wilderness reserves to be protected or restored, covering every major ecological region in the United States, a plan similar to (but less sophisticated than) modern proposals by leading conservation biologists.

In 1985, Wolke met Marilyn Olsen, who eventually became his partner and his wife. In June 1985, Wolke and Olsen were desurveying a new road that the Chevron Oil Company was building in an unprotected roadless area of the Bridger-Teton National Forest. Wolke got caught and was jailed in Pinedale, Wyoming, from February 11 to August 10, 1986. As a guest of the county, Wolke drafted *Wilderness on the Rocks*, which was published in 1991. *Wilderness on the Rocks* was an overview of the wilderness controversy and a critique of the modern conservation movement's efforts to save and restore the American wilderness. Upon his release from jail, Wolke and Olsen moved to their property in the Bitterroot Mountains near Darby, Montana, which they had bought a few months before Wolke's jail term.

In the mid- and late 1980s, Wolke teamed up with Foreman again to coauthor *The Big Outside* (1992), a descriptive inventory of the remaining big roadless wildlands in the United States south of Alaska. It was the first such inventory since Robert Marshall's historic wildland survey of 1936.

By the late 1980s, Earth First! began to lose some of its focus on wildlands, with new people taking it into social activism, animal rights, anarchy, and other realms beyond the wilds. Wolke began to understand the loss of effectiveness that accompanies political ideology, be it leftism, rightism, anarchism, or any other ism with a political ax to grind. Wolke quietly left EF! in 1990, convinced more than ever that saving and restoring wilderness requires conservationists to reach out, educate, and organize a broad spectrum of people spanning a variety of sociopolitical realms. It was the grassroots element again.

Today, Wolke and Marilyn Olsen live on forty-eight forested acres in the Bitterroot Mountains near Darby, Montana. Their two children (Wolke's stepchildren), Joshua and Bridget, are grown and active outdoors in the wilds. Their guide service, Big Wild Adventures, runs wilderness backpacking trips throughout the western United States. Their nonprofit conservation group, Big Wild Advocates, is a nonmembership service organization that works to support the efforts of grassroots conservation groups. Their widely distributed "action reports" cover an array of wilderness-related topics; in fact, Wolke considers his *Roadless Areas and Wilderness!* action report to be one of his most important accomplishments as a conservationist.

In addition, Wolke is associated with various regional and national groups, including Friends of the Bitterroot (steering committee), the Ecology Center (board of directors), the Alliance for the Wild Rockies (advisory council), and the Wildlands Project (correspondent). He also believes that one of his most vital contributions to conservation is the personal contact he has had as a wilderness guide over the years with a wide variety of people. Many have become involved with efforts to save wilderness, wildlife and wild nature in general.

Looking back over many politically volatile years, Wolke can also point to some definable successes, such as efforts to enlarge the Jedediah Smith Wilderness (the west slope of the Grand Tetons) and to designate the Winnegar Hole and Gros Ventre wildernesses and a number of other Wyoming areas protected in 1984. In addition, a handful of wild areas that Wolke helped to protect via administrative appeals and related legal actions remain wild to this day.

Nonetheless, Wolke believes that his most vital function has been to educate people and inspire them to act, for wildland defense must be like a chain letter, always enlisting new people who in turn enlist more into the fracas until a critical mass is created that cannot be ignored. His advice to people who wish to become active in conservation is "just do it!" Join local conservation organizations and volunteer to help. Read up on the issues. Start writing letters to editors, bureaucrats, and politicians. Don't expect to walk into a paying job, at least until you have really proven your effectiveness, so be sure you can make a living at something else. Maintain healthy distrust of those who really love politics, and don't forget to save time for exploring the wilds you are trying to save.

As we begin the so-called twenty-first century humans have created an unprecedented ecological mess. Leading scientists believe that up to half of all existing species may be extinct by the year 2050. They also tell us that the very process of evolution in large terrestrial vertebrates has ground to a halt due to the destruction of wildland habitats and the fragmentation of habitat into increasingly smaller chunks inadequate to support evolutionary processes and healthy, viable plant and animal populations. Wolke is convinced that saving and restoring wilderness is basic to correcting these horrific trends. In addition, he realizes that any future wildland gains will be negated unless human population growth is halted. Finally, in the long run, unless humans learn to treat each other with some measure of dignity, it is unlikely that they will ever learn to treat other forms of life, and indeed the land itself, with much respect.

BIBLIOGRAPHY

Foreman, D., & Wolke, H. (1992). *The Big Outside: A Descriptive Inventory of the Big Wilderness Areas of the United States*. Rev. ed. New York: Harmony Books.

Wolke, H. (1991). *Wilderness on the Rocks*. Tucson, AZ: Ned Ludd Books.

—Howie Wolke

HOWARD ZAHNISER
(1906–1964)

Howard Zahniser is known as the chief architect of and primary lobbyist for the 1964 Wilderness Act, which he largely drafted in his capacity as executive director of the Wilderness Society. That act created the National Wilderness Preservation System, which, in 2000, included 105 million acres of designated wilderness areas on federal public lands. Zahniser was born on February 25, 1906, in Franklin, Pennsylvania, to the Reverend Archibald H. M. Zahniser and his wife Bertha Belle Newton Zahniser. He was the oldest of his siblings to survive childhood. His father's pastorates moved throughout western Pennsylvania every three years when Zahnie—as Zahniser's friends called him—was young. That was then the policy of the Free Methodist church, an evangelical Christian denomination in which all four of Zahniser's uncles also were ordained ministers. When Zahniser entered his teen years, however, his father became a district elder in the denomination, and the family, which included the children Howard, Elizabeth, Harold, and Helen, settled in a two-story frame home on Bridge Street in Tionesta, Forest County, Pennsylvania. Set along the steep valley of the Allegheny River, Tionesta was the community Zahniser would adopt as home. At his death at age fifty-eight in 1964, Zahniser was buried in Tionesta above the river and near his father, mother, and a sibling who had died in infancy.

Many pockets of wildness existed around Tionesta, and Forest County remains of lively interest to hunters, anglers, birders, and other outdoor enthusiasts. Today part of the nearby Allegheny National Forest is protected within the National Wilderness Preservation System. Young Zahniser delighted in the outdoors, especially once he was equipped with eyeglasses and discovered the rich, crisp textures of the natural world. Zahniser later recalled that crisp vision so thrilled him that he ran through fields and vaulted over fences in sheer delight. An elementary-school teacher interested him in birdwatching, which became a lifelong interest. Through that interest, Zahniser eventually broadened his vision of conservation and the protection of nature to encompass the preservation of large tracts of roadless wilderness on federal public lands of the national forests, national parks, and national wildlife refuges. (In the 1970s, the protection of the Wilderness Act was also extended to those federal lands administered by the Bureau of Land Management of the U.S. Department of the Interior.)

Zahniser attended and graduated from Tionesta High School. He often

Photo by Larry Strausbaugh.

joked that he was graduated in the top ten of his high-school class—quickly adding that there were only ten students in the class. He attended Greenville College in Greenville, Illinois, a four-year liberal-arts college affiliated with the Free Methodist church. In high school, Zahniser had contracted osteomyelitis, a bone disease. At the time, it was considered a 50 percent fatal disease, so when Zahniser entered Greenville College on crutches, not sure that he would live four more years, he was allowed to take only those courses that interested him. At the end of four years, Zahniser was still alive, but lacked many required courses for graduation, so he spent a fifth undergraduate year as an upperclassman taking required courses. He often bragged that he "crammed four years of college into five."

At Greenville College, Zahniser majored in English, edited the student newspaper, the *Papyrus*, and was a member of the college debate team. His deep and lively interest in literature and his skills in writing, editing, and public speaking proved pivotal to his later embarking on a federal government career in editing, writing, and public relations. His federal career in turn further enmeshed him in wildlife conservation and introduced him to the mentors who shaped his appreciation for the need to secure legislative protection for wilderness areas.

During two summers of his college years, Zahniser worked on a state roads crew and, in 1927, as a reporter for the Pittsburgh, Pennsylvania,

Press newspaper. The year after his graduation from college, Zahniser taught high-school English in Greenville, Illinois, and worked on the Greenville, Illinois, *Advocate* newspaper. One of his duties was to summarize the sermons preached at various community churches. In 1929, his college chum Paul H. Oehser and his wife Grace Edgbert Oehser convinced Zahniser to take the federal civil service examination for editors. After college, the Oehsers had gone to the Washington, D.C., area, where Paul Oehser worked as an editor for the U.S. Department of Commerce and eventually became chief of publications for the Smithsonian Institution. Zahniser passed the entrance exam and moved to Washington to work for the Department of Commerce, arriving on New Year's Eve in 1929.

Zahniser soon moved from the Department of Commerce to the U.S. Bureau of Biological Survey as an editor, writer, researcher, and broadcast specialist. There he began to work with the naturalists and scientists who were to mentor his growth in conservation and wilderness-preservation interests. He worked with biologists such as Olaus J. Murie, soon to be the foremost expert on North American elk and an outspoken proponent of preserving wildlands. He also worked with the self-taught naturalist Edward A. Preble, Zahniser's primary personal and conservation mentor. Through his connection with Preble, Zahniser began writing a monthly book-review column, "Nature in Print," for *Nature* magazine in 1935. He served as books editor for more than twenty-five years, until *Nature* magazine was folded into *Natural History* magazine. Zahniser's wide and up-to-date reading in natural history and conservation helped make him, as historian Stephen J. Fox (1981) asserts, one of the most literate conservationists of his era. The essay character of "Nature in Print" also enabled him to read and reread the works of earlier nature writers such as Ralph Waldo Emerson, Henry David Thoreau, and John Muir. In it, he also explored the views of nature and a just society expounded by his other principal literary heroes, Dante Aligheri, William Blake, St. Francis of Assisi, and the biblical Book of Job.

On March 3, 1936, Zahniser married Alice Bernita Hayden, and the couple had four children: A. H. Mathias, Esther, Karen, and Edward. His wife's willing abilities as a wilderness cook and trip planner enabled Zahniser to include his family in many wilderness trips, and their firsthand experiences buttressed his witness for conservation and wilderness-preservation advocacy.

The Bureau of Biological Survey eventually became what is now the U.S. Fish and Wildlife Service. Zahniser worked there during its formative years when the forester and ecologist Aldo Leopold was working with Zahniser's boss J. N. "Ding" Darling and his future boss Ira N. Gabrielson on the creation of the federal system of wildlife and game refuges. (Zahniser also worked at the Fish and Wildlife Service briefly with Rachel Carson, who went on to write *The Sea around Us* and *Silent Spring*.) During these same years, the forester, plant physiologist, and recreation planner Robert Marshall was assembling the small cadre of scientists and humanists who formed the nucleus of the Wilderness Society, which was first organized in 1935. Zahniser and Paul Oehser became charter members of the society and attended the organizing meeting of its Washington section with Marshall and other founders in 1937. In 1942, Zahniser became chief of information for the Bureau of Plant Industry of the U.S. Department of Agriculture.

When the United States dropped atomic bombs on Hiroshima and Nagasaki in World War II, Zahniser began to question the ethics of his continued federal employment. He also began to question the notion of technological progress if the end result was the exploitation of basic forces of nature to wreak such destruction on human beings and nature. On the death of the Wilderness Society's only staff member, Robert Sterling Yard, in early 1945, Zahniser had agreed to edit and produce the Wilderness Society's irregular magazine *The Living Wilderness* on a freelance basis while the society's governing council thought through new directions for the struggling young organization. Its founder, Robert Marshall, had died in 1939 at age thirty-eight and had effectively endowed the society with part of his personal estate. In September 1945, Zahniser went to work for the Wilderness Society as executive secretary (later executive director) and editor of the magazine. He quickly established it as a regular quarterly membership magazine and ran the society's Washington office. Olaus Murie became the society's half-time director, working out of his home in Moose, Wyoming.

The society's new twin goals were to build a broad-based, national membership organization and to make it a focus of cooperation among any and all other groups that would support the preservation of wilderness. To help achieve these goals, Zahniser turned *The Living Wilderness* into a showcase of feature articles on wilderness and wildlands and a central clearinghouse for news of the broadest possible definition of conservation organizations and cooperators. Historian T. H. Watkins has said that Zahniser's editing of *The Living Wilderness* set the pattern that most other major conservation magazines, such as *Audubon* and *Sierra*, later followed. Zahniser was also skillful at and worked very hard on developing local and regional leaders for conservation struggles far from Washington, D.C. Many of these grassroots advocates also contributed to the pages of *The Living Wilderness*.

In 1946, Zahniser and the Wilderness Society joined in the campaign to preserve wildlands in the Adirondack State Park of New York State. His association there with the New York conservationist Paul A. Schaefer led to the Zahniser family buying a small, hillside subsistence farm at the end of a road in Warren County. Today the family's cabin Mateskared abuts the Siamese Ponds Wilderness Area of the Adirondack Forest Preserve. The definition of wilderness Zahniser would craft for the 1964 Wilderness Act now also defines more than one million acres of designated state wilderness areas in New York.

In the mid-1940s, the Wilderness Society governing council had decided that the organization should press for national legislation for the preservation of wilderness. By then, the U.S. Forest Service had been pressured into administratively designating several large tracts of roadless wilderness lands for protection. However, during World War II and immediately after, many such wilderness areas were administratively declassified so they could be logged or mined or have roads built into or through them. In 1946, Wilderness Society president Benton MacKaye, founder of the Appalachian Trail, and Zahniser put together proposed legislation to create a Federal Wildlands Project. The goal was to protect a variety of wildlands, especially in river corridors. The proposal went nowhere, but it gave Zahniser his first experience at crafting legislation and trying to get action on it.

In the early 1950s, the U.S. Bureau of Reclamation determined to build a dam on the Green and Yampa rivers in Echo Park, a beautiful canyon inside Dinosaur National Monument. The dam would violate and develop lands of the national park system, of which Dinosaur was a part. The governing bodies of the Wilderness Society and the Sierra Club decided to fight the dam proposal, which was a part of the massive Colorado River Storage Project system of dams. Zahniser and David Brower of the Sierra Club then embarked on building a national coalition of conservation and civic organizations to defeat the dam and defend the principle of the inviolability of national park lands. It was considered a nearly futile effort by many conservationists. However, the coalition was built—the first-ever national conservation coalition—and pressure mounted as citizens across the nation were spurred to contact their members of Congress to protest the proposed dam. Zahniser and Brower also secured independent expert hydrologists and engineers who demonstrated for members of Congress that the Bureau of Reclamation's own engineering planning for the dam was not sound. Congress eventually voted to remove the Echo Park Dam from the overall series of dams on the Colorado River system. Zahniser negotiated the final settlement with the government on behalf of the conservationists. Today the Yampa River, which flows into the Green at Echo Park, is the only free-flowing, undammed river in the entire Colorado River system.

With that unlikely battle won, Zahniser and Brower decided to motivate the coalition they had crafted to work for securing national legislative protection for wilderness on the federal public lands. The first wilderness bill was drafted in 1955 and introduced in Congress in 1956. It met great resistance from mining, logging, and waterpower interests and also from the federal land-managing agencies. Zahniser would see a series of wilderness bills through a series of eleven congressional public hearings and sixty-six drafts or revisions. Many of these dealt with the compromises needed to gain the support of such opponents. Zahniser wrote widely in support of wilderness in the national conservation press and addressed countless citizens' groups to rally support for the legislation. He also worked with newspaper and magazine writers and editors to shape favorable publicity. His tireless, constant advocacy, as Brower later characterized it, finally saw a wilderness bill enacted by Congress in the summer of 1964 and signed into law by President Lyndon B. Johnson on September 3, 1964. Unfortunately, Zahniser had died of a heart attack on May 5, 1964, just weeks after testifying at the final public hearing on the bill. His widow Alice attended the act's signing in the Rose Garden at the White House, along with Margaret E. "Mardy" Murie, the widow of Olaus J. Murie, Zahniser's mentor and close collaborator, who had died in 1963.

Zahniser was a tireless networker within a broad range of conservation and citizen organizations, another key to his effectiveness as a wilderness advocate. In 1948 and 1949, he chaired the Natural Resources Council of America, which he had helped organize in 1946. He was a member of the secretary of the interior's Advisory Committee on Conservation from 1951 to 1954 and an honorary vice president of the Sierra Club from 1952 until his death. He served as vice chairman of the Citizens Committee for Natural Resources in 1955. He helped publicize early efforts to save Thoreau's Walden Pond from development and served as president of the Thoreau Society for the 1956–1957 term. In 1957, he received an honorary doctor of letters degree from Greenville College. He told the convocation he was so far behind in his correspondence that "the degree should be called a Doctor of Postcards!"

Zahniser's writings were mostly published speeches and reviews, articles, and editorials throughout the conservation press, but particularly *Nature* magazine and *The Living Wilderness*. He also drafted many speeches for members of Congress, especially Senators Clinton P. Anderson of New Mexico and Hubert H. Humphrey of Minnesota and Congressman John P. Saylor of Pennsylvania. Many of Zahniser's seminal Adirondack writings and speeches are collected as *Where Wilderness Preservation Began: Adirondack Writings of Howard Zahniser*, edited by Ed Zahniser and published in 1992. Environmental historian Mark W. T. Harvey, author of *A Symbol of Wilderness: Echo Park and the American Conservation Movement*, is preparing a biography of Howard Zahniser.

BIBLIOGRAPHY

Fox, S. R. (1981). *John Muir and His Legacy: The American Conservation Movement*. Boston: Little Brown & Co.

Harvey, M. W. T. (1994). *A Symbol of Wilderness: Echo Park and the American Wilderness Movement*. Albuquerque: University of New Mexico Press.

Oehser, P. H. (1964, November). Howard Zahniser (1906–1964). *Cosmos Club Bulletin, 17,* (11).

Schaefer, P. *Defending the Wilderness: The Adirondack Writings of Paul Schaefer.* Syracuse, NY: Syracuse University Press, 1989.

Snyder, H. Z. (1999, December). Personal interview.

Zahniser, E. (Ed.). (1992). *Where Wilderness Preservation Began: Adirondack Writings of Howard Zahniser.* Utica, NY: North Country Books.

Zahniser, H. (1951, July 28). Conservation and the Family. Address. Issued as a news release by The Wilderness Society, Washington, DC.

—Ed Zahniser

Appendix: Environmental Timeline

1620 to 1860	Erosion becomes a major problem on many American farms. Fields are abandoned. Rivers and streams are filled with silt and mud. The publication of farm journals is initiated by early soil conservationists to improve farming methods.
1748	Jared Eliot, a minister and doctor of Killingsworth, Connecticut, writes the first American book on agriculture to improve crops and to conserve soil.
1824	Solomon and William Drown of Providence, Rhode Island, publish *Farmer's Guide* that discusses erosion and its causes and remedies. A year later, John Lorain, of the Philadelphia Agricultural Society, also publishes a book devoted to the prevention of soil erosion in which he discusses methods such as using grass as an erosion-control crop.
1827	**John James Audubon** begins publication of *Birds of America*.
1830	**George Catlin** launches his great western painting crusade to document Native American peoples.
1838	**John Muir** is born in Dunbar, Scotland. The family emigrates to the United States and settles in Wisconsin. Muir spends much of his youth helping his father on the family farm.
1845	**Henry David Thoreau** begins to write *Walden*, his observations of the fauna and flora of Concord, Massachusetts. The book is published in 1854.
1847	U.S. Congressman George Perkins Marsh of Vermont delivers a speech calling attention to the destructive impact of human activity on the land.
1849	U.S. Department of the Interior (DOI) is established. It is charged with a wide range of responsibilities, including construction of the Washington, DC, water supply system, exploration of the

western wilderness, administration of some hospitals and universities, and management of Native American affairs, public lands, patents, and pensions.

1854   Henry David Thoreau's *Walden; or, Life in the Woods* is published.

1857   Frederick Law Olmsted is appointed to be the first superintendent to develop a city park, New York City's Central Park.

1860   Nicholas T. Sorsby, a physician who farmed in Alabama and Mississippi, writes *Horizontal Plowing and Hillside Ditching*, the first book of its kind devoted to erosion control by plowing methods.

John Muir leaves the Wisconsin farm and enrolls at the University of Wisconsin.

1861   Frederick Law Olmsted writes a series of letters to the *New York Daily Times* that are later reprinted in three volumes as *The Cotton Kingdom*. The letters are considered strong antislavery statements.

1862   President Abraham Lincoln establishes the Department of Agriculture (USDA) as part of the Cabinet. Lincoln calls it the "people's department," since at the time, 90 percent of Americans are farmers who need good seed and reliable information to grow their crops.

1864   Yosemite Valley, California, is reserved as a state park by President Abraham Lincoln.

Man and Nature, by George Perkins Marsh, is published. Marsh's book reports for the first time how human activity is destroying forests and causing erosion that is devastating the land. Marsh has been called "the fountainhead of the conservation movement."

1865   Frederick Law Olmsted returns to New York to begin work on Central Park. Over the next 25 years, he plans, designs, and builds many city parks, including the United States Capitol Grounds and the landscapes for the White House and the Jefferson Memorial.

1866   American Society for the Prevention of Cruelty to Animals is established.

The term "ecology" is coined by German biologist Ernst Haeckel.

1869   John Wesley Powell travels the Colorado River through the Grand Canyon. Between 1880 and 1894, Powell is the director of the U.S. Geological Survey, which becomes one of the largest scientific organizations in the world at that time.

1871	*The New York Tribune* publishes an article titled, "Yosemite Glaciers," by **John Muir**, who hypothesizes that the Yosemite valley was carved up by glaciers.
	The Commission of Fish and Fisheries is established and later becomes the Bureau of Fisheries of the Department of Commerce. The commission is established by a resolution passed by Congress to study the declining number of the nation's ocean and lake food fishes and to recommend ways to reverse the decline.
1872	Yellowstone National Park (Wyoming, Montana, and Idaho) is established and becomes the first national park in the world (today the largest in the lower 48 states).
	General Mining Law is passed.
1873	*Forest and Stream* is started in 1873 and becomes a leading sportsmen's magazine. The magazine is created to help conserve wildlife and forests.
1875	American Forestry Association is established. Its emphasis is on the care and protection of trees rather than on the economics of forestry.
1879	U.S. Geological Survey is established by the U.S. Congress to collect, analyze, interpret, publish, and disseminate information about the natural resources of the nation's public lands.
1880	**George Bird Grinnell** becomes the natural history editor-in-chief of *Forest and Stream* and edits the magazine for 31 years (1880–1911).
1881	Norway tracks signs of acid precipitation on its western coast.
1882	On September 30, the world's first hydroelectric power plant begins operation on the Fox River in Appleton, Wisconsin. The plant, later named the Appleton Edison Light Company, is initiated by Appleton paper manufacturer H. F. Rogers, who is inspired by Thomas Edison's plans for an electricity-producing station in New York.
	American Forestry Congress is organized.
1883	Krakatoa, a small island of Indonesia, is virtually destroyed by a volcanic explosion. The eruption of Krakatoa also results in the local extinction of virtually all island wildlife.
1885	**Gifford Pinchot** decides on a career in forestry. At the time, much of the eastern sections of forests in the United States have been cut down and burned to create farmlands.
	New York State establishes the Adirondack Forest Preserve, to be "kept forever as wild forest lands," a landmark in conservation legislation.

1886 **George Bird Grinnell**, editor of *Forest and Stream*, encourages his readers to join him in forming the country's first bird preservation organization, the Audubon Society. The society is named after the American naturalist and wildlife painter, **John James Audubon**, who lived from 1785 to 1851. In only three months, more than 38,000 people join the society. Overwhelmed by the response, Grinnell has to disband the group in 1888. Grinnell has a great impact on the early national legislation relating to wildlife, parks, and forests.

1887 Boone and Crockett Club, named after two of America's famous hunters, is formally founded by **Theodore Roosevelt, George Bird Grinnell**, and others. It is the first private organization to deal effectively with conservation issues.

1890 **Marjory Stoneman Douglas** is born in Minneapolis, Minnesota. She becomes a well-known activist in preserving the Florida Everglades.

1891 Yosemite National Park is established in California.

 The passing of the Forest Reserve Act allows National Forests (forest reserves) to be established on public domain lands.

 By Proclamation, President Benjamin Harrison sets aside a tract of land in Wyoming as the nation's first forest reservation.

1892 **John Muir**, Robert Underwood Johnson, and William Colby are cofounders of the Sierra Club, in Muir's words, to "do something for wildness and make the mountains glad."

1893 **Gifford Pinchot**'s account of his work at the Biltmore estate in North Carolina is published. The publication is the first to suggest a sustainability approach in forestry by applying forest management skills that call for cutting down trees without damage to the forest.

1895 American Scenic and Historic Preservation Society is founded in response to industrialization's impact on natural resources.

1896 A group of women appalled by the slaughter of birds form the Massachusetts Audubon and refuse to buy or wear hats and clothing decorated with bird plumes or parts.

1898 The first college program in forestry is offered at Cornell University.

 Gifford Pinchot becomes head of the U.S. Division of Forestry (now the U.S. Forest Service) and serves until 1910. Under President **Theodore Roosevelt**, many of Pinchot's ideas become national policy. During his service, the national forests increase from 32 in 1898 to 149 in 1910.

1899	The River and Harbor Act bans pollution of all navigable waterways. Under the act, the building of any wharves, piers, jetties, and other structures is prohibited without congressional approval.
	Frank Chapman, an ornithologist with the American Museum of Natural History in New York, begins publishing *Bird Lore* magazine, which becomes a unifying national forum for the Audubon movement.
1900	Lacey Act is the first federal law that protects game birds and prevents illegally killed game animals to be transported across state boundaries.
	Society of American Foresters is founded.
	Hurricane hits Galveston, Texas. More than 6,000 people are killed and about 3,600 buildings are demolished. The Galveston hurricane remains the worst natural disaster in U.S. history.
1901	**Theodore Roosevelt** becomes president of the United States. During his two terms, Roosevelt creates the National Wildlife Refuge System, expands the National Forests from 42 million acres to 172 million, and establishes 18 areas as national monuments.
	John Muir's *Our National Parks* is published, bringing him to the attention of President **Theodore Roosevelt**.
1902	Bureau of Reclamation is established in the Department of Interior. The Reclamation Act of 1902 creates this bureau within the U.S. Geological Survey (USGS) to help alleviate water problems. At the time, irrigation projects are known as "reclamation" projects because they "reclaimed" arid lands for human use. The bureau manages, develops, and protects water and related resources in an environmentally and economically sound manner.
1903	**John Muir** and **Theodore Roosevelt** spend 3 days on a camping trip through Yosemite Valley.
	President **Theodore Roosevelt** establishes the first federally protected wildlife refuge on Pelican Island. Early Audubon Societies hire wardens to protect many of these first sanctuaries.
1905	Some of the Audubon societies are incorporated into the National Association of Audubon Societies for the Protection of Wild Birds and Animals. National Audubon Society, a nonprofit, national organization, is founded to conserve and restore natural ecosystems, focusing on birds and other wildlife, for the benefit of humanity and the earth's biodiversity. The National Audubon Society is named for **John James Audubon**, renowned ornithologist, explorer, naturalist, and wildlife artist.
	George Washington Carver's *How to Build Up Worn Out Soils* is published.
1906	President **Theodore Roosevelt** is awarded the Nobel Prize for Peace.

1907   The Sierra Club announces its opposition to the proposal to use the Hetch Hetchy Valley, in Yosemite, as a reservoir.

Gifford Pinchot is appointed first chief of U.S. Forest Service.

Inland Waterways Commission is established.

1908   Grand Canyon is set aside as a national monument.

Chlorination is first used extensively at U.S. water treatment plants.

President **Theodore Roosevelt** hosts first Governors' Conference on Conservation.

President Roosevelt sets aside Muir Woods as a national park in California.

1909   **Wallace Stegner**, referred to as "the dean of western writers," is born in Lake Mills, Iowa. During his lifetime, Stegner, a conservationist, lobbies for regulations to prohibit construction of dams in the national parks and raises the country's consciousness about the environment.

Aldo Leopold joins the U.S. Forest Service, and by 1912 he is supervisor of the million-acre Carson National Forest in what is now New Mexico.

1913   President Woodrow Wilson and Congress approve the Hetch Hetchy Valley Dam in Yosemite National Park despite the efforts from **John Muir** and other preservationists to stop the damming of the valley.

1914   Martha, the last passenger pigeon, dies in the Cincinnati zoo.

John Muir dies, leaving a legacy of books and articles to preserve wilderness areas.

Ruth Patrick, age 7, is given her first microscope from her father. She begins to study diatoms, which she studies throughout her life.

Hazel Wolf, age 16, gets her first job in a law office. She receives $25 a month, $15 of which she pays her mother for board and $5.00 of which she uses to pay for a bicycle.

Papago Saguaro National Monument in Arizona is established.

1915   Dinosaur National Monument is established in Utah.

1916   National Park Service is established as an agency of the U.S. Department of Interior. By 1990, the NPS includes more than 350 historical and recreational parks throughout the United States, including Washington, DC, the Virgin Islands, Guam, and Puerto Rico.

1917   Mount McKinley National Park is established in Alaska.

1918   Save-the-Redwoods League is created.

                Migratory Bird Treaty Act, signed by the United States and Canada, restricts the hunting of migratory bird species.

1919   National Parks and Conservation Association is established to protect and improve the U.S. national parks and to educate people about the need to protect and preserve these natural resources.

                Roger Tory Peterson's seventh-grade teacher, Blanche Hornbeck, enrolls her students in the Junior Audubon Club. The experience shapes the 11-year-old boy's interest in birds.

1920   Mineral Leasing Act regulates mining on federal lands.

                The federal Water Power Act is created to establish a commission to oversee waterways and the construction and use of water power projects.

1922   **Will Dilg**, a fisherman and conservationist, is the founder and first president (1922–1926) of the Izaak Walton League of America, the first mainstream conservation organization with mass appeal. The league is formed in the name of Izaak Walton, the world-known English angler and author of *The Compleat Angler*, one of the most famous books in the English language.

1924   Through the work of environmentalist **Aldo Leopold**, Gila National Forest, New Mexico, is established as the first extensive wilderness area.

                Jay Norwood Darling receives a Pulitzer prize for editorial cartooning. He receives his second in 1943 for a 1942 cartoon.

                Marjory Stoneman Douglas, of the *Miami Herald*, writes newspaper columns opposing the draining of the Florida Everglades.

                Bryce Canyon National Park is established in Utah.

1925   **George Bird Grinnell** becomes president of the National Parks Association and receives the Theodore Roosevelt Gold Medal of Honor for distinguished service in promoting outdoor life and conservation causes.

1926   **Margaret** and **Olaus Murie** study declining elk populations in Jackson Hole, Wyoming.

                Great Smoky National Park is established in North Carolina and Tennessee.

Ynes Mexia begins one of her many expeditions to collect and study hundreds of plant specimens. During her career, Mexia will collect more than 100,000 plant specimens and hundreds of new species.

1928   Boulder Canyon Project (Hoover Dam) is authorized to provide combined irrigation, electric power, and flood control system along the border between Arizona and Nevada.

Ansel Adams's photographs appear in the *Sierra Club Bulletin* during the 1920s.

1929   Grand Teton National Park is established in Wyoming.

1930   The 1930s includes severe drought conditions in much of the Midwest known as the "Dust Bowl." Dry weather and over-farming force many to leave the area. **Hugh Bennett** spends time reviewing abandoned farms and soil conditions. During the mid-1930s, Bennett gives numerous speeches on radio and in person at agricultural events and on college campuses to farmers and scientists. He writes hundreds of articles and several books about the threat of soil erosion.

Chlorofluorocarbons are found to be safe refrigerants for use in refrigerators and air conditioners. They are nontoxic and non-combustible.

Robert Marshall defines the term "wilderness" in "The Problem of the Wilderness."

1932   On June 11, **Gloria Anable** sets the women's record of 410 feet for the deepest oceanic descent. She studies deep-sea life from a steel bathysphere.

1933   Tennessee Valley Authority is formed to conduct studies to an-alyze the environmental impact of hydroelectric projects before developing a plan to dam rivers in Tennessee. By 1975 all of the rivers in Tennessee are dammed.

Civilian Conservation Corps employs over two million Ameri-cans in forestry, flood control, soil erosion, and beautification projects.

Hugh Bennett is given the opportunity to put his soil conserva-tion ideas into practice to help reduce soil erosion. He becomes the director of the Soil Erosion Service (SES) created by the De-partment of Interior.

1934   The greatest drought in U.S. history continues. Portions of Texas, Oklahoma, Arkansas, and several other midwestern states are known as the "Dust Bowl."

Taylor Grazing Act regulates livestock grazing on federal lands.

A Field Guide to the Birds, by **Roger Tory Peterson**, is published. Today, there are 52 volumes in the Peterson Field Guide Series.

President Franklin Roosevelt appoints **Jay Norwood Darling** to become chief of the U.S. Biological Survey. He initiates the federal Duck Stamp Program (Migratory Bird Hunting and Conservation Stamp) that is sold to raise funds for land acquisitions and wetlands for waterfowl habitats. Darling also designs the first Duck Stamp.

1935   U.S. Soil Conservation Service is established. **Hugh Bennett**, the father of soil conservation, continues on as head of the new organization.

Wilderness Society is founded by **Aldo Leopold, Robert Marshall**, and Benton MacKaye to protect the wilderness areas of the United States. Later, **Margaret** and **Olaus Murie** join the Wilderness Society and Olaus becomes director from 1945 to 1962.

1936   National Wildlife Federation is founded by **Jay Norwood Darling** to help combat abuse and waste of soil, water, and wildlife.

Rosalie Edge, a preservationist, establishes Hawk Mountain Sanctuary Association as a refuge for hawks in Pennsylvania.

President Franklin Roosevelt's administration outlaws the killing of predators in National Parks.

1937   The Migratory Bird Treaty is extended to include Mexico, Canada, and the United States.

1938   **Sigurd Olson**'s "Why Wilderness?" is published.

1939   **David Brower** creates his first nature film for the Sierra Club called *Sky Land Trails of the Kings*. In the same year, Brower, who is an excellent climber, completes his most famous ascent, Shiprock, a volcanic plug that rises 1,400 feet from the floor of the New Mexico desert.

1940   The Bald Eagle Protection Act is legislated.

Rosalie Edge plays a major role in the establishment of Kings Canyon National Park.

1941   A system of aqueducts begins diverting water from streams feeding into Mono Lake to the city of Los Angeles to help meet the water needs of its residents. Diversion of this water significantly alters the Mono Lake ecosystem. By 1989, the lake's water level drops as much as 33 feet.

The National Park Service commissions noted photographer **Ansel Adams** to create a photo mural for the Department of the Interior Building in Washington, DC. The theme is to be nature as exemplified and protected in the U.S. National Parks, but the project is halted because of World War II and never resumed.

1943   Audubon nature center is opened in Greenwich, Connecticut, and it becomes the model for many other centers that exist today.

1944	Soil Conservation Society of America is founded by **Hugh Bennett** and others.
1945	**Olaus Murie** becomes the director of the Wilderness Society from 1945 and will remain in that post until 1962.
1946	U.S. Bureau of Land Management is created to centralize the administration of lands in the public domain.
	The International Whaling Commission is formed to do research on whale populations.
	Gifford Pinchot, at the age of 81, dies of leukemia in New York City.
	First atomic bomb is exploded in the New Mexico desert. The Atomic Energy Commission is created to oversee the development of peaceful and military uses of nuclear power.
1947	**Marjory Stoneman Douglas** writes a best-selling book, *The Everglades: River of Grass*. The same year, President Truman meets Douglas in Florida to announce the establishment of Everglades National Park.
	Defenders of Wildlife is established.
	Federal Insecticide, Fungicide and Rodenticide Act (FIFRA) is passed and requires the U.S. Department of Agriculture (USDA) to register all pesticides prior to their introduction in interstate commerce.
1948	Air pollution incident in Donora, Pennsylvania, causes 20 deaths and 14,000 related illnesses.
	Federal Water Pollution Control Law is established to look into waste disposal problems.
	Maria Telkes designs the first solar heating system.
	Paul Mueller is awarded a Nobel Prize for his discovery of insect-killing properties of DDT.
1949	*A Sand County Almanac*, written by **Aldo Leopold**, is published a year after Leopold suffers a fatal heart attack.
	Ruth Patrick begins studies of the Savannah River on the future site of what now is the Department of Energy's Savannah River Site. Patrick's involvement with the Savannah River Site continues until 1997 when she steps down from the Savannah River Site's Environmental Committee.
1951	Nature Conservancy is founded. By the year 2000, the conservancy has a membership of 900,000 and has protected over 10 million acres in all 50 states, Latin America, Canada, the Pacific, and the Caribbean.

1952	*Handbook of Turtles*, by **Archie Carr**, is published. Carr is a conservationist who becomes well-known throughout his life as the "turtle man." He is the foremost authority on turtles.

David Brower becomes executive director of the Sierra Club.

About 4,000 people are killed during London's "Killer Smog" over one weekend. |
| 1953 | Radioactive iodine from atomic bomb testing is found in children's thyroid glands in Utah.

Gloria Anable becomes chair of the Mianus River Gorge Conservation Committee of the Nature Conservancy from 1953 to 1988. The Mianus River Gorge is a 555-acre nature preserve in Westchester County in New York. It is the birthplace of the Nature Conservancy's method of direct action conservation.

Barry Commoner begins to study the effects of atomic fallout from nuclear bomb tests.

Eugene Odum's *Fundamentals of Ecology* is published. |
| 1956 | Water Pollution Control Act authorizes water treatment plants.

The proposed Echo Canyon Park Dam, to be built in the Dinosaur National Monument in the upper Colorado River in Utah, is canceled due to the hard lobbying of activists **Wallace Stegner, Rosalie Edge, David Brower,** and others.

United Kingdom's Halley Station in the Antarctica begins to study ozone in the atmosphere. |
| 1959 | The Antarctic Treaty, an international agreement signed in 1959 and amended in 1991, binds its signatories to work together to preserve the natural resources of Antarctica until the year 2041. The treaty states that Antarctica should be used exclusively for peaceful purposes and to promote the exchange of scientific data.

George Schaller begins a project on mountain gorillas in the Congo and later writes a book, *The Year of the Gorilla*. **Dian Fossey** credits Schaller's book for her interest in mountain gorillas. |
| 1960 | **Wallace Stegner** writes *The Wilderness Letter* about the importance of federal protection of wild places. |
| 1961 | World Wildlife Fund (WWF) is founded. WWF is dedicated to protecting the world's wildlife and their habitats and recognized globally by its panda logo.

During the 1960s, a major drought occurs in the vast region of north central Africa known as the *Sahel* that increases world |

awareness of desertification. The drought causes starvation, live-stock death, agricultural land loss, and relocation of large human populations.

1962   *Our Synthetic Environment*, by **Murray Bookchin**, is published.

White House Conservation Conference is held.

Hazel Wolf joins the National Audubon Society in Seattle, Washington, and plays a prominent role in environmental efforts in local, national, and international arenas during her lifetime.

Rachel Carson's *Silent Spring*, a groundbreaking study of the dangers of DDT and other insecticides, is published.

1963   First Clean Air Act authorizes money for air pollution control efforts.

Nuclear Test Ban Treaty between United States and U.S.S.R. stops atmospheric testing of nuclear weapons.

Kirkpatrick Sale begins teaching European History classes at the University of Ghana. He writes a book about his experiences entitled *The Land and People of Ghana*.

1964   **Howard Zahniser, David Brower, Margaret** and **Olaus Murie, Wallace Stegner,** and others lobby for several years to bring about the passage of the 1964 Wilderness Act signed into law by President Lyndon Johnson to create National Wilderness Preservation System.

Hazel Henderson organizes women in the local play park in New York City and starts a group called Citizens for Clean Air. It is the first environmental group east of the Mississippi. She built Citizens for Clean Air from a very small group to a membership of 40,000. Two years later New York City becomes the site where 80 people die from air pollution–related causes during four days of atmospheric inversion.

1965   Water Quality Act gives the federal government power to set water standards in absence of state action.

Solid Waste Disposal Act is passed.

The Freedom of Information Act is passed.

The Land and Water Conservation Fund is established to acquire land for National and State Parks.

Unsafe at Any Speed, by **Ralph Nader**, is published.

1966   Eighty die in New York City from air pollution–related causes.

William Niering writes a book, *The Life of the Marsh: The North American Wetlands*. He becomes a well-known authority on both freshwater and saltwater wetlands.

Barry Commoner establishes the Center for the Biology of Natural Systems.

1967   Environmental Defense Fund is formed. The group leads the effort to save the osprey from DDT.

Dian Fossey establishes Karisoke, a research camp in Rwanda to study mountain gorillas.

Robert MacArthur and **Edward O. Wilson**'s *The Theory of Island Biogeography* is published.

1968   Wild and Scenic Rivers Act is enacted. It is designed to protect certain rivers from being dammed or damaged in other ways.

The Population Bomb, by **Paul Ehrlich**, is published.

The Grand Canyon dam project is removed from the Colorado River bill due to the efforts of **David Brower** and other activists. Their activism saves sections of the canyon from being flooded.

Joan Wolfe starts the West Michigan Environmental Action Council (WMEAC) that works with various organizations including churches, social clubs, PTA groups, labor unions, and student groups on environmental issues. It may be the first large, broad-based environmental organization of its kind in the United States.

Redwood National Park is established in California.

Garrett Hardin's "The Tragedy of the Commons" is published.

Zero Population Growth organization is formed by **Paul Ehrlich**.

Edward Abbey's *Desert Solitaire* is published.

Margaret Wentworth Owings cofounds Friends of the Sea Otter.

1969   Major oil spill from offshore wells causes beach pollution.

National Environmental Policy Act (NEPA) is passed by the U.S. Congress and requires the government to consider the consequences of decisions involving projects that could significantly affect the quality of the human environment. It establishes White House Council on Environmental Quality.

William Niering is instrumental in the passage of Connecticut's Tidal Wetlands Act.

Friends of Earth is founded by **David Brower**. It now has offices in 66 countries.

John Todd, Nancy Jack Todd, and Bill McLarney are the cofounders of the New Alchemy Institute in Cape Cod, Massachusetts. The institute begins to pioneer a new way of treating sewage and other wastes.

Lester Brown establishes the Overseas Development Council with Jim Grant.

1970   **Ruth Patrick** is elected to the National Academy of Sciences and is only the twelfth woman at that time to be so honored.

Denis Hayes is the National Coordinator of the first Earth Day, celebrated on April 22.

The Clean Air Act (CAA) is passed; amendments are enacted in 1977 and 1990. The federal law regulates air emissions from sources that are stationary, such as power plants and factories, as well as those that are mobile, such as automobiles.

Occupational Safety and Health Act (OSHA) is founded. Its mission is to save lives, prevent injuries, and protect the health of the country's workers.

Environmental Protection Agency (EPA) is established in response to growing public concern about unhealthy air, polluted rivers and groundwater, unsafe drinking water, endangered species, and hazardous waste disposal.

Friends of the Everglades is founded by **Marjory Stoneman Douglas**. One of the organization's first activities is to defeat a proposal to build an airfield in the Florida Everglades.

Concerns about mercury in tuna lead the U.S. Food and Drug Administration (FDA) to order a national recall of canned tuna because tests on some fish show mercury levels above the acceptable standard of 0.5 parts per million.

Lee Botts helps organize the Lake Michigan Federation in 1970. One of their major actions is to prevent the building of 20 to 24 nuclear power plants around the lake. She and her group also help in getting the first Clean Water Act passed in 1972.

The Natural Resources Defense Council is founded.

Time magazine calls **Barry Commoner** "The Paul Revere of Ecology."

1971   *The Closing Circle*, by **Barry Commoner**, is published.

Greenpeace, an international organization dedicated to bringing an end to activities that harm the environment and to protecting endangered species, is founded in Vancouver, British Columbia, Canada, by members of the Don't Make a Wave Committee to oppose nuclear testing at Amchitka Island in Alaska. Greenpeace Foundation is established with special emphasis against radioactive and toxic waste dumping, nuclear weapons, and acid precipitation.

1972   EPA phases out DDT in the United States. The pesticide, technically known as dichlorodiphenyltrichoroethane, is banned in the United States but will still be used in many developing nations for non-agricultural pest control.

The Clean Water Act (Federal Water Pollution Act) is passed. The Act restores polluted waters for recreational use and eliminates discharges of pollutants into navigable waterways. **Lee Botts** and others are instrumental in getting the act passed. She organizes local participation in Wisconsin, Illinois, Indiana, and Michigan to review the discharge permits that industries and sewage treatment plants are now required to obtain under the Clean Water Act.

Oregon passes the first bottle-recycling law.

Coastal Zone Management Act, Federal Environmental Pesticide Control Act, and Ocean Dumping Act pass.

Marine Mammal Protection Act legislation is passed to protect and manage marine mammal species by making it illegal to take, import, possess, harass, buy, sell, or offer to buy or sell marine mammals, their parts, or their products.

1973   The Convention on International Trade in Endangered Species of Wild Fauna and Flora (CITES) is signed by over 80 nations.

Energy crisis is felt in the United States from the OPEC oil embargo.

Congress approves the 1,300-kilometer pipeline from Alaska North Slope oil field to Port of Valdez.

Small Is Beautiful, by E. F. Schumacher, is published. Schumacher is a German-born English economist and philosopher who proposes small-scale technologies for economic development.

Planet Drum Foundation is founded by **Peter Berg** and **Judy Goldhaft**. The organization's emphasis is on bioregionalism—regional self-reliance, sustainability, and community self-determination.

Endangered Species Act is legislated and requires the conservation of threatened and endangered species and the ecosystems upon which they depend.

Friends of the River is founded by **Mark Dubois** for river preservation and restoration.

1974   Safe Drinking Water Act sets standards to protect nation's drinking water.

Karen Silkwood, a chemical technician at the Kerr-McGee nuclear fuels production facility in Crescent, Oklahoma, is critical of allegedly inadequate safety provisions. She dies in a suspicious, one-car accident in November.

World population reaches 4 billion.

Lester Brown creates the Worldwatch Institute.

Edwin Way Teale's *A Naturalist Buys an Old Farm* is published.

1975   Unleaded gas goes on sale. The equipment of cars with anti-pollution catalytic converters to reduce tailpipe pollution emissions begins.

Asbestos insulation is banned in new buildings.

The Kerr-McGee nuclear fuel production plant in Crescent, Oklahoma, closes. **Karen Silkwood**'s estate files a civil suit against Kerr-McGee.

The last undammed river in Tennessee is dammed.

Atlantic salmon return to the Connecticut River to spawn for the first time since 1875.

Edward Abbey's *The Monkey Wrench Gang* is published.

Edward O. Wilson's *Sociobiology* is published.

1976   National Academy of Sciences reports that chlorofluorocarbon gases from spray cans are damaging the ozone layer.

Resource Conservation and Recovery Act empowers EPA to regulate the disposal and treatment of municipal solid and hazardous wastes.

Seabrook anti-nuclear demonstrations take place in New Hampshire.

Toxic Substances Control Act legislation is enacted by Congress that requires testing, regulation, and screening to determine potential adverse effects of all chemicals produced or imported into the United States to the health of humans, other organisms, and the environment.

Wes and Dana Jackson create an agricultural school, The Land Institute, in Salinas, Kansas, to do prairie plant research.

1977   Department of Energy is established by the federal government to regulate and manage the energy policy of the United States. The Department is a world leader in the research and development of programs and technologies that generate energy from fossil fuels, nuclear fuels, and alternative energy resources such as solar energy, wind power, and biofuels.

First created as the Water Pollution Control Act in 1948 and renamed the "Clean Water Act" in 1972, the law is amended with changes that give further protection to surface water resources. The 1977 amendment sets the basic structure for regulating discharges of pollutants into waters of the United States.

Surface Mining Control and Reclamation Act is enacted. This federal legislation requires companies involved in surface mining operations to restore mined lands back to their natural conditions after the mining operations cease.

1978   Rainfall in Wheeling, West Virginia, is measured at a pH of 2, the most acidic yet recorded. The water is 5,000 times more acidic than normal rainfall.

Robert Bullard reports a case study, "Cancer Alley," about an African-American community in Triana, Alabama, located near a local stream that is highly contaminated with DDT from a plant that was shut down in 1970.

Aerosols with fluorocarbons are banned in the United States.

A National Wildlife Refuge in Florida is dedicated to the memory of Jay Norwood Darling.

David Brower and Paul Ehrlich are nominated for a Nobel Peace Prize. They are nominated again in 1979 and 1998.

Amoco Cadiz oil spill off the coast of France occurs. Oil is deposited along about 200 miles of coastline, often to depths of 12 inches. As many as 30,000 seabirds die.

David Gaines, his wife Sally, and several friends form the Mono Lake Committee. The 15,000-person membership supports the continued protection, restoration, and education about Mono Lake.

Lois Gibbs and others begin a door-to-door petition drive to demand the evacuation, restitution, and clean up of a toxic waste dump in their Love Canal neighborhood in Niagara Falls, New York. The evacuation of pregnant women and children begins.

1979   Three Mile Island Nuclear Power Plant in Pennsylvania experiences near-meltdown. The nuclear reactor's Unit 2 (TMI-2) is the site of the worst U.S. nuclear power plant accident. On the morning of March 28, equipment failure, design deficiencies, and worker error led to severe damage to the TMI-2 reactor core and small off-site releases of radiation.

Gaia: A New Look at Life on Earth, by James E. Lovelock, is published.

U.S. Environmental Protection Agency (EPA) suspends the domestic use of 2,4,5,-T, one of the herbicides in Agent Orange. The suspension is later changed to a permanent ban of the substance.

Robert Bullard and his wife Linda McKeever Bullard represent a group of African Americans opposing a plan to build a municipal landfill near their neighborhood in Houston, Texas. Their work results in a lawsuit, the first of its kind in the United States, that charges environmental discrimination in a waste facility siting under the civil rights laws to challenge environmental discrimination.

Denis Hayes receives the national Jefferson Medal for Greatest Public Service by an individual under 35 years old.

1980   President Jimmy Carter declares a health emergency in Love Canal, New York, and all of the families are evacuated.

The Superfund (Comprehensive Environmental Response, Compensation, and Liability) Act is passed with the mission to clean up the worst hazardous and toxic waste sites areas on land and water that constitute threats to human health and the environment.

Debt-for-nature swap idea is proposed by Thomas E. Lovejoy, whereby nations could convert debt to cash that would then be used to purchase parcels of tropical rainforest to be managed by local conservation groups.

Earth First! is founded by **Dave Foreman, Howie Wolke** and Mike Roselle. Earth First! is originally designed to strengthen the conservation movement, to make it easier for groups like the Sierra Club to take stronger, more effective positions and to advocate far more wilderness than most conservationists are willing to publicly support.

George Schaller becomes the first foreigner allowed to study the panda in its native habitat in China's Sichuan Province.

Low Level Radioactive Waste Act is passed.

Global Report 2000 to the President addresses world trends in population growth, natural resource use, and the environment by the end of the century and calls for international cooperation in solving problems.

Dian Fossey's *Gorillas in the Mist* is published and is later made into a motion film starring Sigourney Weaver.

Richard Moore becomes executive director of Southwest Organizing Project of New Mexico, a citywide and statewide grassroots organization in Albuquerque and throughout New Mexico. Its tasks include addressing issues of health care, child care, water rights, land, housing, and unemployment.

Susan Witt begins to serve as executive director of the newly-founded E. F. Schumacher Society. The Schumacher Society is a national, nonprofit organization that promotes and supports grassroots initiatives to develop more self-reliant regional economies.

John Todd and **Nancy Jack Todd** establish Ocean Arks International. Its purpose is to disseminate methods and practices of ecologically sustainable development throughout the world.

1981   **Lois Gibbs** founds Citizens' Clearinghouse for Hazardous Waste, later named Center for Health, Environment, and Justice (CHEJ).

Making Things Happen: How to Be an Effective Volunteer, by **Joan Wolfe**, is published. Wolfe shows volunteers how to attract new members and provides tips on how to raise funds and initiate action based on her successful environmental initiatives in Michigan.

The Quebec Ministry of Environment informs U.S. officials that about 60 percent of the sulfur dioxide polluting Canada's air and water is derived from industries in the United States.

1982   Earth Island Institute is founded by **David Brower**.

Warren County, North Carolina, plans to build a landfill for polychlorinated byphenyl (PCB) wastes in a predominately African-American neighborhood. Many groups think the site is racially motivated and poses an environmental health problem for the residents. As a result, there are widespread protests including marches. Reverend Benjamin F. Chavis, Jr. is one of many who participate in a protest march. More than 500 demonstrators are arrested. Although the protesters are unsuccessful in blocking the PCB landfill, they bring national attention to waste facility siting inequities and galvanize African-American church and civil rights leaders' support for environmental justice. Many believe that the march is a major impetus for the environmental justice movement in the United States.

The U.S. Congress passes the Nuclear Waste Policy Act that directs the Department of Energy to find a site to safely dispose of spent nuclear fuel. Although sites in Texas and the state of Washington are considered, Congress selects Yucca Mountain in Nevada.

World Resources Institute is founded as a research organization to help public and private groups pursue sustainable development.

The British Antarctica Survey station located at Halley Bay, Antarctica, reveals decreases in the extent of Earth's ozone layer.

Nuclear Waste Policy Act is passed.

1983   Asian Immigrant Women Advocates is founded. Its mission is to empower disenfranchised, low-income, Asian immigrant women who work as garment workers, electronics assemblers, hotel room cleaners, and janitors in the greater San Francisco–Oakland–Santa Clara County Bay Area. **Helen Kim** joins in 1991.

Randy Hayes's film, *The Four Corners, A National Sacrifice Area*, wins the Student Academy Award for the Best Documentary. The film documents the tragic effects of uranium and coal mining on Hopi and Navajo Indian lands in the American Southwest.

The residents of Times Beach, Missouri, are ordered to evacuate their community. Investigations of Times Beach disclose the fact that oil contaminated with dioxin, a highly toxic substance, had been used to treat the town's streets.

1984   Native Action is founded by Gail Small for social justice for American Indians.

David Haenke is the coordinator of the first North American Bioregional Congress (NABCI), held in May.

Major accident of a Bhopal chemical plant in India causes a disaster. Toxic gases leak from a chemical manufacturing plant, killing an estimated 3,000 people and injuring many thousands more. Unofficial estimates suggest even higher casualties.

Lester Brown inaugurates the *State of the World* series of annual environmental assessments.

1985   *Dwellers in the Land: The Bioregional Vision*, by **Kirkpatrick Sale**, is published.

International Rivers Network is founded by **Phil Williams**. One of its major goals is to demolish dams and to oppose the building of new ones throughout the world.

Randy Hayes becomes Executive Director of Rainforest Action Network. Hayes believes that rainforest problems are the most important issue today. The Rainforest Action Network has more than 40,000 members.

Concerned Citizens of South Central Los Angeles becomes one of the first people of color environmental groups in the United States. Juanita Tate is the present Executive Director. The organization aims toward providing a better quality of life for the residents of this Los Angeles community. **Maria Perez, Nevada Dove,** and **Fabiola Tostado** will later join the group and be known as the Toxic Crusaders.

Huey D. Johnson becomes the founder and president of the Resource Renewal Institute (RRI), a nonprofit organization in California. Johnson suggests that green plans are the path that countries should take to respond to environmental decline. Green plans treat the environment as it really exists—a single, interconnected ecosystem that can only be safeguarded for future generations through a systemic, long-range plan of action.

The ozone "hole" is discovered by scientists of the British Antarctica Survey. The hole appears during the Antarctica spring and is caused by the chlorine from CFCs.

The Greenpeace ship *Rainbow Warrior* is bombed on July 10 while docking in Auckland Harbour, New Zealand. It is later discovered that French intelligence agents are responsible for the bombing of the ship and the death of a Greenpeace photographer on board.

A section of the Food Security Act, popularly known as the swampbuster provision, is designed to protect wetland areas of the United States from conversion into farmland as a means of preserving these unique ecosystems and the species that live there.

The wild population of the largest flying bird species in North America, the California Condor, falls to only nine individuals.

Dian Fossey is murdered in her cabin in Karisoke, Rwanda. The murder case is never solved.

The Emergency Planning and Community Right-to-Know Act is passed.

1986   A nuclear power plant in Chernobyl, Ukraine, suffers a catastrophic accident. The fire and explosion of the power plant create nuclear fallout and radiation that contaminates large areas of northern Europe, particularly the United Kingdom, Finland, and Sweden.

Randy Hayes spearheads the first environmental demonstration against the largest financial institution on the planet—the World Bank. In 1987 he leads the first nonviolent civil disobedience against that institution.

Safe Drinking Water Act is amended to direct the EPA to set standards for 83 specified contaminants and to ban the use of lead pipes in new drinking water systems.

1987   The Mobro, a Long Island garbage barge, travels 9,600 kilometers in search of a place to dump its trash.

The Montreal Protocol on Substances that Deplete the Ozone Layer proposes to cut in half the production and use of chlorofluorocarbons. In 1990, an amended protocol is accepted by 93 nations agreeing to phase out five key chlorofluorocarbons. The Montreal Protocol is adopted by governments in 1987 and is modified five times as of 2000. Its control provisions are strengthened through four adjustments to the Protocol adopted in London (1990), Copenhagen (1992), Vienna (1995), Montreal (1997), and Beijing (1999). The Protocol aims to reduce and eventually eliminate the emissions of man-made ozone-depleting substances.

Juana Gutiérrez becomes president and founder of Mothers of East Los Angeles, Santa Isabel Chapter (Madres del Este de Los Angeles–Santa Isabel) (MELASI). Its mission is not only to fight against the toxic dumps and incinerators but to take a proactive approach to community improvement.

Groundwater Contamination in the United States, coauthored by **Ruth Patrick**, is published.

Charles Lee is one of the principal authors of the first national study on the demographic patterns associated with location of hazardous waste sites. It is considered a landmark study in the field and is a seminal study that gives rise to a lot of national attention. The report is called *Toxic Wastes and Race in the United States: A National Report on the Racial and Socioeconomic Characteristics of Communities with Hazardous Waste Sites*.

1988   Plastic Pollution Research and Control Act bans ocean dumping of plastic materials.

EPA studies report that indoor air can be 100 times as polluted as outdoor air. Radon is found to be widespread in U.S. homes.

West Harlem Environmental Action is founded by **Peggy Shepard,** Vernice Miller, and Chuck Sutton. The West Harlem Environmental Action group focuses on a local sewage treatment plant and public health issues. The organization is also involved in economic development and historical preservation issues in the West Harlem neighborhood.

Max Cordova becomes president of the Truchas Land Grant, one of the few surviving Spanish land grants of northern New Mexico.

Beaches on the east coast of the United States and lakeshore beaches on Lake Erie and Lake Michigan are closed because of contamination by medical waste washed on shore.

James Hansen, a NASA scientist, reports to Congress the problems of global warming that include droughts and polar ice melts.

Worst United States drought since the 1930s occurs.

Global temperature is the highest in 130 years.

Ocean Dumping Ban legislates international dumping of wastes in the ocean.

Plastic ring six-pack holders are required to be made biodegradable.

Vienna Convention for the Protection of the Ozone Layer is held.

Paul Hawken's book, *The Ecology of Commerce*, is voted as the number-one college text on business and the environment by professors in 67 business schools.

1989   New York Department of Environmental Conservation reports that 25 percent of the lakes and ponds in the Adirondacks are too acidic to support fish.

Exxon Tanker *Valdez* runs aground on Prince William Sound, Alaska, spilling 11 million gallons of oil in one of the world's most fragile ecosystems.

Concern about the future of the Mono Lake ecosystem results in a court order that forbids further diversion of water from its feeder streams.

Whatever Happened to Ecology?, by **Stephanie Mills**, is published. Mills's memoir discusses the worsening ecological crisis and how bioregionalism is a key to saving Earth.

1990   U.N. report forecasts that world temperatures could rise 2 degrees Fahrenheit within 35 years because of greenhouse gas emissions.

The Arctic ozone hole is discovered.

Ocean Robbins, age 16, and Ryan Eliason, age 18, are the co-founders of YES!, Youth for Environmental Sanity. The goal of YES! is to educate, inspire, and empower young people to take positive action for healthy people and healthy planet. Ocean serves as director for five years and is now the organization's president. As of 2000, the program has reached 600,000 students in 1,200 schools in 43 states through full school assemblies.

Southwest Network is founded with **Richard Moore** as coordinator. The organization is formed to bring together people of color from across the Southwest to broaden regional strategies and perspectives on environmental degradation and other injustices.

Clean Air Act amendments include requirements to control the emission of sulfur dioxide and nitrogen oxides into the air. They are intended to meet unaddressed or insufficiently addressed air pollution problems such as acid rain, ground-level ozone, stratospheric ozone layer depletion, and air toxics. The amendments also mandate the application of Best Available Technology (BAT) to reduce air toxics.

Earth Day 1990. **Denis Hayes** is international chairman and **Mark Dubois** is the international coordinator.

The publication of **Chellis Glendinning**'s *When Technology Wounds: The Human Consequences of Progress* is the first book to present an overview of technology survivors that have been harmed by toxic wastes. Her book is nominated for a Pulitzer Prize in general nonfiction in 1991.

Sylvia Earle is appointed Chief Scientist of the National Oceanic and Atmospheric Administration.

The Oil Pollution Act is passed.

The Pollution Prevention Act is passed.

1991   Persian Gulf War in Kuwait results in an extensive number of oil wells being set on fire. The area becomes a pollution disaster.

Defending the Earth, by **Murray Bookchin** and **Dave Foreman**, is published.

Confessions of an Eco-Warrior, by **Dave Foreman**, is published.

The United States accepts an agreement that prohibits activities relating to mineral resources, protects native species of flora and fauna, and limits tourism and marine pollution in Antarctica.

Helen Kim becomes the coordinator of the Asian Immigrant Women Advocates (AIWA). From 1992 to 1996, Kim coordinates national support network for the Garment Workers Justice Campaign.

The First National People of Color Environmental Leadership Summit is held in Washington, DC. More than 600 members attend the meeting.

Dave Foreman becomes chairman of the Wildlands Project.

1992   **Juana Gutiérrez** is honored for her community work by citations from Pope John Paul II and later in 1995 by President Bill Clinton.

The United Nations Conference on Environment and Development (the Earth Summit) is attended by delegates from 108 nations who work together for the development and adoption of several major agreements designed to benefit the global environment. Among the major agreements of the summit was a document known as Agenda 21, which provided a plan of action for achieving sustainable development.

Cathrine Sneed is the founder and director of The Garden Project in San Francisco. The Garden Project is a horticulture class for inmates of the San Francisco County Jail, using organic gardening as a metaphor for life change. The United States Department of Agriculture calls the Project "one of the most innovative and successful community-based crime prevention programs in the country."

1993   Sugar producers and U.S. government agree on restoring the Florida Everglades.

George B. Schaller's book, *The Last Panda*, is published. The book relates Schaller's four years in the highlands of China working to protect both panda habitat and the few pandas that remained.

"Developing Working Definitions of Urban Environmental Justice" is written by **Charles Lee** and published in *Earth Island Journal*.

Denis Hayes becomes president of the Bullitt Foundation, Seattle, Washington.

Marjory Stoneman Douglas is awarded the Presidential Medal of Freedom by President Bill Clinton for her work to save the Florida Everglades.

1994   Once endangered in all of the lower 48 states, the bald eagle's status is upgraded to "threatened."

Dumping in Dixie: Class and Environmental Quality, 2nd Edition, by **Robert Bullard**, is published. The book reports on five environmental justice campaigns in states ranging from Texas to West Virginia.

United Nations Framework Convention on Climate Change is held.

California Desert Protection Act is passed.

1995   Government reintroduces endangered wolves to Yellowstone Park.

Stephanie Mills authors *In Service of the Wild: Restoring and Reinhabiting Damaged Land*, a book that provides a common sense argument for rehabilitating damaged landscapes.

1996   **Danny Kennedy** becomes the founder of Project Underground in California. The organization has the technological, legal, and scientific capacity to work on environmental issues related to mines and oil drilling operations.

Paul Hawken becomes the co-chairman and co-founder of The Natural Step International. Its purpose is to teach and support environmental systems thinking in corporations, cities, government, unions, and academic institutions.

Hazel Henderson shares the Global Citizen Award with Nobelist A. Perez Esquivel of Argentina.

Ralph Nader and **Winona LaDuke** run for president/vice president on the Green Party ticket.

Adam Werbach, age 23, is elected as the 46th president of the Sierra Club. In his two years as the youngest president, the Sierra Club grows to 600,000 members.

1997   *Dying from Dioxin: A Citizen's Guide to Reclaiming Our Health and Rebuilding Democracy*, by **Lois Gibbs,** is published.

Stephanie Mills edits *Turning away from Technology: A New Vision for the 21st Century*, which is published by the Sierra Club.

Max Cordova's work with the Forest Service earns him the Al Gore Hammer Award for cutting government red tape and making government more efficient.

Kyoto Protocol is adopted.

1998   For her lifelong commitment to conservation, President Bill Clinton awards **Margaret Murie** the nation's highest civilian honor, the Medal of Freedom.

In honor of her 100th birthday, the Audubon Society creates a **Hazel Wolf** "Kids for the Environment" fund to foster environmental appreciation among young people.

El Niño causes the deaths of thousands of southern sea lions and South American fur seals along the coasts of Chile, Peru, Ecuador, and the Galapagos.

Marjory Stoneman Douglas passes away at 108 years of age.

Melodie Dove of Concerned Citizens of South Central Los Angeles hires **Maria Perez** to work on a pilot program that focuses on environmental and workplace rights. The south central district has the highest number of children inflicted with lead poisoning in Los Angeles County.

1999   *Beyond Globalization: Shaping a Sustainable Global Economy*, by **Hazel Henderson,** is published.

Paul Hawken co-authors *Natural Capitalism: Creating the Next Industrial Revolution.*

Off the Map (An Expedition Deep into Imperialism, the Global Economy, and Other Earthly Whereabouts), by **Chellis Glendinning,** is published.

Scientists report that the human population of Earth now exceeds six billion people.

Peregrine falcon is removed from the U.S. Endangered Species List.

New Carissa runs aground off coast of Oregon, leaking some oil into Coos Bay. The tanker is later towed into the open ocean and sunk.

Julia Butterfly Hill, age 23, comes down out of a 180-foot California redwood tree after living there for two years to prevent the destruction of the forest. A deal is made with the logging company to spare the tree as well as a three-acre buffer zone.

2000   **Denis Hayes** is the coordinator and **Mark Dubois** is the international coordinator of Earth Day 2000.

Ralph Nader and **Winona LaDuke** run for U.S. President and Vice President on the Green Party ticket.

In January, **Hazel Wolf** passes away at the age of 101.

The Chernobyl nuclear power plant is closed in December.

Anthropologists for the Wildlife Conservation Society in New York announce that a type of large West African monkey is extinct, making it the first primate to vanish in the 20th century.

The study by National Park Trust, a privately funded land conservancy, found more than 90,000 acres within state parks in 32 states are threatened by commercial and residential development and increased traffic, among other things.

A bone-dry summer in north central Texas breaks the Depression-era drought record as the Dallas area marks 59 days without rain.

In November, **David Brower** passes away at 88.

Massachusetts announces that the state will spend $600,000 to see if petroleum pollution in largely African-American city neighborhoods contributes to lupus, a potentially deadly immune disease.

Half of the world's 6 billion people now live in cities.

The Environmental Protection Agency announces new regulations that will cut air pollution from heavy-duty trucks and buses by 90 percent over the next decade.

The World Food Program Agency estimates that by 2020, nearly half of the world's population will live in countries with water shortages.

Index

Page numbers for main entries are in **bold**.

About the Editors and Contributors

JOHN MONGILLO is a science writer and coauthor of *Encyclopedia of Environmental Science* and the series *Reading About Science*.

BIBI BOOTH is an environmental education specialist with the U.S. Department of the Interior, Bureau of Land Management.

PETER BERG is founder and director of Planet Drum Foundation in San Francisco.

JIM BERRY is at the Roger Tory Peterson Institute.

MAKEDA BEST is at The Garden Project in San Francisco.

LEE BOTTS serves on the boards of directors or as an advisor for several environmental groups.

MIKHAIL DAVIS is director of the Brower Fund, Earth Island Institute.

NEVADA DOVE is a member of Concerned Citizens of South Central Los Angeles.

MARK DUBOIS is international coordinator for Earth Day 2000.

LOIS GIBBS is executive director of the Center for Health, Environment, and Justice.

CHELLIS GLENDINNING is a psychologist and founding member of the neo-Luddite Jacques Ellul Society in Washington, DC.

JUDY GOLDHAFT is managing director of Planet Drum Foundation in San Francisco.

CHARLES R. GOLDMAN is a professor of limnology in the Department of Environmental Science and Policy at the University of California, Davis.

GABRIEL GUTIÉRREZ is a professor at California State University, Northridge.

DENIS HAYES is president of the Bullitt Foundation in Seattle.

RANDY HAYES is founder and president of the Rainforest Action Network in San Francisco.

GLENN S. JOHNSON is an assistant professor of Sociology in the Environmental Justice Resource Center at Clark Atlanta University.

HELEN SUNHEE KIM is a community organizer of the Asian Immigrant Women Advocates (AIWA).

STEPHANIE MILLS is an author, editor, lecturer, and bioregional activist.

MONO LAKE COMMITTEE is located in Lee Vining, California.

CATHERINE NIERING is the wife of William Albert Niering.

MARIA PEREZ is a member of Concerned Citizens of South Central Los Angeles.

JEFF PRIEST is at the University of South Carolina.

KIRKPATRICK SALE is a writer concerned with decentralism and active in the bioregional movement.

GEORGE SCHALLER is a field biologist and the director of science at the Wildlife Conservation Society in New York.

JOHN TODD is president of Ocean Arks International and a research professor and distinguished lecturer at the University of Vermont.

NANCY JACK TODD is vice president of Ocean Arks International.

FABIOLA TOSTADO is a member of Concerned Citizens of South Central Los Angeles.

PHIL WILLIAMS is founder of the International Rivers Network.

SUSAN WITT is the executive director of the E. F. Schumacher Society.

JOAN LUEDDERS WOLFE started the West Michigan Environmental Action Council.

HOWIE WOLKE runs the guide service, Big Wild Adventures, and the nonprofit conservation group, Big Wild Advocates.

ED ZAHNISER is a research associate of the Zahniser Institute of Environmental Studies, Greenville College, Illinois, and works for the Division of Publications of the National Park Service in Harpers Ferry, West Virginia.